Time-correlated Single Photon Counting

Time-correlated Single Photon Counting

Desmond V. O'Connor
David Phillips

The Royal Institution, London, UK

1984

ACADEMIC PRESS
(Harcourt Brace Jovanovich, Publishers)
London Orlando San Diego San Francisco New York
Toronto Montreal Sydney Tokyo

ACADEMIC PRESS INC. (LONDON) LTD.
24-28 Oval Road,
London NW1 7DX

United States Edition published by
ACADEMIC PRESS, INC.
Orlando, Florida 32887

British Library Cataloguing in Publication Data

O'Connor, Desmond V.
Time-correlated single photon counting.
1. Fluorescence 2. Photon correlation
I. Title II. Phillips, David
535'.35 QC477

ISBN 0-12-524140-2
LCCCN 83-73318

PRINTED IN THE UNITED STATES OF AMERICA

84 85 86 87 9 8 7 6 5 4 3 2 1

Preface

Of all the techniques available to obtain information on the rates of depopulation of excited electronic singlet states of molecular species, monitoring of fluorescence provides in principle the simplest and most direct measure of concentration. The time evolution of fluorescence, on a time-scale from hundreds of nanoseconds to tens of picoseconds, can be measured by a variety of techniques, but the one which has become most popular over the last 12 years is time-correlated single photon counting. The level of sophistication accompanying the use of these methods has developed rapidly, and several instruments are now available commercially. It is perhaps arrogance which causes us to believe that new users of the technique do not always appreciate fully the advantages and pitfalls of the equipment and curve-fitting routines necessarily used with the hardware. We have written this monograph in the sincere hope that by relating our own experiences with a variety of different single photon counting systems, we may provide a useful service to users and potential users of this formidably sensitive technique.

March 1984

Desmond V. O'Connor
David Phillips

Acknowledgements

We are grateful to colleagues from the Departments of Chemistry at the University of Western Ontario, London, Ontario, Canada, and at the University of Southampton, England, and latterly, at the Davy–Faraday Laboratory of The Royal Institution, for sharing with us their expertise in the use of time-correlated single photon counting for fluorescence decay measurements in a variety of applications. D.O'C acknowledges an especial debt of gratitude to Professor W. R. Ware. At the same time we emphasize that any shortcomings in this book are entirely our own responsibility.

Contents

Time-correlated Single Photon Counting

1 Fluorescence, its Time Dependence and Applications

It is hardly conceivable that a purchaser of this volume will not be fully conversant with molecular fluorescence in all its aspects, but for the sake of completeness, and to provide an introduction, this chapter will outline in cameo some aspects of fluorescence.

1.1 Fluorescence Intensities

Fluorescence is defined simply as the electric dipole transition from an excited electronic state to a lower state, usually the ground state, of the same multiplicity. Mathematically, the probability of an electric-dipole induced electronic transition is proportional to R_{if}^2 where R_{if}, the transition moment integral between initial state i and final state f, is given by Equation 1.1. In this equation ψ_{e} represents the electronic wavefunction, ψ_{n} the vibrational wavefunction, $\hat{\mathbf{M}}$ is the electronic dipole moment operator, and the Born–Oppenheimer principle of separability of electronic and vibrational wavefunctions has been invoked. The first integral involves only the electronic wavefunctions of the system, and the second term, when squared, is the familiar Franck–Condon factor.

$$R_{\mathrm{if}} = \int_{-\infty}^{+\infty} \psi_{\mathrm{ef}} \hat{\mathbf{M}} \psi_{\mathrm{ei}} \, \mathrm{d}\tau_{\mathrm{e}} \int_{-\infty}^{+\infty} \psi_{\mathrm{nf}} \psi_{\mathrm{ni}} \, \mathrm{d}\tau_{\mathrm{n}}. \tag{1.1}$$

To a good approximation the electronic integral is zero unless the states i and f are of the same spin; hence electronic absorption in most organic molecules results in the formation of an excited singlet state from the ground singlet state. There are also symmetry restrictions on transitions between states i and f imposed by the necessity of the electronic integral being totally symmetric if it is not to become of zero value upon integration. For a transition which satisfies this requirement (symmetry allowed), the molar decadic absorption coefficient will have a maximum value of the order of 10^5 dm^3 cm^{-1} mol^{-1}. The corresponding value for the rate constant for radiative decay, k_{R}, of the excited state via the spontaneous fluorescence process is given approximately by Equation 1.2, and more exactly by Equation 1.3 (Strickler and Berg, 1962).

$$k_R(s^{-1}) \simeq 10^4 \varepsilon_{max}(dm^3\,cm^{-1}\,mol^{-1}) \quad (1.2)$$

$$k_R = 2.88 \times 10^{-9} n^2 \langle \bar{\nu}_F^{-3} \rangle_{Av}^{-1} \int \varepsilon \, d\bar{\nu}/\bar{\nu} \quad (1.3)$$

where $\langle \nu_F^{-3} \rangle^{-1}$ is a measure of the average frequency of the fluorescence, $\int \varepsilon \, d\bar{\nu}/\bar{\nu}$ is the area under the absorption curve, and n is the refractive index of the medium in which the experiment is carried out. Use of Equation 1.2 for a symmetry-allowed transition gives a value of k_R of $10^9\,s^{-1}$. In the absence of any other depopulation process, therefore, the fluorescence decay time of such an excited electronic state would be $1/k_R = 10^{-9}$ s, or 1 ns. This quantity, $1/k_R$ is termed the natural or mean radiative lifetime, τ_R. In practice, because of competing processes, the actual or measured decay time, τ_F, is for complex polyatomic molecules invariably less than the mean radiative lifetime.

For a transition which does not satisfy the symmetry restrictions imposed by Equation 1.1, the transition moment integral can be non-zero if a second-order mechanism is invoked which necessitates the excitation of a vibration which is not totally symmetric in one or other of the electronic states. This has the effect of reducing the magnitude of the transition moment integral compared with that for a symmetry-allowed transition, and also leads to the absence of the absorption and emission spectral features corresponding to transitions between the vibrationless ground and excited electronic states.

In addition to symmetry restrictions on electronic transitions there are restrictions caused by the necessity of overlap in space of molecular orbitals for the electron in its initial and final states. Where this is small, for example in $n \rightarrow \pi^*$ transitions in carbonyl compounds, the electronic integral in Equation 1.1 is diminished. The "allowedness" of a transition, F, expressed as an oscillator strength, can be summarized as in Equation 1.4 by a series of factors, f, which relate in turn to spin, (s), overlap (o), p (parity), sy (symmetry), and which have the following approximate values $f_s = 10^{-5}$, $f_o = 10^{-2}$, $f_p = 10^{-1}$, $f_{sy} = 10^{-1}$ to 10^{-3}.

$$F = f_s f_o f_p f_{sy} F_A \quad (1.4)$$

F_A is the oscillator strength of a fully allowed transition. For fluorescence then ($f_s = 1$), values of k_R extend from $10^9\,s^{-1}$ for a fully allowed transition (typical say of dyestuffs), to $10^7\,s^{-1}$ for symmetry forbidden transitions (typical say of aromatic molecules), down to $10^5\,s^{-1}$ and less (for say carbonyl compounds).

It should be noted that the radiative rate constant referred to above depends upon the refractive index of the medium, n, in which it is measured. Hirayama and Phillips (1980) have reviewed the approach of several workers to the dependence of k_R upon n, and showed that all treatments lead to the expression

$$k_R \propto n^2 \int \frac{\varepsilon(\nu)}{\nu}\, d\nu \qquad (1.5)$$

where the integral is the area under the absorption spectrum. Controversy has arisen over the unknown dependence of this integral on n, but recent work on the substituted anthracene molecule I has shown that k_R has an overall n^2 dependence.

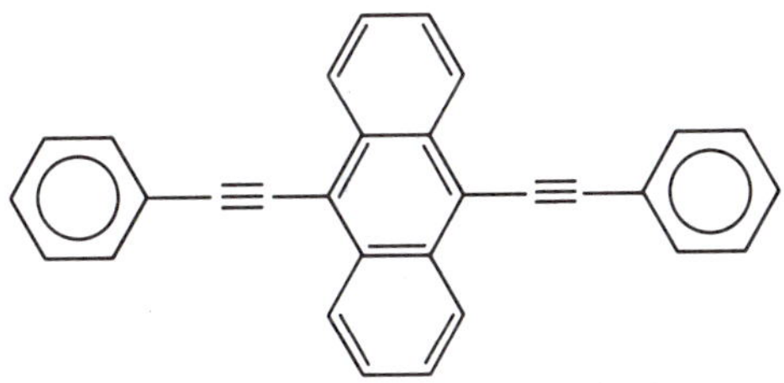

I

Table 1.1 demonstrates that for the fluorescence of the activator I, in xylene as a function of temperature, an n^2 rather than an n^3 dependence certainly gives a better fit to the observations.

Table 1.1 Dependence of τ_F for compound I upon n, n^2, n^3 in xylene

T/°C	n	τ_F/ns	$\frac{n(T)}{n(21)}\tau_F$	$\left[\frac{n(T)}{n(21)}\right]^2\tau_F$	$\left[\frac{n(T)}{n(21)}\right]^3\tau_F$
21	1.4990	3.17	3.17	3.16	3.16
39	1.4910	3.20	3.18	3.16	3.14
64	1.4785	3.27	3.22	3.17	3.13
95	1.4630	3.31	3.23	3.15	3.07
127	1.4470	3.40	3.28	3.16	3.05

The intensity of fluorescence from any excited molecule, although clearly dependent upon the magnitude of k_R as outlined above, is strongly dependent upon internal competing processes. These are shown schematically in the familiar Jablonskii diagram, Fig. 1.1. The unimolecular processes competing with fluorescence are intersystem crossing to the triplet manifold, internal conversion to the ground state, and photochemical reaction. In condensed media vibrational relaxation occurs on a picosecond timescale, and thus only chemical processes with rate constants in excess of $10^{12}\ s^{-1}$ will compete with vibrational relaxation. Subsequent to excitation therefore, vibrational relaxation is usually complete before electronic relaxation. Internal conversion is

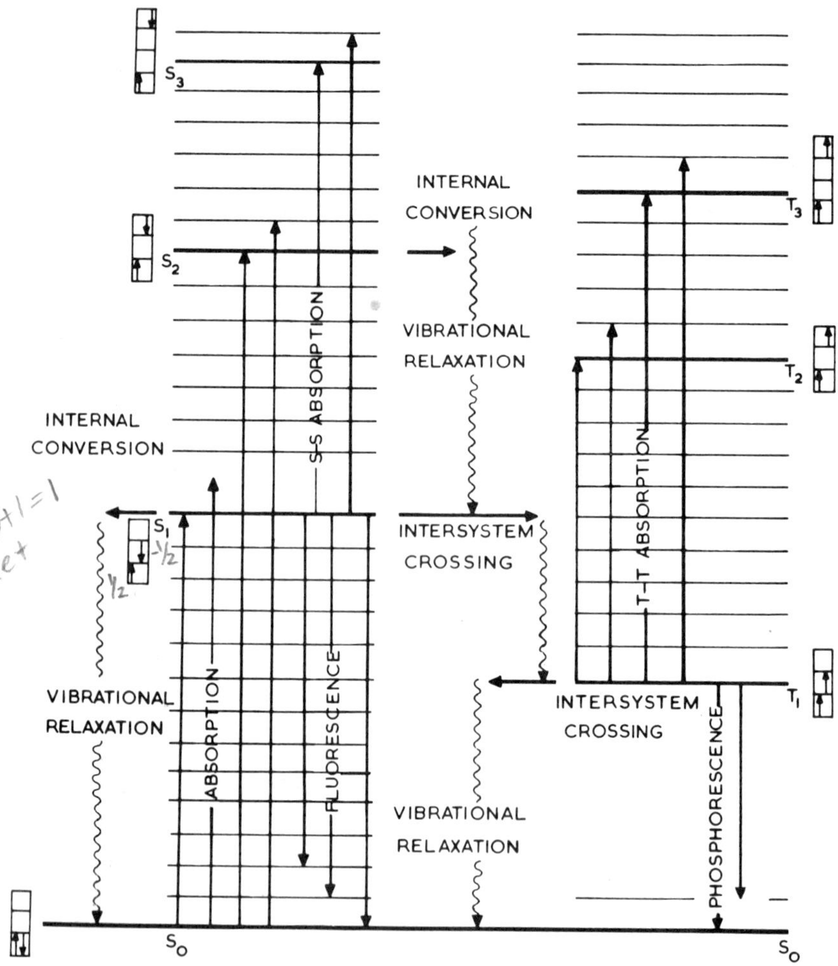

Figure 1.1 Jablonskii diagram showing fates of photoexcited complex polyatomic molecule.

usually faster at higher excess energies, but is not of great importance for lower-lying vibrational levels of the first excited singlet state. The principal process competing with fluorescence is therefore intersystem crossing to the triplet manifold of levels. Writing a simple kinetic scheme permits definition of the quantum yield of fluorescence, Φ_F, and decay time τ_F in terms of first-order rate constants (see Fig. 1.1).

	Rate constant	
$M + h\nu \rightarrow {}^1M^*$	I_a (absorption)	(1.6)

$$^{1}M^{*} \rightarrow M + h\nu \qquad k_R \text{ (fluorescence)} \tag{1.7}$$

$$^{1}M^{*} \rightarrow {}^{3}M^{*} \qquad k_{ISC} \text{ (intersystem crossing)} \tag{1.8}$$

$$^{1}M^{*} \rightarrow M \qquad k_{IC} \text{ (internal conversion)} \tag{1.9}$$

$$^{1}M^{*} \rightarrow \text{products} \qquad k_D \text{ (dissociation)} \tag{1.10}$$

From a steady-state analysis of the scheme, the quantum yield of fluorescence Φ_F is given by

$$\Phi_F = \frac{k_R}{(k_R + k_{ISC} + k_{IC} + k_D)}. \tag{1.11}$$

The fluorescence decay time, τ_F, is given by Equation 1.12 as:

$$\tau_F = (k_R + k_{ISC} + k_{IC} + k_D)^{-1}. \tag{1.12}$$

The intensity of fluorescence seen from any molecule, even before consideration of *bimolecular* interactions, depends upon the magnitude of the rate constant k_R relative to the sum (called Σk) of k_{ISC}, k_{IC} and k_D. In some molecules, for example carbonyls, k_R is very small with respect to Σk and these molecules are thus only weakly fluorescent. Aromatic molecules, where k_R is usually of the same order of magnitude as Σk, are strongly fluorescent, and consequently very useful as probes. It is very clear from Equation 1.11 that rationalization of the photophysics and photochemistry of any singlet state molecular species in terms of the *absolute* magnitudes of rate constants for the various competing decay processes cannot be obtained with the sole knowledge of quantum yields, which are merely rate constant ratios. However, a knowledge of fluorescence decay times together with quantum yields does, in principle, provide the absolute rate information required, since

$$k_R = \frac{\Phi_F}{\tau_F} \tag{1.13}$$

$$k_{ISC} = \frac{\Phi_{ISC}}{\tau_F}, \tag{1.14}$$

and so on.

Simple decay time and quantum yield measurements have been widely used in the development of our understanding of the role of specific vibrational mode excitation in the radiative and non-radiative decay of some complex molecules. We show in Table 1.2, as an example, the fluorescence decay times and quantum yields of selected vibronic states of $^{1}B_{2u}(S_1)$ benzene in collision-free vapour, taken from the paper by Spears and Rice (1971). Each pair of quantum yield and lifetime values corresponds to a specific vibronic state, the identity of which has been omitted for simplicity.

Table 1.2 Lifetimes and quantum yields of single vibronic states in $^1B_{2u}$ benzene (adapted from Spears and Rice, 1971)

Energy above origin/cm^{-1}	Observed lifetime/ns	Quantum yield	Radiative lifetime/ns	Non-radiative lifetime/ns
0	100	0.22	455	128
521	79	0.27	290	108
764	77	0.26	295	104
923	83	0.21	395	105
1007	66	0.23	285	86
1042	72	0.29	250	101
1170	52	0.16	325	62
1444	71	0.21	340	90
1470	65	0.19	340	80
1547	51	0.18	285	62
1687	61	0.16	380	73
1691	52	0.24	215	68
1846	62	0.16	380	74
1930	49	0.20	245	61
1965	61	0.22	280	78
2070	55	0.06	920	59
2367	55	0.17	320	66
2393	39	0.26	150	53
2610	42	0.08	525	46
2614	42	0.06	700	45
2769	49	0.07	700	53
2888	47	0.03	1470	48

1.2 Bimolecular Interactions

In the simple kinetic scheme discussed so far bimolecular interactions other than collisional removal of vibrational energy have been neglected. Clearly in condensed media, bimolecular interactions, which may take place on a picosecond timescale, must influence the fate of the excited electronic state. The effect of the many collisions with surrounding inert medium, such as solvent, will merely be to produce a Boltzmann distribution of levels emitting fluorescence. There are, however, many specific interactions which can greatly influence the kinetics of decay. These will now be discussed briefly.

One obvious consequence of increasing the concentration of molecules M in any fluorescence experiment is to increase the probability that energy can migrate from one molecule to another by non-radiative processes, thus influencing the decay characteristics of the fluorescence:

$$^1M^* + M \rightarrow M + {}^1M^*. \qquad (1.15)$$

The energy migration may occur through a series of near neighbour random hopping processes, or may occur over long distances through an induced dipole mechanism (Forster, 1946), the probability of which is given by Equation 1.16, which requires overlap of absorption spectrum of acceptor, (here M) and emission spectrum of donor (here $^1M^*$), and depends inversely on the sixth power of the distance separating the chromophores.

$$k_{ET} = \frac{K}{\tau_R R^6} \int_0^\infty f_D(\bar{\nu})\varepsilon_A(\bar{\nu})\, d\bar{\nu}/\bar{\nu}^4 \tag{1.16}$$

τ_R here is the mean radiative lifetime of the donor, and K is a constant given by Equation 1.17, where κ^2 is an orientation factor arising from the induced dipole nature of the transfer, which takes values between 0 and 4, a value of 2/3 being appropriate for a random distribution of molecules:

$$K = \frac{9000\kappa^2 c^4 \ln 10}{128\pi^5 n^4 N}. \tag{1.17}$$

Energy migration may reveal itself as a fast component in the decay of fluorescence of $^1M*$, being often in the picosecond region, but possibly in some synthetic polymer systems on the nanosecond timescale. It gives rise to a non-exponential form of the donor fluorescence decay law, $I(t)$, given by

$$I(t) = I(0)\exp\{-t/\tau_0 - 2B\gamma(t/\tau_0)^{1/2}\} \tag{1.18}$$

in which τ_0 is the donor fluorescence decay time in the absence of acceptor,

$\gamma = \frac{[A]}{[A]_0}$, the ratio of concentrations of acceptor,

$B = \left(\frac{1 + 10.87x + 15.5x^2}{1 + 8.743x}\right)^{\frac{3}{4}}$, where

$x = D\alpha^{-1/3}t^{2/3}$, and

$\alpha = (R^0/\tau_0)^6$ where R^0 is the value of R such that the rate of energy transfer is equal to the rate of spontaneous decay of the donor.

The interaction of an excited molecule $^1M^*$ with a ground state may in some instances lead to electronic quenching of the excited state:

$$^1M^* + M \rightarrow M + M. \tag{1.19}$$

This concentration quenching in fluid media leads to the inclusion of the term $k_M[M]$ in the expression for decay time now given by Equation 1.20. In rigid media, where diffusional kinetics do not pertain, other models are appropriate (see for example Beavan *et al.*, 1979).

$$\tau_F^{-1} = k_R + k_{ISC} + k_{IC} + k_D + k_M[M]. \tag{1.20}$$

Inclusion of the term $k_M[M]$ yields the familiar relationship for quantum yield of fluorescence relative to that at a reference concentration,

$$\frac{(\Phi_F)_0}{\Phi_F} = 1 + k_M\tau_0[M] \tag{1.21}$$

a particular form of the Stern–Volmer relationship which here refers to self-quenching of the excited state. τ_0 in this equation is the fluorescence decay time at the reference concentration.

In many systems concentration quenching is accompanied by the appearance of a new emission band to the red of the fluorescence of the uncomplexed molecule $^1M^*$ which is attributable to the formation of an excited dimer, termed an *excimer* (Equation 1.22). Excimer fluorescence is characteristically broad, as in Fig. 1.2, and decay characteristics are complex, since excimer formation is reversible

$$^1M^* + M \rightleftharpoons {}^1(MM)^*. \tag{1.22}$$

A complete kinetic scheme is shown in Fig. 1.3, after Birks (1970). Analysis of this scheme leads to the following relationships for the decay of monomer (M) and excimer (D) fluorescence, respectively.

$$i_M(t) = A_1 \exp(-\lambda_1 t) + A_2 \exp(-\lambda_2 t) \tag{1.23}$$

$$i_D(t) = A_D[\exp(-\lambda_1 t) - \exp(-\lambda_2 t)], \tag{1.24}$$

where

$$\begin{aligned}\lambda_{1,2} = \tfrac{1}{2}\{&(k_{IM} + k_{FM} + k_{DM}[M] + k_{ID} + k_{FD} + k_{MD}) \\ &\pm [(k_{DM} + k_{FM} + k_{DM}[M] - k_{ID} - k_{FD} - k_{MD})^2 \\ &+ 4k_{DM}[M]k_{MD}]^{1/2}\}.\end{aligned} \tag{1.25}$$

The monomer decay should thus be represented by the sum of two exponential terms; that of the excimer by the difference between the same two exponential terms. These kinetics are obeyed by many simple molecular systems, but in the case of synthetic polymers, in which excimer-forming moieties are attached covalently to the backbone of, for example, a vinyl chain, the kinetics are more complex. Thus for example in a copolymer of 1-vinylnaphthalene and methylmethacrylate, the fluorescence decay of monomeric naphthalene can only be described adequately by three components. Typical data are shown in Table 1.3.

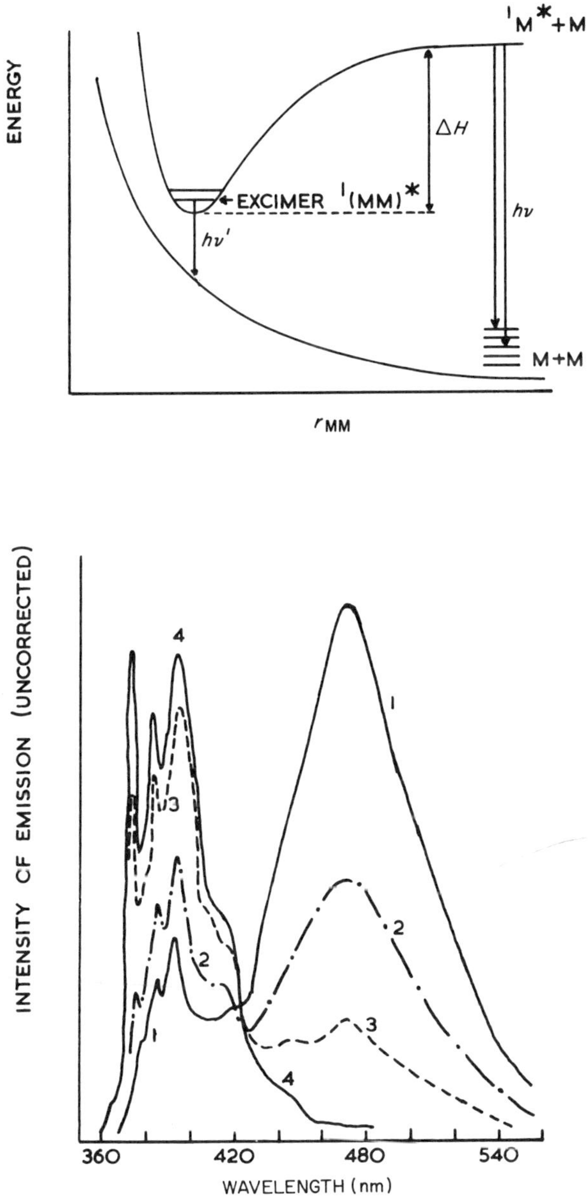

Figure 1.2 Excimer fluorescence characteristics: exaggerated potential energy diagram for excimer and resulting fluorescence spectra typified by the case of pyrene in fluid solution at concentrations (1) 3×10^{-3} M, (2) 10^{-3} M, (3) 3×10^{-4} M, (4) 2×10^{-6} M (after Parker and Hatchard, 1963).

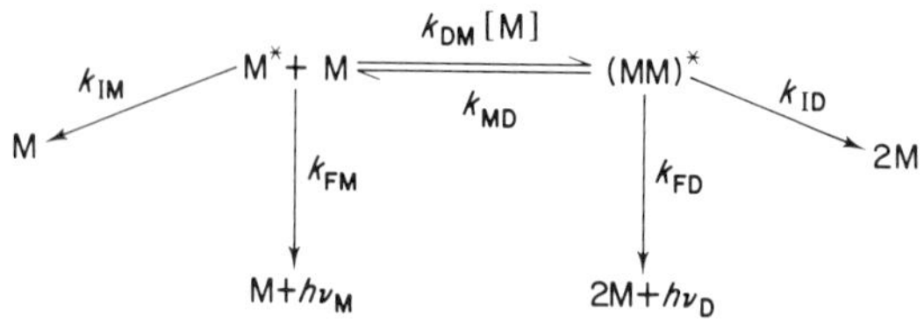

Figure 1.3 Kinetic scheme for excimer formation and decay (after Birks, 1970).

Table 1.3 Triple exponential decay data for 1-vinylnaphthalene methylmethacrylate copolymers (Roberts *et al.*, 1981)

f_n*	A_1	τ_1/ns	A_2	τ_2/ns	A_3	τ_3/ns
0.17	0.005	73.55	0.023	12.65	0.112	40.06
0.27	0.011	67.60	0.041	15.31	0.092	34.17
0.38	0.027	59.13	0.057	11.88	0.074	26.44
0.47	0.021	59.50	0.075	8.46	0.066	23.09
0.58	0.025	53.34	0.100	6.41	0.069	19.52
0.66	0.025	53.24	0.115	4.18	0.065	16.36
0.75	0.032	48.83	0.149	3.57	0.072	14.41
0.83	0.029	47.07	0.257	1.87	0.072	13.52

*f_n is mole fraction of naphthalene polymer.

Such multiple component fitting represents a somewhat hazardous excercise, with extreme care needed to ensure that parameters derived from it are meaningful, but this is now done routinely and will be described later in this volume. A rationale for the three components is given by the kinetic scheme shown in Fig. 1.4, in which two types of monomeric naphthalene are

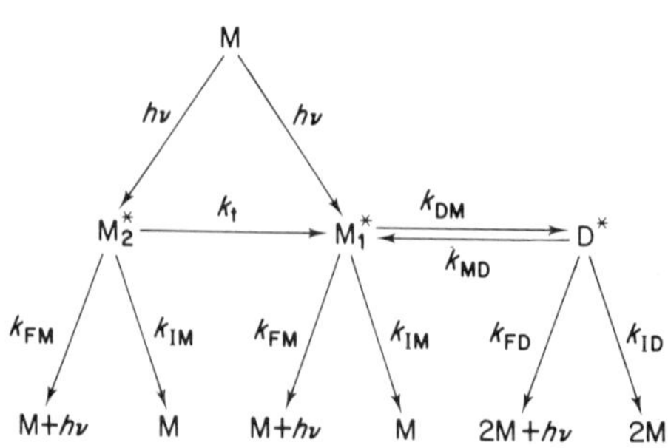

Figure 1.4 Kinetic scheme for excimer formation in vinyl aromatic polymers. In this scheme, M_1 and M_2 are two spectrally identical, but kinetically distinct monomer species, differing only in the extent to which they can participate in energy migration and thereby excimer formation.

postulated, having the same spectral distribution of fluorescence, but different kinetic behaviour. M_1 can rapidly and reversibly form excimers through energy migration and molecular motion, whereas M_2 behaves essentially as an isolated chromophore which is weakly coupled to M_1 through long range energy transfer. Such schemes are discussed widely elsewhere (see for example Roberts *et al.*, 1981).

1.2.1 Impurity quenching

In the presence of additive or impurity molecules, Q, electronically excited states may be quenched with a consequent decrease in the observed fluorescence decay time:

$$^1M^* + Q \rightarrow \text{products}. \tag{1.26}$$

Quenching may result from electronic energy transfer, often by the induced dipole mechanism referred to previously, from chemical reaction, from enhancement of non-radiative decay brought about by paramagnetic species or molecules with atoms of high nuclear charge, or by complex formation. In all cases, if the concentration of absorbing molecule M is kept constant, the Stern–Volmer relationship becomes

$$(\Phi_F)_0/\Phi_F = 1 + k_Q\tau_0[Q], \tag{1.27}$$

where $(\Phi_F)_0$ and τ_0 now refer to quantum yield and decay time of $^1M^*$ in the absence of Q. In the case of quenching by complex formation, a new fluorescence may be observed which resembles excimer fluorescence in that it is red-shifted and structureless, but is a result of charge transfer to form an excited complex (exciplex). The kinetics of this are exactly comparable to excimer kinetics.

Thus for example, the time dependence of monomer and complex concentrations, $[^1M^*](t)$ and $[^1MQ^*](t)$, are given by

$$[^1M^*](t) = -A_1e^{-\lambda_1 t} + A_2e^{-\lambda_2 t} \tag{1.28}$$

$$[^1MQ^*](t) = A_3e^{-\lambda_1 t} - A_4e^{-\lambda_2 t} \tag{1.29}$$

where the rate constants are identified in Fig. 1.5, and

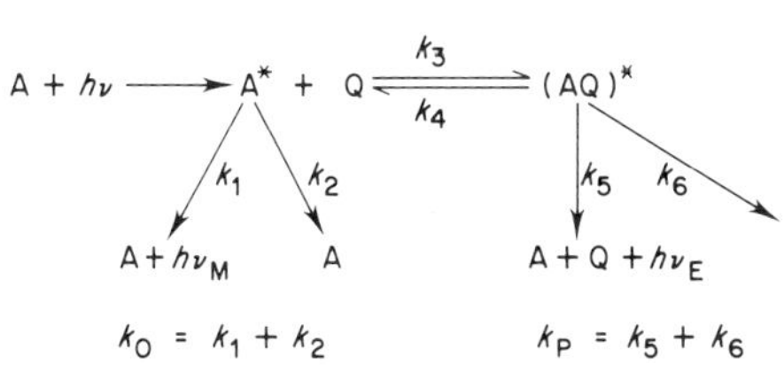

Figure 1.5 Kinetic scheme for exciplex formation.

$$A_1 = \left(\frac{k_0 + k_3[\mathrm{Q}] - \lambda_2}{\lambda_1 - \lambda_2}\right)[{}^1\mathrm{M}^*]_0 \tag{1.30}$$

$$A_2 = \left(\frac{\lambda_1 - (k_0 + k_3[\mathrm{Q}])}{\lambda_1 - \lambda_2}\right)[{}^1\mathrm{M}^*]_0 \tag{1.31}$$

$$A_3 = -A_4 = \frac{k_3[\mathrm{Q}]}{\lambda_1 - \lambda_2}[{}^1\mathrm{M}^*]_0 \tag{1.32}$$

and

$$\lambda_{1,2} = \tfrac{1}{2}\{k_0 + k_3[\mathrm{Q}] + k_4 + k_\mathrm{p} \pm [(k_0 + k_3[\mathrm{Q}] - k_4 - k_\mathrm{p})^2 + 4k_4k_3[\mathrm{Q}]]^{1/2}\}. \tag{1.33}$$

Equation 1.32 indicates that the A factors for the grow-in A_3 and decay A_4 of exciplex fluorescence should be of the same magnitude but opposite sign. That this relationship can be obeyed almost exactly is shown by the last two columns of Table 1.4, which contains data measured for a gas-phase exciplex system (O'Connor *et al.*, 1982).

In a few cases, weak association between a molecule M in the ground state and an additive molecule may occur.

$$\mathrm{M} + \mathrm{Q} \rightleftharpoons \mathrm{MQ} \tag{1.34}$$

Excitation of the associated complex MQ then produces an excited state $^1(\mathrm{MQ}^*)$ which may resemble closely a true exciplex. However, the kinetics of fluorescence in such a system are in principle distinguishable, since in the case of the ground-state complex excitation is instantaneous, and the complex fluorescence does not then exhibit a "growing in" period. Without giving a full analysis, in cases where such *static* quenching is of importance, measure-

Table 1.4 Fluorescence decay time data for 1-cyanonaphthalene–triethylamine exciplex in the gas phase at 188°C (O'Connor *et al.*, 1982).

	Monomer			Exciplex			
Concentration of amine $10^{-3}M$	τ_1/ns	τ_2/ns	$\frac{a_1}{(a_1+a_2)}$	τ_1/ns	τ_2/ns	a_3	a_4
0		24.1					
0.222	8.17	11.96	0.64	8.64	12.05	−0.46	0.46
0.530	5.41	10.13	0.82	5.80	11.60	−2.69	2.68
0.837	4.07	10.50	0.90	4.16	11.11	−3.68	3.70
2.13	2.00	11.23	0.98	1.96	10.97	−2.33	2.33
3.36	1.31	10.35	0.98	1.31	10.38	−1.93	1.92

ment of relative quantum yields of fluorescence $(\Phi_F)_0/\Phi_F$ and relative decay times of fluorescence, $(\tau_F)_0/\tau_F$ can be used to obtain a form of the equilibrium constant for ground-state complex formation through the following equation:

$$[(\Phi_F)_0/\Phi_F]/[(\tau_F)_0/\tau_F] = 1 + K'_{eq}[Q]. \tag{1.35}$$

Since molecular oxygen is an ubiquitous impurity, the processes by which it may quench excited states have been widely studied. Quenching of fluorescent singlet states may occur through Equations (1.36) or (1.37), with very high efficiencies.

$$^1M^* + O_2(^3\Sigma\bar{g}) \rightarrow {}^3M^* + O_2(^3\Sigma\bar{g}) \tag{1.36}$$

$$\rightarrow {}^3M^* + O_2(^1\Delta g) \tag{1.37}$$

In a heterogeneous emitting system with two (or more) fluorescence decay components, a "chemical timing" method of achieving time resolution of fluorescence spectra has been developed which depends upon the selective quenching of the longer-lived molecules by molecular oxygen.

In fluid media, the various quenching processes we have discussed, with the exception of dipole-induced dipole electronic energy transfer, yield decay curves which can be described by an exponential or a sum of exponentials. In some circumstances however, diffusional kinetics can yield an additional, observable decay term which has a $t^{-1/2}$ dependence. For a diffusion-controlled quenching reaction the continuum model predicts the fluorescence decay law to be (Nemzek and Ware, 1975)

$$I(t) = I_0 \exp(-at - 2bt^{1/2}), \tag{1.38}$$

where

$$a = 1 + 4R'D_{AQ}N'[Q] \tag{1.39}$$

and

$$b = 4(R')^2(\pi D_{AQ})^{1/2}N[Q]. \tag{1.40}$$

R' is related to the encounter distance, and D_{AQ} is the mutual diffusion coefficient of A and Q, N' is 6.023×10^{20}, and τ_0 is the decay time at $[Q] = 0$. The ratio of intensities of fluorescence under *steady* illumination is predicted to be

$$(I_0/I) = [1 + 4R'D_{AQ}N'[Q]\tau_0]Y^{-1}, \tag{1.41}$$

where

$$Y = 1 - (b/a^{1/2})\pi^{1/2}\exp(b^2/a)\operatorname{erfc}(b/a^{1/2}) \tag{1.42}$$

and

$$\mathrm{erfc}(x) = 2\pi^{-1/2} \int_x^{\infty} \exp(-u^2)\,du. \tag{1.43}$$

That the decay law is indeed represented by Equation 1.38 is established in Fig. 1.6 for the quenching of 1,2-benzanthracene by CBr_4 (Nemzek and Ware, 1975). However, a test of the consistency of the model is to deconvolve decay curves according to Equation 1.38, use resulting values of a and b to compute R' and D, and *calculate* values of $(I_0)/I$. These can then be compared with experiment. However in no case reported to date is there good agreement, and a large contribution of static quenching is thus invoked to explain the observed results (see for example Nemzek and Ware, 1975; Beddard *et al.*, 1978).

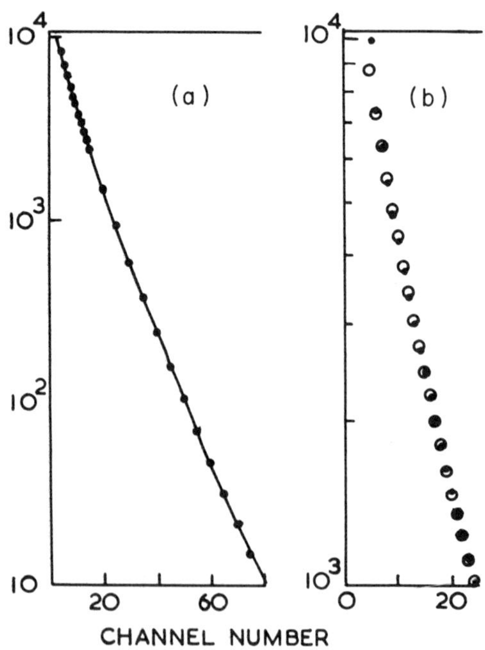

Figure 1.6 Experimental data (a: line, b: open circles) for decay of 1,2-benzanthracene fluorescence quenched by CBr_4 and computed decay (filled circles) for $I(t)$ using Equation 1.38 (from Nemzek and Ware, 1975). (a) Three decades; (b) one decade decay.

1.3 General Environmental Effects

There are effects other than specific electronic quenching effects which may influence the time-resolved behaviour of fluorescence of molecules. We have already discussed the effect of refractive index upon the radiative rate constant, but there are general dielectric effects of the medium which are responsible for *spectral* shifts, and which may also influence the nature of the emitting state, and hence its decay characteristics.

The spectral distribution of fluorescence is dictated by the Franck–Condon factors defined as the square of the second term in Equation 1.1 and the position of the centre of gravity of the fluorescence depends upon any geometry changes between ground and excited states. The latter point is illustrated in Fig. 1.7 from which it can be seen that the most probable transition in absorption is to higher energies than that for fluorescence if the potential surface of the excited state undergoes some non-zero displacement with respect to the ground state, and assuming that vibrational relaxation is complete prior to electronic relaxation. If vibrational relaxation is incomplete, then a time-dependent fluorescence spectrum results. However, since in condensed media vibrational relaxation occurs on a picosecond timescale, time-dependent spectra due to this phenomenon can only be observed in picosecond experiments. Solvent relaxation occurs on a longer timescale, and can result in the observation of nanosecond time-resolved spectra.

The spectral position of absorption and fluorescence is influenced by the dielectric properties of the medium in which observations are made. Figure 1.8 shows that the vapour phase O–O bands in absorption and fluorescence of a molecule are identical, whereas in solution with solvent of static dielectric constant ε and refractive index n, the bands are no longer coincident. The differences can be rationalized as follows. From Onsager theory, a solute molecule of dipole moment μ in a spherical cavity of radius a polarizes the dielectric of the solvent, producing a reaction field. This field, R_0, is given for the ground state of the solvent molecule (of dipole moment μ_0), by:

$$R_0 = \frac{2\mu_0(\varepsilon - 1)}{a^3(2\varepsilon + 1)} \tag{1.44}$$

Upon excitation, and invoking the Franck-Condon principle, the electronic excitation is much more rapid than the dielectric relaxation time of the solvent, so the reaction field of the Franck–Condon excited state R_1' is given by Equation 1.45, in which the solvent still experiences the ground state solvent dipole moment, and the high frequency dielectric constant ($=n^2$) is used in place of ε:

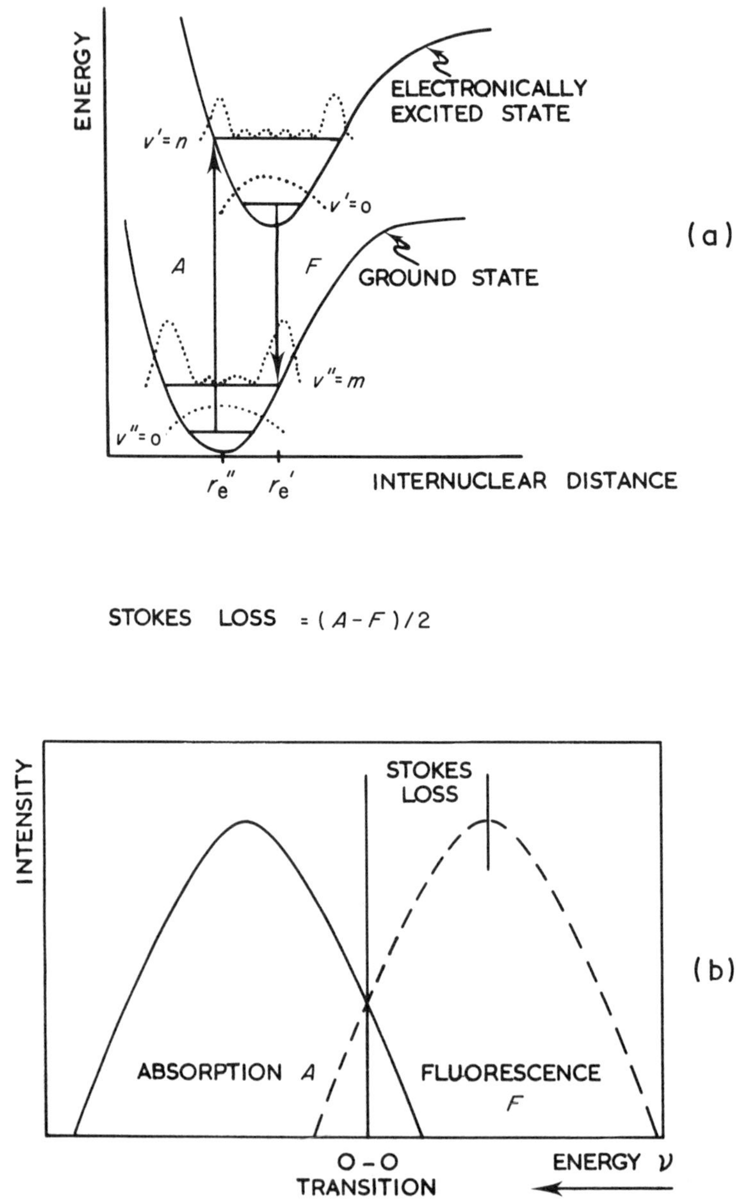

Figure 1.7 Stoke's loss. (a) Potential energy surfaces for pseudo-diatomic molecule showing most probable transitions. (b) Resulting absorption and emission spectra. (No vibrational structure is shown, although this can be encountered in many polyatomic molecules.)

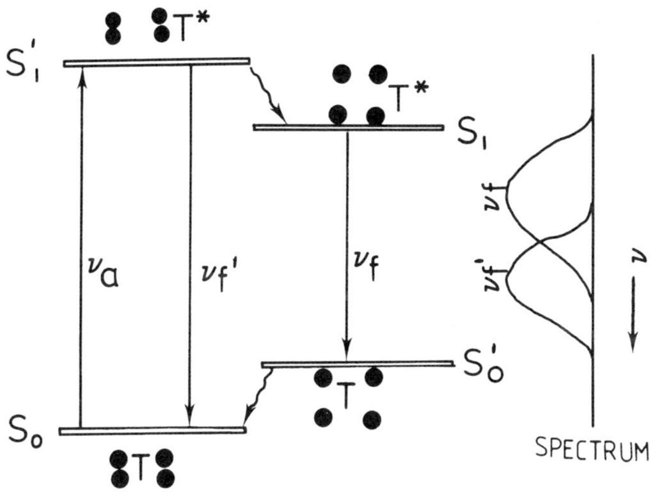

Figure 1.8 Schematic of solvent relaxation, with resulting fluorescence spectral shift for solute molecule T.

$$R_1' = \frac{2\mu_0}{a^3}\frac{(n^2-1)}{(2n^2+1)}. \tag{1.45}$$

If the decay time of the excited state of the solvent exceeds the dielectric relaxation time, then the equilibrated excited state system of dipole moment μ', has a reaction field given by:

$$R_1 = \frac{2\mu'}{a^3}\frac{(\varepsilon-1)}{(2\varepsilon+1)}. \tag{1.46}$$

Following fluorescence, the non-equilibrium ground state has a reaction field R_0', given by:

$$R_0' = \frac{2\mu'}{a^3}\frac{(n^2-1)}{(2n^2+1)} \tag{1.47}$$

Manipulation of these equations shows that the difference in energy of the O–O transitions in absorption and emission is given by Equation 1.48.

$$\Delta\bar{\nu} = \frac{2(\mu'-\mu_0)^2}{hca^3}\left[\frac{(\varepsilon-1)}{(2\varepsilon+1)} - \frac{(n^2-1)}{(2n^2+1)}\right]. \tag{1.48}$$

For non-polar solvents $\varepsilon \simeq n^2$, and no shifts are observed. For polar solvents shifts can be large. Therefore fluorescence spectral position can be very sensitive to the environment of the fluorescent molecule, and such observations are for example widely used in biological studies to identify different sites in heterogeneous systems such as cell membranes. Where solvent relaxation is slow compared to the decay time of the fluorescent molecule, time-dependent fluorescence spectra will be observed. For example, it is shown in Fig. 1.9 that the time-resolved (gated) spectra of the toluidino-naphthalene sulphonate (TNS) molecule (much used as a probe in biological membranes) are markedly dependent upon the time elapsing after excitation, clearly demonstrating the timescale of solvent relaxation.

Due to the nature of some excited and ground states, *specific* solvent effects may sometimes be observed. For example in states arising from $n \rightarrow \pi *$ excitations, the absorption bands undergo blue shifts in polar solvents relative to those observed in non-polar solvents, due to the binding of the non-bonding electrons to the solvent. Where there are two excited states of differing electronic configuration in close proximity, change of solvent polarity can invert the ordering of the excited states, leading to a change in the nature of the fluorescence, which can clearly influence spectral profile, decay time and yield. In mixed solvents, or in heterogeneous systems, level inversion can lead to time-dependent spectra, and complex fluorescence decays.

An example is the aminonaphthalene molecule, widely used as the basic chromophore of many probes in biochemistry, for which time-resolved spectra are shown in Fig. 1.10 (Meech *et al.*, 1983). Results can only be

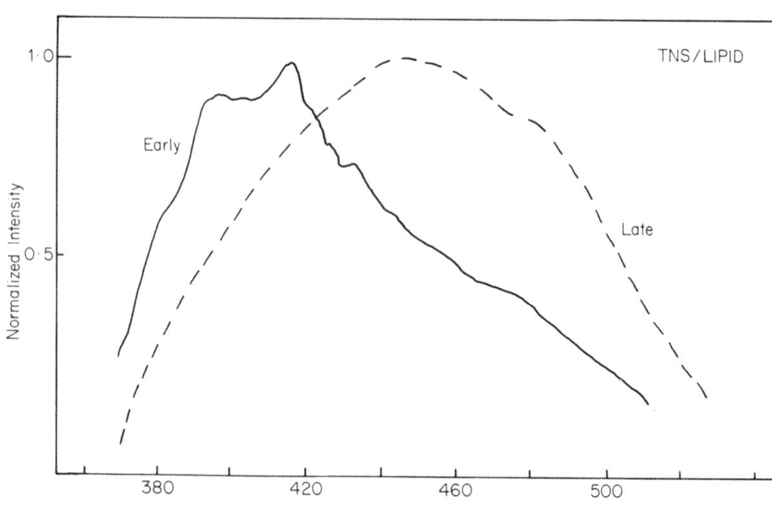

Figure 1.9 Time-resolved spectra of the probe molecule TNS (see text) in a lipid bilayer of egg phosphatidylcholine. Early spectrum at $t = 0$ gate width 2.5 ns, late spectrum at $t = 31$ ns, gate width = 2.5 ns (after Phillips *et al.*, 1979).

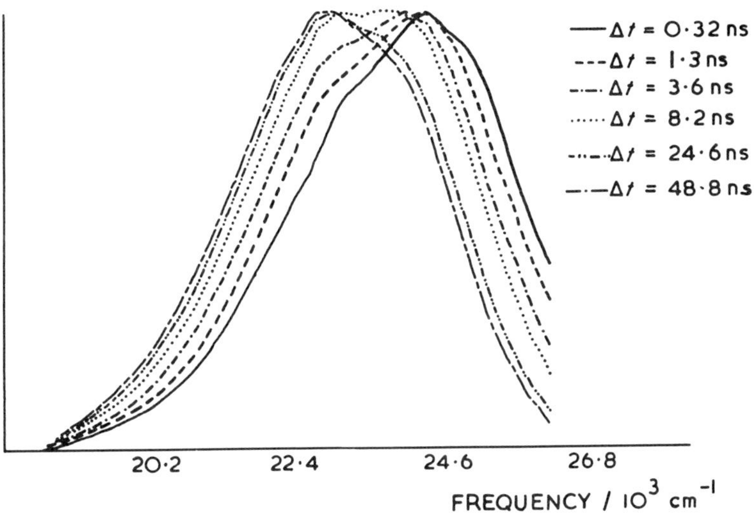

Figure 1.10 Time-resolved spectra of amino-naphthalene in butanol at 190 K. "Gated" spectra taken after delays shown (after Meech *et al*., 1983).

interpreted in terms of two relaxation mechanisms, one an intramolecular electronic reorganization, increasing the charge-transfer nature of the emitting state in polar media (see Fig. 1.11), and the second, a *solvent* dipole relaxation about the excited state dipole moment. The relative positions of the lowest two states of this chromophore as a function of solvent, time and viscosity are summarized in Fig. 1.11 (Meech, 1983). That such a complex system should yield non-exponential fluorescence decays is not perhaps surprising.

It must not be forgotten that the viscosity of the medium can play a role in determining decay characteristics. In the case of anthracene and 9-methylanthracence, for instance, which exhibit only small spectral changes in hydrocarbon solvents of varying viscosity, measured decay times vary as shown in Figs 1.12 and 1.13 (Blatt *et al*., 1981). The dependence in each case is different, but pronounced, and is attributable to different viscosity dependences of Franck–Condon factors for non-radiative decay of the molecules.

As a further specific example of solvent effects upon measured decay times, we can cite one example of the very widely met problem of acid–base equilibria in β-carboline (Sakurovs and Ghiggino, 1982). Various acid–base equilibria, as shown in Fig. 1.14, are possible. The measured total fluorescence decay times vary from $\tau = 22.0$ ns at pH $= 3.6$ to $\tau = 1.60$ ns at pH $= 14$ corresponding to emission from the cation, $\lambda_{max} \sim 450$ nm at pH $= 3.6$ to

(a)

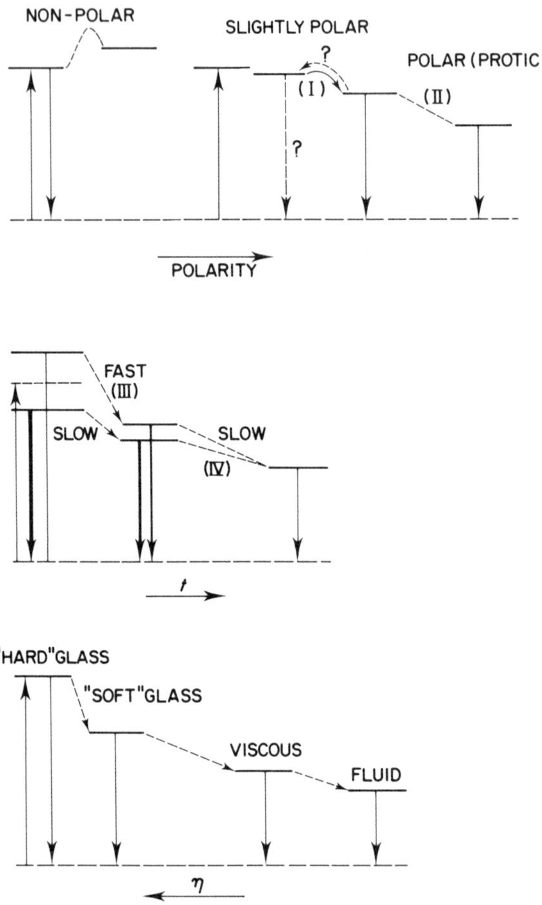

Figure 1.11 State diagrams for aminonaphthalene in solvents as function of (a) solvent, (b) time, and (c) viscosity (after Meech *et al.*, 1983).

that from the Zwitterion, $\lambda_{max} = 510$ nm at pH = 14. Clearly in such complex systems identification of emitting species must be approached with care.

1.4 Polarization of Electronic Transitions

The transition dipole moment operator $\hat{\mathbf{M}}$ contained in Equation 1.1 is a vector quantity; in other words the change in position of the electron in the transition is in a fixed direction with respect to some system of coordinates in which the molecular frame is fixed. The dipole moment operator is resolvable

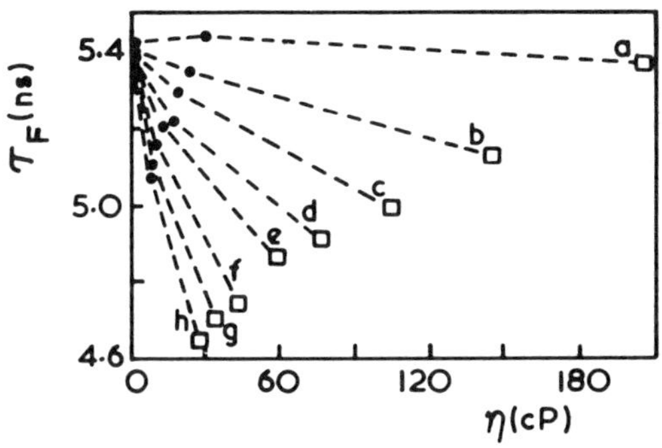

Figure 1.12 Viscosity dependence of fluorescence decay time of anthracene in hexane. ●, oil A; and □, oil B at (a) 20°C, (b) 25°C, (c) 30°C, (d) 35°C, (e) 40°C, (f) 45°C, (g) 50°C and (h) 55°C. [Reprinted with permission from Blatt *et al.* (1981). Copyright 1981 American Chemical Society.]

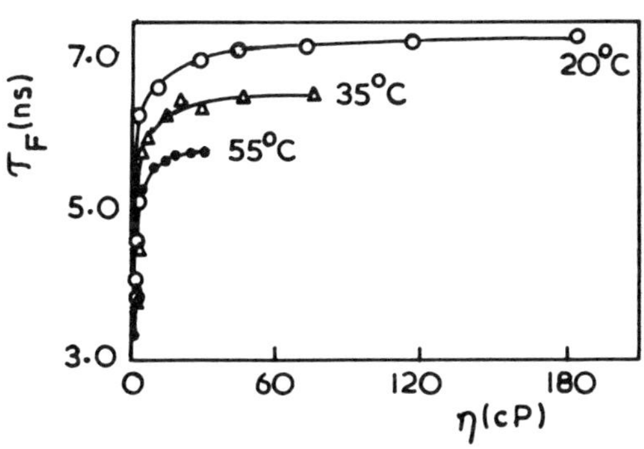

Figure 1.13 Viscosity dependence of 9-methylanthracene fluorescence decay times. [Reprinted with permission from Blatt *et al.* (1981). Copyright 1981 American Chemical Society.]

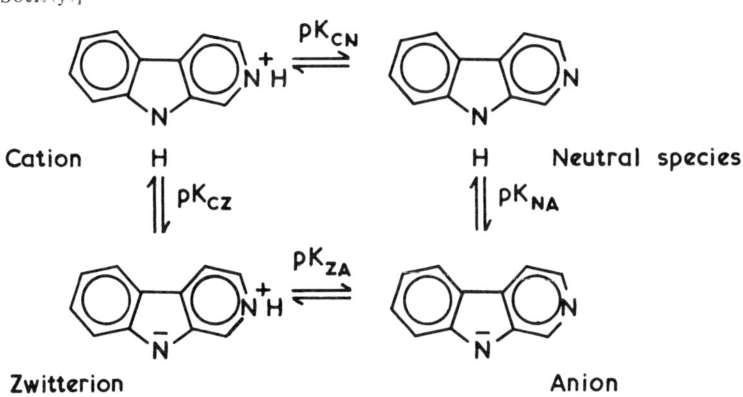

Figure 1.14 Acid–base equilibria and pKa's for B-carboline (after Sakurovs *et al.*, 1982).

in three directions such that, for any transition, only one component will be finite.

$$\hat{\mathbf{M}}^2 = \hat{\mathbf{M}}_x^2 + \hat{\mathbf{M}}_y^2 + \hat{\mathbf{M}}_z^2 \tag{1.49}$$

Hence if a perfectly ordered system such as a fixed molecular single crystal absorbs plane-polarized light, there will generally be one orientation of the crystal axes with respect to the plane of the polarization which maximizes the absorption probability. Normally fluorescence involves the transition between the same two states as are observed in absorption (i.e., the light emitted as the same directional properties with respect to the electric vector as has the light absorbed).

In two-dimensional representation the fluorescence intensity component parallel to the plane of polarization of the electric vector of the incident radiation (the reference direction), $I_{\parallel}$ and that perpendicular to the reference direction $I_{\perp}$ varies with θ, the angle between the reference direction and direction of the optical transition moment on a laboratory scale, since the molecule is rigid. The exact magnitude of $I_{\parallel}$ and $I_{\perp}$ as a function of θ can be traced out in the circular diagram shown in Fig. 1.15.

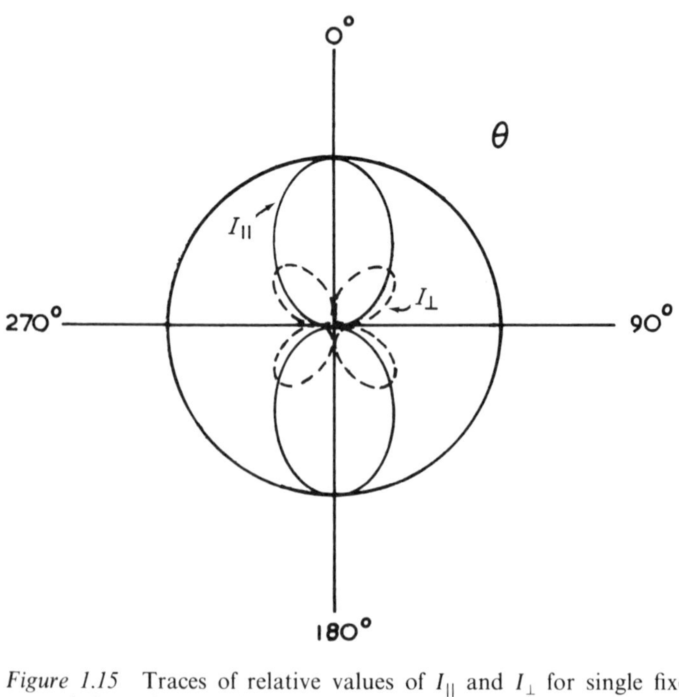

Figure 1.15 Traces of relative values of $I_{\parallel}$ and $I_{\perp}$ for single fixed chromophore rotated through angle θ with respect to vertically polarized exciting light. At $\theta = 0$, transition moment is parallel to the plane of exciting light.

It should be noted that this diagram has been constructed for a fixed single molecule, but is clearly also appropriate for a perfectly ordered fixed system with transition moments of all absorbing chromophores parallel. If, as usual, the degree of polarization P is defined as

$$P = \frac{I_{\parallel} - I_{\perp}}{I_{\parallel} + I_{\perp}}, \tag{1.50}$$

it can be seen that it will vary with θ for this system between the limits of $+1$ and -1 between $\theta = 0$ and $\theta \to 90°$. For a perfectly random array of absorbing chromophores, it is easily shown that for a parallel (fluorescent) emission, P has the value of $+0.5$ independent of θ, whereas for a perpendicular transition, $P = -0.33$.

In fluid media, this analysis is only appropriate if the average rotational relaxation time τ_r is very long compared with the fluorescence decay time, τ_F, ($\tau_r \gg \tau_F$). If the opposite is true, i.e., $\tau_F \gg \tau_r$, all anisotropy introduced in the sample upon excitation is lost through the rotational relaxation prior to emission. If τ_r and τ_F are comparable, then partial relaxation will occur on the same timescale as emission of fluorescence, and analysis of the time-dependence of the fluorescence anisotropy can yield information concerning the rotational motion. The emission anisotropy r is defined as

$$r(t) = \frac{I_{\parallel}(t) - I_{\perp}(t)}{I_{\parallel}(t) + 2I_{\perp}(t)} \tag{1.51}$$

and at time zero, r_0 is given by

$$r_0 = \frac{1}{5}(3\cos^2\delta - 1), \tag{1.52}$$

where δ is the angle between absorption and emission dipoles, which for fluorescence, is usually zero. In general,

$$r(t) = r_0 M_2(t) = r_0 \frac{\langle 3\cos^2\theta(t) - 1\rangle}{2}, \tag{1.53}$$

where $M_2(t)$ is the orientation autocorrelation function, and $\theta(t)$ the angle through which the emission transition moment turns between time zero and t.

For a rigid sphere, Einstein showed that $r(t)$ is a single exponential function.

$$r(t) = r_0 e^{-6D/\tau}, \tag{1.54}$$

where D is the rotational diffusion coefficient.

The time-averaged value of $r(t)$ is

$$\bar{r} = \frac{\int_0^\infty r(t) I(t)\,\mathrm{d}t}{\int_0^\infty I(t)\,\mathrm{d}t}. \tag{1.55}$$

This leads to the Perrin relationship:

$$\frac{1}{r} = \frac{1}{r_0}\left(1 + \frac{3\tau_F}{\tau_r}\right), \tag{1.56}$$

or, expressed in terms of degree of polarization:

$$\left(\frac{1}{P} - \frac{1}{3}\right) = \left(\frac{1}{P_0} - \frac{1}{3}\right)\left(1 + \frac{3\tau_F}{\tau_r}\right). \tag{1.57}$$

Therefore if the decay time τ_F of a probe molecule is known, a measure of the single rotational relaxation parameter τ_r can be made if the time averaged degree of polarization is measured (using continuous excitation), and P_0 is known. In practice P_0 is not known, and use is made of the relationship $\tau_r = 3V\eta/kT$ where η is the viscosity of the medium. Measurement of P in solvents of different viscosity then permits evaluation of V and P_0. A typical result of such an experiment is shown in Fig. 1.16, for a copolymer system tagged with anthracene.

There are, however, many uncertainties in the procedure, and it seems clear that direct measurement of the time dependence of emission anisotropy through $I_{\parallel}(t)$ and $I_{\perp}(t)$ is preferable, particularly since the general form of $r(t)$ will only rarely conform to a single exponential component. This area is discussed fully in Chapter 8 of this volume, but a simple example is shown in Fig. 1.17 of measurement of $I_{\parallel}(t)$ and $I_{\perp}(t)$, from which $r(t)$ can be constructed through Equation 1.51. Unusually, in this case, $r(t)$ is not strongly time-

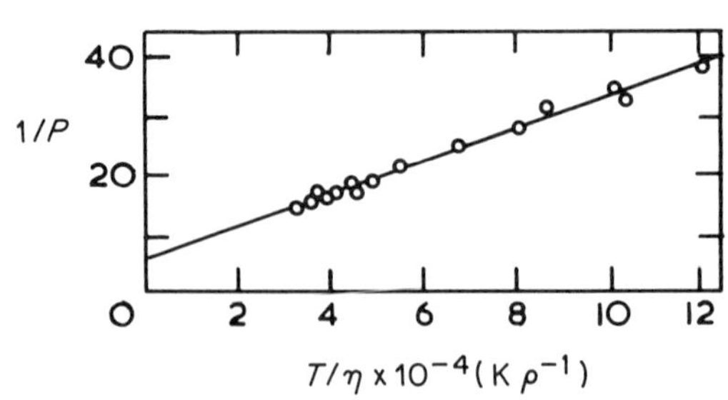

Figure 1.16 Perrin plot of reciprocal fluorescence polarization against T/η for a copolymer (P-chlorostyrene-9-vinylanthracene) in toluene, where P is the degree of polarization T is the temperature in degrees Kelvin, and η is the viscosity of the medium (after North and Soutar, 1972).

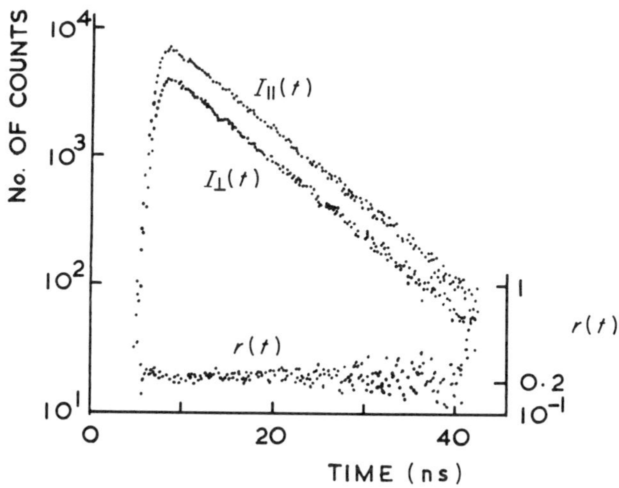

Figure 1.17 Parallel ($I_{\|}(t)$) and perpendicular ($I_{\perp}(t)$) polarized components of the fluorescence decay for DPH dispersed in a nematic liquid crystal. The emission anisotropy $r(t)$ is also shown (after Palmer *et al.*, 1974).

dependent, indicating the substantial order in the system under study, a nematic liquid crystal, and the lack of freedom of motion of the probe.

1.5 Measurement of Fluorescence Decay Times

The preceding discussion will have served, it is hoped, to illustrate some of the many needs for accurate measurement of fluorescence decay times. We shall now outline briefly methods available for such measurement.

1.5.1 Modulation methods

The first method employed directly for the measurement of decay times was that of phase-modulation, described as long as fifty years ago. In this technique, a sample is excited by a light source, modulated (usually sinusoidally) at between 10 and 50 MHz, and reliance is placed upon the measurement of the phase angle between detected fluorescence and the exciting light, δ, and the amplitude of the modulated signal. Such a system is illustrated in Fig. 1.18, shown for use in measuring polarized components of fluorescence. For an exponentially decaying signal, the relationship between δ, and the modulation frequency, v, is

$$\tan \delta = 2\pi v \tau. \tag{1.58}$$

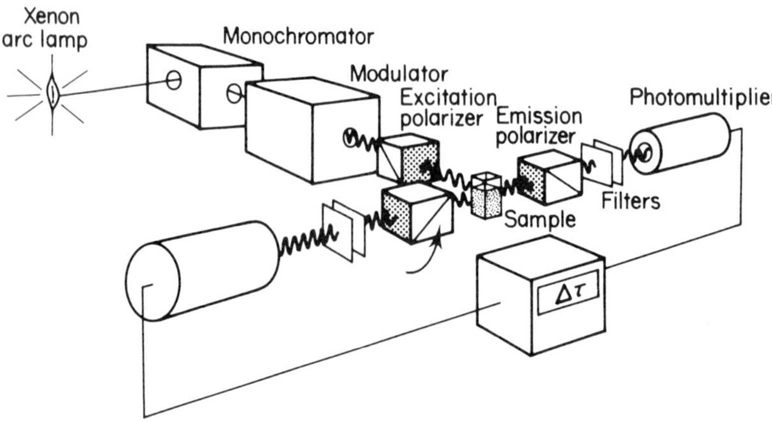

Figure 1.18 Schematic diagram of phase-modulation fluorimeter, used in this arrangement for anisotropy measurements (after Cossins, 1981).

This deceptively simple expression forms the basis of commercial instruments, such as that manufactured by SLM Inc., and the method has gained a considerable number of users. Its drawbacks include the stringent requirement that spurious modulations be excluded, and the fact that, until recently, analysis for anything other than a single component was almost impossible. However, there have been some remarkable improvements recently, which render the analysis of multiple components much more simple, and also permit the recording of time-resolved fluorescence spectra. We refer the reader to a system developed by Gugger and Calzaferri (1980), based upon laser excitation, which is capable of high resolution, and in which decay times as short as 10 ps could in principle be measured. The apparatus is shown in Fig. 1.19, and is essentially a dual beam instrument of high resolution. It may well be that in future years more reliance will be placed upon such phase-modulation techniques, but at present, it seems that pulsed techniques still command the loyalty of the majority of users.

A method related to phase-modulation has been used to demonstrate that lifetimes as short as 200 ps can be measured using the mode noise in a free running argon ion laser to produce variations in the excited state population of a fluorophore. Measurement of the r.f. power spectrum of the resulting fluctuations yield the excited state lifetime. Mode noise contains very high frequency fluctuations which the excited state population cannot follow because of its finite lifetime, and therefore these components will be absent from the r.f. spectrum of the fluorescence fluctuations. The fluorescence process thus acts like a low-pass exponential filter, and comparison of the

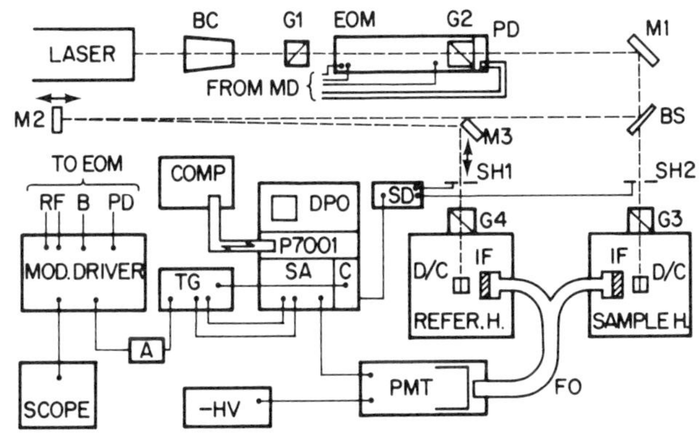

Figure 1.19 Experimental layout for laser-excited modulation experiment for fluorescence decays (after Gugger and Calzaferri, 1980). ---, laser beam; ○———○, electrical connections. Optics: BC, beam collimator; G1–G4, Glan prism polarizers; M1–M3, mirrors; BS, beam splitter; IF, interference filter; FO, fibre optics. Electronics: DPO, digital processing oscilloscope; P7001, processor of the DPO; COMP, minicomputer; SA, spectrum analyser plug in; C, counter plug in; TG, tracking generator; A, amplifier; SD, shutter driver; PMT, photomultiplier; HV, high voltage supply. Modulator: EOM, electro-optical modulator; PD, photodiode assembly; MD, modulator driver; RF, output and termination of r.f. driving voltage; B, bias voltage. Housings: REFER. H., reference chamber; SAMPLE H., sample chamber; D/C, diffuser or cuvette.

fluorescence power spectrum with that of the source provides the decay time data, as the following treatment demonstrates.

For a fluorophore, the intensity decaying exponentially as in Equation 1.59 has a spectral power density in the frequency domain

$$I(t) = I_0 \exp(-t/\tau), \tag{1.59}$$

found by Fourier transformation to be

$$P_F(\omega) = \tau^2/(1 + \tau^2\omega^2). \tag{1.60}$$

The frequency-domain signal is measured in this case at the discrete frequencies at which laser mode beats occur. The time domain representation of these beat frequencies is

$$A(t) = \sum_n A_n \cos(\omega_n t), \tag{1.61}$$

where A_n is the amplitude of the beat at frequency ω_n. Fourier transformation of this yields

$$P_s(\omega) = \sum_n A_n \delta(\omega - 2\pi f_0 n). \tag{1.62}$$

The observed noise spectrum of a fluorescence signal excited by a laser with the beat frequency pattern of Equation 1.61 is then the product of Equations 1.60 and 1.62, or

$$P'_F(\omega) = \sum_n A_n^2 \tau^2 \delta(\omega - 2\pi f_0 n)(1 + \omega^2\tau^2)^{-1}. \tag{1.63}$$

The time-averaged beat frequency of a laser is stable over long periods of time, so that it is possible to obtain a spectrum (from scatter) proportional to Equation 1.62 (Fig. 1.20(a)). Division of the observed fluorescence fluctuation spectrum of Fig. 1.20(b) by such a pattern yields the frequency domain lifetime directly through Equation 1.63. This is shown for Rhodamine B in Fig. 1.21 to yield a decay time of 3.2 ns.

There are limits on this method which are imposed by the large spacing of

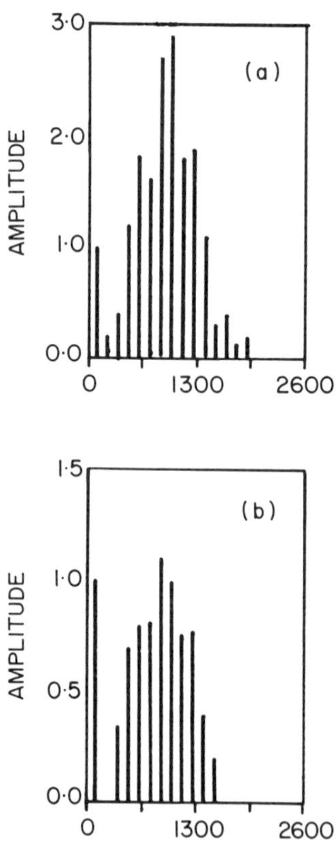

Figure 1.20 Time-resolved power spectrum of (a) fluctuations in scattered radiation and (b) fluctuations in fluorescence of Rhodamine B, normalized to the 130 MHz beat of the exciting Ar^+ in laser (after Hieftje *et al.*, 1977).

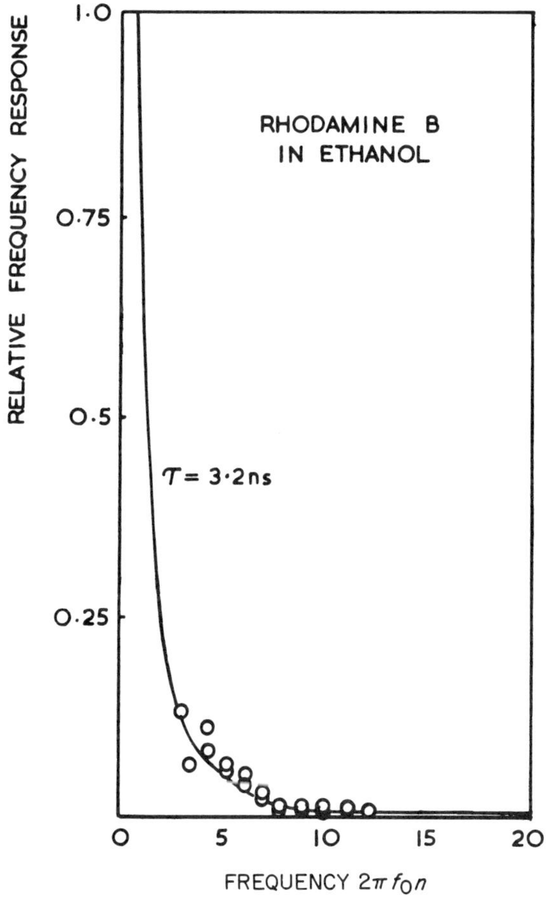

Figure 1.21 Frequency response of fluorophore Rhodamine B normalized at the first beat frequency (after Hieftje *et al.*, 1977).

laser mode-noise peaks (130 MHz), setting an upper decay time limit of 7 ns, by detector noise background setting a lower limit of 200 ps, of cost (of a spectrum analyser) and of difficulties in analysis of multicomponent fluorescence. However, as any source could in principle be used to measure a noise spectrum, the method is worth further consideration.

1.5.2 Pulse-sampling methods

Early techniques for fluorescence decay measurements used a gaseous discharge lamp, of a few nanoseconds duration to excite the sample, and

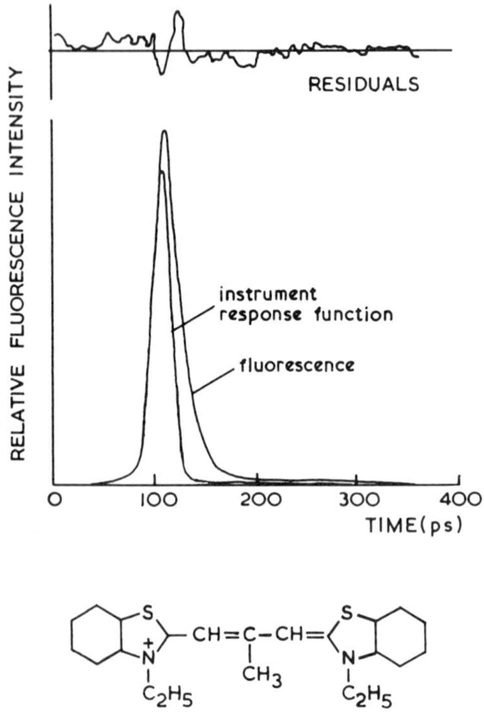

Figure 1.22 Fluorescence decay of dye shown measured by the second harmonic of Nd^{3+} laser excitation, using streak camera detection (from Winkworth *et al.*, 1982).

detected fluorescence was monitored by a photomultiplier for direct display on a fast sampling oscilloscope (Meserve, 1971; North and Soutar, 1972). Although simple, the technique is insensitive, and prone to difficulties when attempts are made to measure decay times below about 10 ns. A more recent and elegant extension of the technique is to use mode-locked picosecond solid-state lasers for excitation with streak-camera detection of fluorescence. With these methods, decay times from 10 ps to, perhaps, 1 ns can be measured, and Fig. 1.22 shows a typical decay.

Although very high temporal resolution is possible, the apparatus is costly, the range limited, and measurements are still far from routine. The advent of a new generation of synchroscanning streak cameras, while not diminishing the costs of these measurements, promises to make experimental techniques easier, and to render accessible both longer and shorter timescales, particularly the latter when used in conjunction with sub-picosecond mode-locked dye laser systems for excitation.

1.5.3 Time-correlated single photon counting

The first use of a time-to-amplitude converter as an approach to the measurement of fluorescence decay seems to have been that of Koechlin (1961). By the 1970s, single photon counting had become standard practice although refinements continue to this day. The method has been the subject of a number of reviews, those of Ware (1971), Yguerabide (1972), Poultney (1972), Knight and Selinger (1973), Isenberg (1975), Badea and Brand (1979), and Birks and Munro (1967) being especially worthy of note. Since the remainder of this volume is concerned exclusively with the technique, no further discussion will be given here.

1.5.4 Pump and probe techniques

The ultimate time resolution of streak cameras may be sub-picosecond, and they thus represent a costly but effective means of detection of very fast decays. Phase-modulation methods may ultimately be used down to about 10 ps, although at present detection limits for reliable measurements are rather longer. Time-correlated single photon counting, being based upon a photomultiplier detection, has a limit perhaps of about 20 ps (Ghiggino *et al.*, 1980).

Improvements over this may be achieved using laser excitation in some pump and probe configuration. An original early example of this technique is shown in Fig. 1.23. Here fluorescence is excited by the second, or fourth

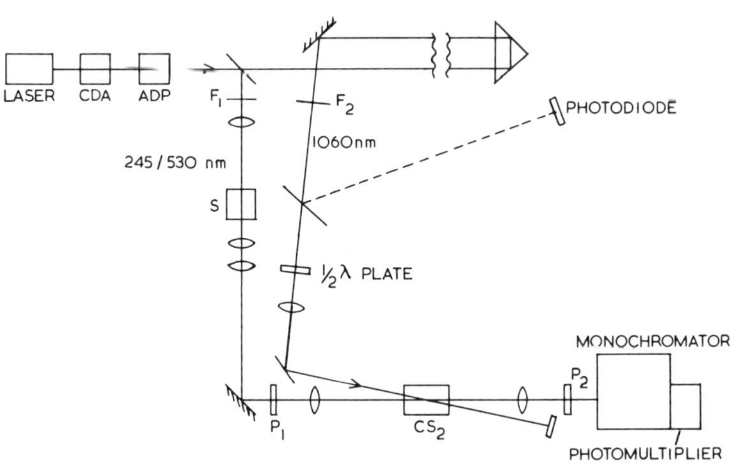

Figure 1.23 Laser arrangement for picosecond measurements of fluorescence decay using CS_2 shutter (Porter *et al.*, 1974).

harmonic of a neodymium in glass laser, and a Kerr cell shutter, (CS_2 solution) is opened by a delayed fundamental pulse from the same laser. Timing is thus achieved by the delay of the interrogating pulse, and this can be achieved accurately, the monochromator and photomultiplier combination being used solely to monitor total intensity at the time the shutter is opened. The Kerr effect consists of rotation of the plane of polarization of light through 90° in the presence of a high electric field. In this experiment, the laser fundamental is of sufficient intensity to achieve this, and by placing the Kerr cell between crossed polarizers, the shutter can be made to operate.

A recent very elegant pump and probe method termed up-conversion has been devised which permits picosecond fluorescence decay and anisotropy measurements to be made. Up-conversion refers to "the non-linear phenomenon of sum-frequency generation in anisotropic crystals", and the apparatus due to Beddard *et al.* (1981) is illustrated in Fig. 1.24. The method relies upon the generation in the anisotropic crystal of a signal of frequency ω_3, the sum frequency of fluorescence ω_2 and laser pulse, ω_1. ω_3 can be distinguished spatially and in frequency, and by delaying the interrogating laser pulse in time, a profile of fluorescence decay can be built up.

The instantaneous power of the sum frequency in a crystal of length L is given by

$$P_\omega(L) \approx L^2 P_1 P_2 \frac{\omega_3}{\omega_1}, \tag{1.64}$$

where the subscripts 1, 2 and 3 refer to the gating pulse, the fluorescence and the sum frequency (given by $\omega_3 = \omega_2 \pm \omega_1$), respectively. By appropriate phase matching the difference frequency can be generated which can be used to measure fluorescence in the ultra-violet.

The fluorescence intensity I_s generated by the exciting pulse has a time profile given by

$$I_s(t) = \int_{-\infty}^{t} I_f(t')L(t-t')\,dt', \tag{1.65}$$

where t' is the time delay, $L(t)$ is the laser pulse shape and $I_f(t)$ is the molecular fluorescence intensity at time t. The signal $S(t)$ after passage through the crystal is given by

$$S(t) = \int_{-\infty}^{\infty} I_s(t')L(t-t')\,dt', \tag{1.66}$$

which combined with Equation 1.65 gives

$$S(t) = \int_{-\infty}^{t} I_f(t')A(t-t')\,dt', \tag{1.67}$$

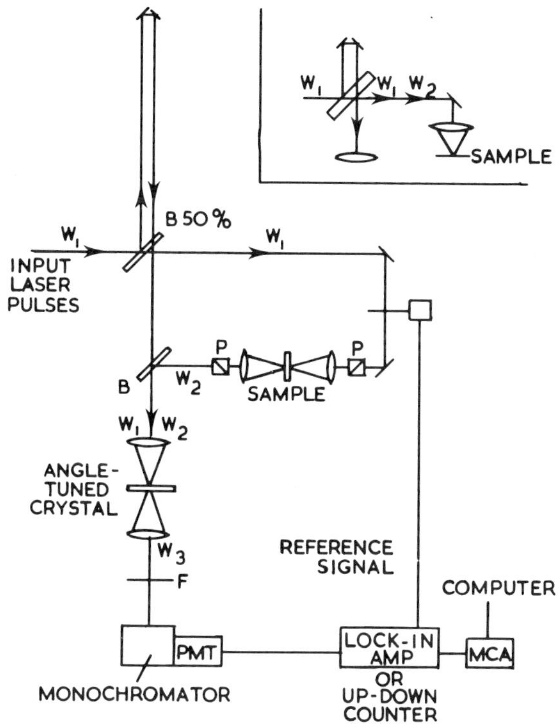

Figure 1.24 Experimental arrangements for time-resolved fluorescence up-conversion: F, filters; P, polarizers; C, a sectored disc chopper connected to a lock-in amplifier or photon counter; W_1, laser beam; W_2, fluorescence. The crystal is $LiIO_3$ (path length, 1 nm). The sample is contained in a glass cell of path length 1 nm mounted perpendicularly to the exciting beam and spun about an axis parallel to the beam, and the fluorescence is collected from the front face of the cell along the exciting beam axis. In an alternative method (inset) the sample is flowed through a cell or pumped through a nozzle to form a jet, and fluorescence is collected at 180° to the exciting beam. (After Beddard *et al.*, 1981.)

where $A(t)$ is the measured autocorrelation function of the laser pulse. As the autocorrelation function is readily measured, the signal can be fitted to the desired decay function at the shortest times with no *a priori* knowledge of the laser pulse shape. This offers a significant advantage over transient absorption methods where it is necessary to make assumptions about the laser pulse shape to facilitate curve fitting. When used to measure fluorescence decays, it offers several advantages over time-correlated single photon counting. As well as greatly improved time resolution, it is possible to detect fluorescence in the far red and infra-red without using special red-sensitive photomultipliers; for example fluorescence at 900 nm can be mixed with laser light

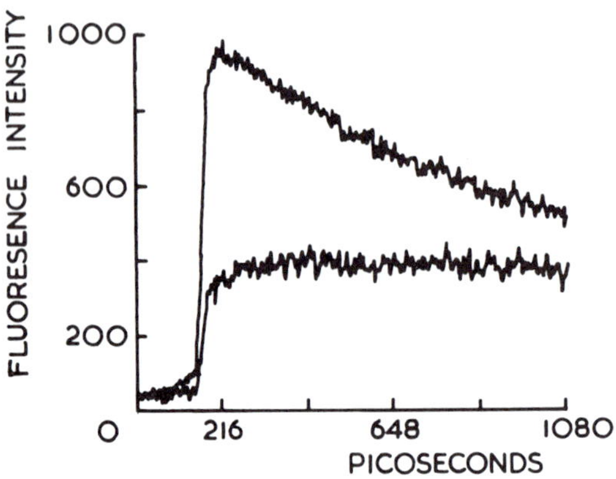

Figure 1.25 $I_{\parallel}$ (upper), and $I_{\perp}$, (lower) curves for fluorophore rotating in alcohol solvent, up-conversion experiment (after Beddard *et al.*, 1981).

at 600 nm to generate a sum frequency at 360 nm. This advantage, however, bestows necessarily a corresponding disadvantage; that is up-conversion of fluorescence towards the blue end of the spectrum is rendered difficult owing to the absence of efficient crystals operating in the far ultra-violet.

The method can easily be adapted to make measurements of fluorescence anisotropy, and the quality of data produced is exemplified by Fig. 1.25. The precision of decay times measured using this arrangement is set by the optical delay time, which for convenience is of the order of 1 ps or less. The lower limit is set by the excitation pulse profile, which with present generation lasers is about 2 ps.

References

Badea, M. G. and Brand, L. (1979). *Meth. Enzymol.*, **61**, 378–425.

Beddard, G. S., Carlin, S., Harris, L., Porter, G. and Tredwell, G. S. (1978). *Photochem. Photobiol.* **27**, 433–438.

Beddard, G. S., Doust, T. and Porter, G. (1981). *Chem. Phys.* **61**, 17–23.

Beavan, S. W., Hargreaves, J. S. and Phillips, D. (1979). *Adv. Photochem.* **11**, 207–303.

Birks, J. B. (1970). "Photophysics of Aromatic Molecules", pp. 301–305. Wiley-Interscience, New York.

Birks, J. B. and Munro, J. H. (1967). *Prog. Reac. Kinet.* **4**, 239–303.

Blatt, E., Treloar, E. F., Ghiggino, K. P. and Gilbert, R. G. (1981). *J. Phys. Chem.* **85**, 2810–2813.

Cossins, A. R. (1981). In "Fluorescent Probes" (Beddard, G. S. and West, M. A., eds), pp. 39–80. Academic Press, London.

Forster, Th. (1946). *Naturwiss.* **33**, 166–171.

Ghiggino, K. P., Roberts, A. J. and Phillips, D. (1980). *J. Phys. E.* **13**, 446–449.
Gugger, H. and Calzaferri, G. (1980). *J. Photochem.* **13**, 295–307.
Hieftje, G. M., Haugen, G. R. and Ramsey, J. M. (1977). *Appl. Phys. Lett.* **30**, 463–465.
Hirayama, S. and Phillips, D. (1980). *J. Photochem.* **12**, 139–146.
Isenberg, J. (1975). In "Biochemical Fluorescence Concepts", Vol. 1 (Chen, R. F. and Edelhoch, H., eds) pp. 43–77. Marcel-Dekker, New York.
Knight, A. E. W. and Selinger, B. K. (1973). *Aust. J. Chem.* **26**, 1–27.
Koechlin, Y. (1961). Thesis, passim. University of Paris.
Meech, S. R. (1983). Thesis. University of Southampton.
Meech, S. R., O'Connor, D., Phillips, D. and Lee, A. G. (1983). *J. Chem. Soc. Faraday Trans. II*, **79**, 1563–1584.
Meserve, E. J. (1971). In "Excited States of Proteins and Nucleic Acids" (Steiner, R. F. and Weinryb, J., eds) p. 57. Plenum, New York.
Nemzek, T. L. and Ware, W. R. (1975). *J. Chem. Phys.* **62**, 477–489.
North, A. M. and Soutar, I. (1972). *J. Chem. Soc. Faraday Trans. II* **68**, 1101–1106.
O'Connor, D. V., Chewter, L. A. and Phillips, D. (1982). *J. Phys. Chem.* **86**, 3400–3404.
Palmer, T. F., Cehelnik, E. D., Cundall, R. B. and Lockwood, J. R. (1974). *J. Chem. Soc. Faraday Trans. II* **70**, 244–252.
Parker, C. A. and Hatchard, V. (1963). *Trans. Faraday Soc.* **59**, 284–290.
Phillips, D., Ghiggino, K. P. and Lee, A. G. (1979). In "Synchrotron Radiation Applied to Biophysical and Biochemical Research" (Castellani, A. and Quercia, U. F., eds) pp. 171–182. Plenum, New York.
Porter, G., Reid, E. S. and Tredwell, C. J. (1974). *Chem. Phys. Lett.* **29**, 469–472.
Poultney, S. K. (1972). *Adv. Elect. Elect. Phys.* **31**, 39–117.
Roberts, A. J., O'Connor, D. V. and Phillips, D. (1981). *Ann. N.Y. Acad. Sci.* **366**, 109–124.
Sakurovs, R. and Ghiggino, K. P. (1982). *J. Photochem.* **18**, 1–8.
Spears, K. G. and Rice, S. A. (1971). *J. Chem. Phys.* **55**, 5561–5581.
Strickler, S. J. and Berg, R. A. (1962). *J. Chem. Phys.* **37**, 814–822.
Ware, W. R. (1971). In "Creation and Detection of the Excited State", Vol. 1A (Lamola, A. A., ed.) pp. 213–326. Marcel-Dekker, New York.
Winkworth, A. C., Osborne, A. D. and Porter, G. (1982). In "Picosecond Phenomena III" (Eisenthal, K. B., Hochstrasser, R. M., Kaiser, W. and Lauterean, A., eds) p. 228. Springer-Verlag, Berlin.
Yguerabide, J. (1972). *Meth. Enzymol.* **26**, 498–578.

2 Basic Principles of the Single Photon Counting Lifetime Measurement

2.1 The Standard Experiment

A schematic diagram of a conventional single photon counting instrument is shown in Fig. 2.1. The single photon counting measurement relies on the concept that the probability distribution for emission of a single photon after

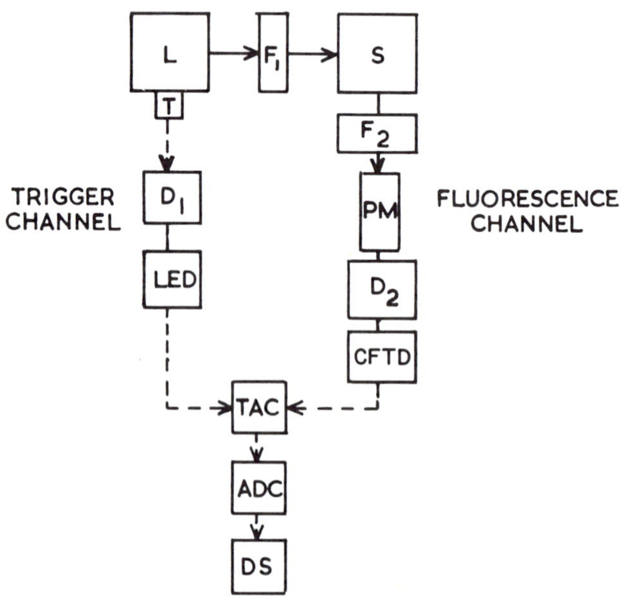

Figure 2.1 Block diagram of a conventional single photon counting apparatus. —— — optical signal; ------ electronic signal; L, excitation source; T, trigger [Antenna, (fibre optic) and photomultiplier tube, etc.]; S, sample holder; F_1,F_2, filter or monochromator; PM, fast photomultiplier tube; D_1,D_2, delay lines; LED, leading edge timing discriminator; CFTD, constant fraction timing discriminator; TAC, time-to-amplitude converter; ADC, analogue-to-digital converter; DS, data store (multi-channel analyser or computer).

an excitation event yields the actual intensity against time distribution of all the photons emitted as a result of the excitation. By sampling the single photon emission following a large number of excitation flashes, the experiment constructs this probability distribution.

With reference to this diagram, the experiment is carried out as follows: the trigger T, which could be a photomultiplier, an antenna pick-up, or a logical sync. pulse from the electronics pulsing the excitation source, generates an electrical pulse at a time exactly correlated with the time of generation of the optical pulse. The trigger pulse is routed, perhaps through a discriminator, to the start input of the time-to-amplitude converter, (TAC) which initiates charging of a capacitor. In the meantime the optical pulse excites the sample, which subsequently fluoresces. An aperture has been adjusted so that at most one photon is "detected" for each exciting event. The signal resulting from this photon stops the charging ramp in the TAC, which puts out a pulse, the amplitude of which is proportional to the charge in the capacitor, and hence to the time difference between START and STOP pulses. The TAC output pulse is given a numerical value in the analogue-to-digital converter and a count is stored in the data storage device in an address corresponding to that number. Excitation and data storage are repeated in this way until the histogram of number of counts against address number in the storage device represents, to some required precision, the decay curve of the sample. If deconvolution is necessary, the time profile of the excitation pulse is collected in the same way by replacing the sample by a light scatterer.

2.2 Details

2.2.1 Statistics

In an ideal time-correlation experiment each photon emitted by the sample as a result of a given excitation flash would be timed and recorded. In fact the response time of the detection equipment and the mode of operation of the TAC render it necessary to time only the first photon in a given time interval after the occurrence of the flash. The statistics of general photon counting have been discussed by Morton (1968); a simpler discussion is given here following Pfeffer *et al.* (1962) and Wahl (1975).

During one excitation cycle an average number of photons, z_i, impinge on the cathode of the fluorescence photomultiplier during the time interval $t_{i-1/2}$ to $t_{i+1/2}$, corresponding to the time width of address i (usually a channel in a multichannel analyser) in the data store. These photons eject an average number w_i of photoelectrons from the photocathode, thus

$$w_i = qz_i, \tag{2.1}$$

where q is the quantum efficiency of the photocathode. The probability of emission of l photoelectrons in the ith time interval is given by the Poisson distribution (Mandel and Wolf, 1965):

$$p_l(i) = \frac{(w_i)^l}{l!} e^{-w_i} \tag{2.2}$$

with

$$\sum_{l=0}^{\infty} p_l = 1 \tag{2.3}$$

Therefore (Koechlin, 1961):

$$p_0(i) = e^{-w_i} \tag{2.4}$$

$$p_1(i) = w_i e^{-w_i} \tag{2.5}$$

$$\begin{aligned} p_{l>1}(i) &= 1 - p_0(i) - p_1(i) \\ &= 1 - e^{-w_i} - w_i e^{-w_i} \\ &= 1 - (1 + w_i)e^{-w_i}. \end{aligned} \tag{2.6}$$

After a large number of excitation cycles (N_E), the number of anode pulses in the ith time interval, N_A, is

$$N_A = N_E[p_1(i) + p_{l>1}(i)]. \tag{2.7}$$

If $w_i \ll 1$, then

$$p_1(i) = w_i \tag{2.8}$$

and

$$p_{l>1}(i) = w_i^2 \ll w_i.$$

Therefore the number of anode pulses, N_A, is proportional to the intensity of the fluorescence at time t_i since

$$\begin{aligned} N_A &\approx N_E(w_i + w_i^2) \\ &\approx N_E w_i \\ &= N_E q z_i. \end{aligned} \tag{2.9}$$

Because the TAC detects only the first photon in a given time interval for a given excitation cycle, N_A is not the number of counts in the ith channel. This number, N_i, is related to N_A by the equation (Coates, 1968):

$$N_A = \frac{N_i}{1 - \dfrac{1}{N_E} \sum_{j=1}^{i-1} N_j} \tag{2.10}$$

Since

$$\sum_j N_j \leqslant N_D,$$

the number of *detected* anode pulses, if $N_D \ll N_E$ it follows that $N_i = N_A$. Consequently the count in channel i is a measure of the fluorescence intensity at time t_i.

Generally, N_D is measured at the output of the TAC and N_D/N_E kept below a certain limit. If N_D is not very much less than N_E, data can be corrected using Equation 2.10 provided that $w_i \ll 1$ (see Section 6.4). Collection at high N_D/N_E ratios need not lead to distorted curves if pile-up inspection is performed (see Section 5.2.5(b)). However, it is simpler and probably just as efficient, when data transfer and analysis are taken into account, to keep the ratio N_D/N_E below a certain value.

It is clear that choice of a value for $N_D/N_E(=F_D)$ will depend on the level of distortion that can be accepted. Yguerabide (1972) has discussed this point; using numerical calculations with Equation 2.10 he concluded that an F_D as high as 0.05 was tolerable since at this count rate distortions in the early channels are negligible, while they amount to only a few per cent in the late channels. However, empirical tests in our laboratory indicate that significant distortions are still present when F_D is as high as 0.05. In fact most workers keep F_D lower than 0.01 or 0.02. As far as pile-up distortions are concerned, the lower the value of F_D the better; therefore, if collection times are short (as for instance in instruments with a cavity dumped laser as excitation source) an F_D of 0.005 or even 0.002 is more usual. It should be pointed out that excitation sources in SPC instruments are comparatively weak in intensity and in many experiments the low level of detected photons, N_D, renders discussion of tolerable F_D values somewhat academic. In conclusion, therefore, when pulse pile-up distortions are to be avoided by limiting the number of detected photons a value of F_D in the region of 0.001 or lower is to be preferred.

2.2.2 Sensitivity, precision and time resolution

When the fluorescence photomultiplier (the PM) is operated in the single photon mode, many of the distortions normally accompanying photomultiplier detection and associated recording circuitry are avoided, since the PM is used merely to time the arrival of a photon at the photocathode. In addition, amplitude jitter in the exciting light pulse is irrelevant provided that the pulse shape remains constant. The SPC technique therefore has all the advantages of digital signal processing and in addition, at least for decay time measurements (see Chapter 7 for spectroscopic measurements), is concerned only with the times of occurrence of events and not with intensities.

The effect of PM noise is greatly reduced in many experiments by the mode of operation of the TAC. When the TAC has received the START pulse it remains "dead" for some fixed time before the voltage sweep (see Section 5.2.5) is initiated. During the voltage sweep, which can continue for a time set by the operator (called the TAC range), it can accept at most one signal resulting from either noise or fluorescence. At the end of the TAC range, or after receiving the STOP signal, it remains "dead" until the next START signal arrives. Since the TAC range is usually set to span the maximum in the photon distribution function, data collection occurs at times of enhanced signal-to-noise ratio. Therefore, a PM noise count rate of say 1000 Hz after the CFTD discriminator may be reduced by a factor of 100 after the TAC when the trigger pulse is used to start the voltage sweep. Otherwise (see Section 2.2.6) noise reduction will be less. Generally background noise offers no serious difficulties since the PM is chosen with low noise characteristics and correction for background is quite straightforward (Section 2.2.5).

Sensitivity in SPC instruments has sometimes been reported in terms of the reciprocal of the product (optical density in 1 cm path length × quantum yield) for the sample, the decay time of which has been measured. Figures as high as 10^6 and 10^7 have been reported for flash lamp (Ware, 1971; Yguerabide, 1972) and laser-based systems (Harris *et al.*, 1977). Unless the degree of light attenuation in monochromators and filters is specified, this figure is virtually meaningless. Another approach is suggested by Knight and Selinger (1973) who used a concrete example. They detected (TAC conversions) 10 photons per second on a TAC range of 80 ns when the vapour over crystalline anthracene at 20°C ($P = 10^{-4}$ Torr) was excited at 337.1 nm. Slit widths and mode of filtering, however, were not specified. Haugen and Lytle (1981) have evaluated the sensitivity of their instrument as (counts actually measured)/(counts that should be measured), the latter calculated from known quantities such as number of excitation photons per second, optical density of sample, quantum yield, f-number of optics, reflections and air glass interfaces, spatial apertures, quantum efficiency of PM photocathode and transmission of filters at the appropriate wavelengths. (They achieved 2.0×10^{13} counts $\text{M}^{-1}\,\text{s}^{-1}$ out of a possible 2.8×10^{13}.) This approach, while rigorous, is difficult to follow in some experiments. We suggest, therefore, that if sensitivity is to be specified it should be done in terms of optical density and quantum yield with *full* details of slit widths and filters also specified.

In a fluorescence decay curve collected by the SPC technique there are two distinct types of noise. One, resulting from photomultiplier dark counts, is a constant background and is easily treated (see Section 2.2.5). The other, a counting error, varies from channel to channel and is of more fundamental importance. In the data store resulting from an SPC experiment there is, in a given time interval (channel), a finite non-zero chance of observing any integral number of counts. The probability of observing any specific number of counts is given by the Poisson probability function (Bevington, 1969), with a mean μ and variance $\sigma^2 = \mu$. Every count in every channel represents an estimate of the mean of a Poisson distribution of counts for that channel. Hence the variance σ^2 equals the number of counts in that channel, N_i, and the uncertainty in the number of counts σ_i is given by

$$\sigma_i = \sqrt{N_i}. \tag{2.11}$$

The implications of this counting error in connection with data analysis are discussed in Chapter 6. Here we can make use of this uncertainty to estimate the number of detected counts that are necessary to record a decay curve with a given precision (Yguerabide, 1972).

If we wish to have a precision of 5% in the number of counts, N_i, in channel i, where the curve has decayed to 1% of its maximum value, then $0.05 = 1/\sigma_i = 1/\sqrt{N_i}$ and $N_i = 400$. Consequently the number of counts in the channel of maximum counts should be 40 000. This figure represents a very rough estimate. Experimentally it will probably be sufficient to accumulate 1 or 2 $\times 10^4$ counts in the maximum for a single exponential decay. The requirements for multiple exponential decays are more severe and depend on the lifetime values and the intensities of each component. For such decays a peak count of at least 3×10^4 should probably be accumulated, but frequently the choice of the acceptable number of counts will be a compromise between high precision and avoidance of distortions resulting from long-term instability in the excitation source. For measurements of the decay of fluorescence anisotropy, peak counts of the order of 10^5 or 10^6 may be advisable because the decay times extracted depend on the subtraction of two decay curves (see Chapter 8).

Many factors contribute to the uncertainties in the lifetimes that are derived from an SPC measurement. These include excitation pulse instability, spread in the photomultiplier transit time, jitter in the electronics and uncertainty in the channel calibration. It is important to realize that the width of the pump pulse profile is irrelevant if the shape of the profile remains absolutely constant. However, no excitation source is perfectly stable and instabilities are usually proportional to the pulse width. Hence it is desirable to have as narrow a pulse width as possible. Similarly, transit time spread will in general be smaller in a PM tube with a shorter transit time; hence the search for faster and faster detection devices. It must also be realized that the SPC experiment is, essentially, an averaging technique with the consequence that the transit time spread (typically 1 to 2 ns for modern tubes) does not lead to the same spread in the measured lifetime. They are proportional but the latter can be made as small as required to a certain limit, given perfect stability in the other experimental conditions, by increasing the number of collected counts. The determination of this limit is basically empirical.

Some authors set it on the basis of the shortest lifetimes they have measured with confidence. Thus Ware (1971) reported a limit of 0.8 ns while Yguerabide (1972) reported 0.5 ns. Others prefer to estimate the time resolution as some fraction of the full width at half maximum (FWHM) of the instrument response function. This function is the curve that results when the pump pulse profile is measured in the way described in Section 2.1 and is a convolution of the true pump pulse profile with the response function of the detector. West (1979) puts this fraction as one tenth for single exponentials and one fifth for double exponentials, whereas one fifteenth is recommended by Grinwald (1976) and Cramer and Spears (1978). On pulsed laser based SPC equipment, instrument response functions can be as low as 75 ps (Murao *et al.*, 1982); lower limits of about 5 ps on measurable lifetimes are therefore set by the one fifteenth criterion. We believe that this figure is unrealistic and that errors of at least 100% should probably be quoted for lifetimes up to 60 ps if the result of a single measurement is given.

The figures for time resolution given in the preceding paragraph are rules of thumb based on authors' confidence in the lifetimes they have measured. If a more exact estimate is preferred, the following experiment (Zimmerman and Cutler, 1975) may be performed. Two pump pulse profiles are measured, a length of time corresponding to the time needed to collect an ordinary decay curve being allowed to elapse between the two measurements. Deconvolution of one pump pulse profile with the other will yield a "decay" time, which is a measure of the time resolution of the instrument. This experiment has yielded values of 18 ps (Zimmerman and Cutler, 1975) for instruments with flash lamp excitation. Koester and Dowben (1978), using a method based on the optical properties of the cavity-dumper in their dye laser excitation source, determined a timing resolution of 25 ps whereas for a similar excitation source using the method proposed by Zimmerman we estimate the resolution at 50 ps.

In spite of these low values, quoted in recent publications, the prospective user should not expect to be able to measure lifetimes less than 500 ps until each component of his instrument has been properly optimized. Even then, constant checks on excitation source stability and electronic jitter must be performed if decay times less than 100 ps are to be believed. Short decay times in two-component decay curves present even more difficulties. These decays should be measured a number of times in order to compensate for random errors. The detection of systematic errors can best be achieved by comparison of the results of SPC analysis with results of independent experiments. We strongly caution against the adoption of models based on extracted lifetimes from two-component decays of less than 100 ps unless the model is justified on other grounds.

2.2.3 Convolution

If the flash of light that excites the sample were infinitely narrow, and if the response of the detection system were infinitely fast, the observed decay curve would represent the true decay, or δ-pulse response, of the sample. We shall refer to this function as $G(t)$.

The form of the observed decay, $I(t)$, when the excitation function, $E(t)$, is not a δ-function can be deduced from the theory of impulse functions and leads to the convolution concept. Convolution, or folding together, occurs because molecules excited by photons at early times are decaying while others are being excited by photons in the tail of the excitation pulse. A simple deduction of the convolution equation is based on the diagram in Fig. 2.2. The pump pulse is assumed to be a sum of δ-pulses of amplitude $E(t')$ at any time t'. Since the number of sample molecules excited at time t' is proportional to $E(t')$ the number at any later time $x-t'$ is proportional to $E(t')G(x-t')$. The total number of excited state molecules at time x, written $[A*](x)$ is then a sum over all times t' preceding time x or, for an infinite sum,

$$[A*](x) \propto \int_0^x E(t')G(x-t')\,dt'. \tag{2.12}$$

We have so far neglected distortions introduced by the detection system.

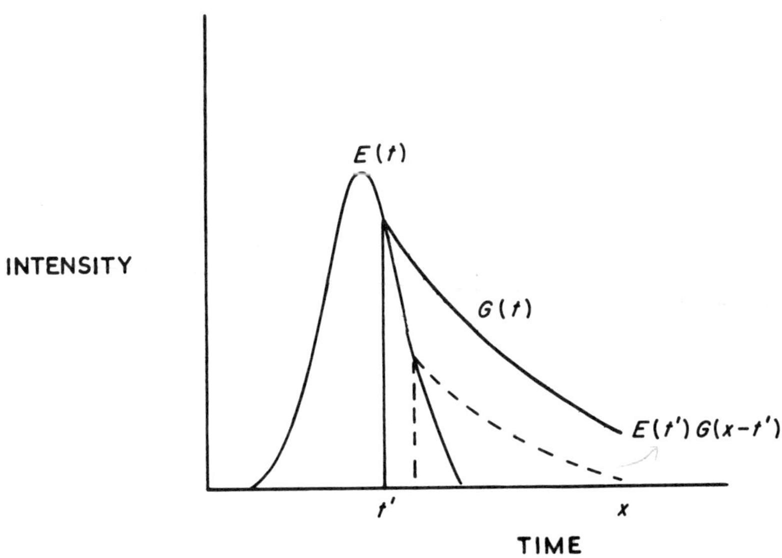

Figure 2.2 Schematic representation of the effect of convolution. $E(t)$ idealized pump pulse profile; $G(t)$ decay law (assumed single exponential) of sample.

Their effect can be evaluated through the use of Laplace transforms. Suppose that $H(t)$ is the δ-pulse response of the detection system, and $P(t)$ the measured time profile of the pump pulse, i.e., the instrument response function. $P(t)$ is a convolution of $E(t)$ and $H(t)$, which we shall write as

$$P(t) = E(t) \otimes H(t). \tag{2.13}$$

Writing the Laplace transform of a function $X(t)$ as $x(s)$ with

$$x(s) = L[X(t)] = \int_0^\infty e^{-st} X(t)\,dt,$$

it follows that

$$p(s) = e(s).h(s). \tag{2.14}$$

Similarly

$$a(s) = e(s).g(s) \tag{2.15}$$

and

$$i(s) = a(s).h(s). \tag{2.16}$$

Therefore

$$\begin{aligned} i(s) &= e(s).g(s).h(s) \\ &= p(s).g(s) \end{aligned} \tag{2.17}$$

and

$$\begin{aligned} I(t) &= P(t) \otimes G(t) \\ &= \int_0^t P(t')G(t-t')\,dt', \end{aligned} \tag{2.18}$$

where $I(t)$ represents the δ-function response of the sample distorted by convolution with both the pump pulse and the detector response, i.e., the measured decay curve.

Equation 2.18, the convolution integral, can be solved for $G(t)$ if $I(t)$ and $P(t)$, measured under the same conditions of instrumental distortion, are known. Methods for solving the convolution integral, or variants thereof, are discussed in Chapter 6.

Convolution will obviously become less significant as $P(t)$ becomes narrower with respect to $G(t)$. Errors attendant upon failure to deconvolve, in the case of instrument response functions of varying FWHM, have been estimated (Shaver and Cline Love, 1975), but only for single exponential decays.

Before a decision is reached that the assumption $I(t) = G(t)$ is valid it must be borne in mind that $P(t)$ represents the instrument response function and not simply the excitation pulse. Many instrument response functions, while having quite a narrow FWHM (as short as 75 ps in some instruments), have an extended tail, usually with a secondary peak resulting from photomultiplier effects (Lewis *et al.*, 1973). If this tail cannot be eliminated it is meaningless to consider only the FWHM of $P(t)$ in deciding whether deconvolution is necessary or not. In addition, the sample decay function may contain a short-lived component which may be overlooked if convolution is not taken into account. For instance, Lopez-Delgado *et al.* (1974) state that with an instrument response function of FWHM 1.4 ns, lifetimes greater than 4 ns can be derived directly from the measured $I(t)$ without deconvolution. As an example they report the decay time of quinine bisulphate in 0.1 N H_2SO_4 as 18.8 ns; significantly, there is now strong evidence that a single exponential function is an inappropriate model for the decay of this fluorescence (Harris and Selinger, 1979; Meech *et al.*, 1982). We therefore suggest that all decay curves should be deconvolved, irrespective of the lifetime values.

When both $P(t)$ and $I(t)$ are employed in data analysis it is usually assumed that the counting errors in $P(t)$ are negligible. Experiments in our laboratory have confirmed the validity of this assumption, instrument response functions collected with varying numbers of counts yielding identical results. Provision for including the errors in $P(t)$ in the uncertainty in $I(t)$ is therefore unnecessary but can be made (see Chapter 6).

2.2.4 Measurement of the instrument response function

The validity of the convolution integral requires that the functions $I(t)$ and $P(t)$ should be measured under the same conditions of distortion. In practice one tries to scatter from the same volume as that from which fluorescence is viewed, thereby equalizing the illuminated areas of the PM tube photocathode, upon which the response function, $H(t)$ depends. Colloidal suspensions such as Ludox*, glycogen (Weber and Teale, 1957), barium sulphate (Bailey and Rollefson, 1953) or simply milk (Millar *et al.*, 1980) can be substituted for the sample and, at the appropriate concentration, will closely, but not exactly, mimic the emission geometry (Demas and Crosby, 1971). Some of these suspensions, however, fluoresce under short wavelength excitation (300 nm) (Eastman, 1967) and are therefore to be avoided at these wavelengths. Aluminium or silver foil or, better, roughened quartz, may be used as a diffuse reflector. A screen coated with magnesium oxide can also

*Ludox is a Trade name for an aqueous suspension of silica manufactured by E. I. du Pont de Nemour Co., Wilmington, Delaware, USA.

serve as a scatterer. A monochromator between the sample and PM tube renders the choice of scatterer or reflector less important, but if the PM tube photocathode views the fluorescing or scattering area directly the correct method should be determined empirically using a standard sample (Section 2.4) and deconvolution.

As might be expected, the photomultiplier response, $H(t)$, is dependent on the energy of the incident radiation. Since fluorescence is normally Stokes shifted from excitation it may not be possible to measure $I(t)$ and $P(t)$ at the same wavelength. Pulses from hydrogen and deuterium flash lamps and from storage ring radiation are thought to be wavelength invariant, so that the instrument response function for such sources can be measured at the wavelength of the fluorescence. Measurement of $P(t)$ for other excitation sources is discussed in Section 2.3.

Long-term drift in the time profile of the excitation source can be treated either experimentally or in the data analysis. The former is probably preferable. A slight elaboration of the simple experiment entails the collection of one instrument response function, then of the decay curve, and then of a second instrument response function (these counts can be simply added to the counts in the first instrument response function in the data store). An instrument that automatically switches data collection between instrument response function and decay curve after a certain number of excitation cycles has been described by Hazan *et al.* (1974). A special cell holder carries the scattering solution and the sample one above the other. The holder is moved up and down by a pneumatic piston and the entire operation of decay curve and instrument response function collection, and background subtraction is controlled electronically and repeated until the required precision has been reached. For long collection times this procedure has much to recommend it and should lead to an improvement in measured decay times. More elegant devices than a pneumatic piston for alternating between scattering solution and sample have been described (Easter *et al.*, 1976; Rayner *et al.*, 1976). Short-term drift or jitter in the pump pulse profile is less easily corrected for and should be eliminated by optimization of the operating conditions in the excitation source.

2.2.5 Correction for constant background

In a good SPC instrument there is usually a very low noise level arising from leaked room light and thermionic noise in the photomultiplier tube. The reason is that the fast PM tubes chosen for single photon counting have very good dark noise characteristics. Most u.v.-sensitive tubes do not need to be cooled but it is advisable to operate red sensitive tubes at a low temperature. All measured curves must be corrected for dark noise since long low-intensity lifetimes are sometimes difficult to distinguish from the background. There are three commonly used methods for background correction.

Simplest and most widely used is the displacement of the decay curve in the data storage channels so that the first ten or twenty channels are used to accumulate only background counts. This background level is then estimated from a recount of the data and subtracted from the curve channel by channel. A justification for this subtraction has been given by Knight and Selinger (1973) who showed that the observed decay function $I(t)$ is given by

$$I(t) = C_1 I_0(t) + C_2, \tag{2.19}$$

where C_2 is the constant background component.

An automated version of background subtraction has been described by Hazan *et al.* (1974). In this version the analyser is run in subtract mode with an excitation shutter closed, for exactly the same number of excitation cycles as are used with the analyser in add mode and the shutter open. This procedure is also justified on statistical grounds but involves rather precise timing and is perhaps most suitable for a fully automated system such as that described by the authors.

More questionable is the inclusion of the constant background as a variable parameter in least-squares fitting procedure (Robbins *et al.*, 1980). There appears to be no advantage in increasing the number of fitting parameters in this way, especially in multi-exponential decays when a variable shift parameter is also included.

Subtraction of the background before deconvolution is therefore quite satisfactory. Care should be taken, if background is estimated from channels corresponding to a non-linear region of the TAC, that sufficient channels to give proper averaging are used. In addition the timescale should be such that in the channels before the curve the probability of collecting signal photons resulting from previous excitation events is negligible.

2.2.6 Operation at high excitation repetition rate

When the repetition rate of the excitation source is so high that the TAC has not time to reset between the occurrence of successive excitation pulses, distortions will be present in the decay curve if the data collection is performed in the conventional fashion. The present generation of TACs can operate at 1 or 2×10^5 Hz START rate (Haugen *et al.*, 1979) and therefore can be operated in conventional mode with most flash lamp excitation sources and with laser sources that employ a Pockels cell. Mode locked ion lasers, cavity dumped dye lasers and storage rings usually operate optimally at repetition rates in the MHz region. Consequently, for these excitation sources, the standard mode of data collection is not suitable.

In one variant of the conventional technique (Wild *et al.*, 1977) an inhibit function is used in such a way that data collection is blocked during periods of TAC recovery. Full advantage of the high repetition rate of excitation is, therefore, not taken. A more common variant, and one that also offers some advantages for low repetition rate excitation sources, is operation of the TAC in "reverse" mode (Swords, 1977; Lopez-Delgado *et al.*, 1974). Fluorescence signals are routed to the START input of the TAC and the trigger signals to the STOP input. It is preferable to arrange the delays (D_1 and D_2 in Fig. 2.1) so that the TAC sweep is stopped by a trigger signal corresponding to the excitation cycle in which the START signal occurred. If the delays are not so arranged the inter-pulse separation in the excitation source must be very stable.

The advantage of this technique is that, for high repetition rate sources, many more of the fluorescence signals (after suitable reduction to avoid pulse pile-up) are processed by the TAC. In a similar way, however, a correspondingly larger number of the dark noise signals are processed, and the need for a low noise fluorescence PM tube is therefore greater. In all other respects it is similar to the conventional mode except that the curves are collected in the data store with time increasing from higher to lower channel numbers. An example of a pump pulse and a decay curve collected with the TAC in "reverse" mode is shown in Fig. 2.3. It is usual to adjust data such as these to the more conventional arrangement before data analysis.

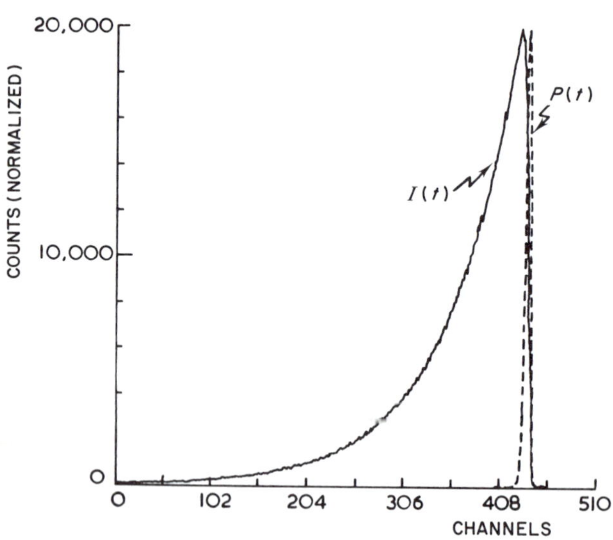

Figure 2.3 Decay curve of 1-methylindole in methylcyclohexane measured with the TAC operated in "reverse" mode. 1 channel = 0.082 ns.

2.3 Experiments to Correct for Wavelength-dependent PM Distortion

In the derivation of the convolution integral, Equation 2.18, wavelength-dependent effects on pump pulse shape and detector response were neglected. When they are taken into account we should write for the measured instrument response function $P(\lambda_E,t)$ or $P(\lambda_e,t)$

$$P(\lambda_E,t)=E(\lambda_E,t)\otimes H(\lambda_E,t) \tag{2.20}$$

$$P(\lambda_e,t)=E(\lambda_e,t)\otimes H(\lambda_e,t), \tag{2.21}$$

where λ_E signifies the wavelength of excitation and λ_e the wavelength at which the emission is observed. The measured decay curve should be written

$$I(\lambda_e,t)=E(\lambda_E,t)\otimes G(\lambda_e,t)\otimes H(\lambda_e,t). \tag{2.22}$$

If the pump pulse shape is wavelength independent, $E(\lambda_E,t)=E(\lambda_e,t)$ and when the instrument response function is measured at the wavelength of emission observation Equation 2.22 is soluble since

$$\begin{aligned} I(\lambda_e,t)&=E(\lambda_e,t)\otimes G(\lambda_e,t)\otimes H(\lambda_e,t)\\ &=P(\lambda_e,t)\otimes G(\lambda_e,t). \end{aligned} \tag{2.23}$$

The wavelength dependence of pump pulse profiles is discussed in Chapter 3. Here it may be noted that only hydrogen and deuterium flash lamps and storage ring radiation have pulse shapes that are believed to be constant over a wide wavelength range. For other sources, the pump pulse shape must be measured at the excitation wavelength and some sort of correction applied.

It has been suggested (Ricka, 1981) that in certain experiments in which two or more decays are measured at the same emission wavelength, having been excited at the same excitation wavelength, there is no need to measure an instrument response function since, if the decay functions are related (e.g., in a fluorescence depolarization measurement), the decay curves can be manipulated and the desired information extracted. While it is indeed possible to analyse some kinds of decay data in this way, we would strongly suggest that all decay curves should be deconvolved in order to ascertain whether the data are distortion free. Mathematical treatment of this wavelength-dependent distortion is discussed in Chapter 6. In general, it is preferable to eliminate experimental distortions by instrumental techniques rather than correct for them in data analysis. Hence, some effort has been made by workers in the SPC field to develop experiments by which the effect of a wavelength-dependent PM response may be counteracted.

The first such experiment appears to be that of Wahl *et al.* (1974), who proposed the use of a sample with a well-known single exponential decay time. In essence their method rests on the solution of Equation 2.22 for the reference compound. The convolution product $(E(\lambda_E,t)\otimes H(\lambda_e,t)$ is then calculated, and may be substituted in Equation 2.22 when $I(t)$ is measured for an unknown decay. The disadvantages of the experiment are that reference compounds must be found for each emission wavelength and that the errors involved in the two deconvolutions as well as the small number of compounds with a definitely known single exponential decay must lead to large errors in the recovered $G(t)$.

Britten and Lockwood (1976) modified the method just described so that a knowledge of the exact lifetime of the reference compound was not required. Instead, two reference compounds, the decays of which are known to be single exponential, but the lifetimes of which need not be known accurately, are used to determine the function $E(\lambda_E,t)\otimes H(\lambda_e,t)$. Since this method entails preparation of two highly pure samples at each emission wavelength of interest its practicability is highly questionable. Wong and Halpern (1976) found that, by adjusting the measured function $P(\lambda_E,t)$ so that all counts at times longer than 3.6 ns after the function maximum were reduced by 20%, a function was obtained that yielded successful results when used in the deconvolution of decay curves of certain samples. Such an entirely empirical technique is not recommended since it could seriously mask real non-exponentiality in the decay curves of interest.

Probably the least objectionable of the techniques involving reference compounds was proposed by Szabo and co-workers (Rayner *et al.*, 1976, 1977). Consider a compound, R, which absorbs at a wavelength λ_a, shorter than the wavelength of excitation of the sample, and that emits both at this excitation wavelength and at the wavelength at which the sample's emission is observed. In addition, assume that the decay of this reference compound is wavelength independent [i.e., $G^R(\lambda_E,t)=G^R(\lambda_e,t)$]. One measures four functions: $P(\lambda_E,t)$ (Equation 2.20); $I(\lambda_e,t)$ (Equation 2.22); $I^R(\lambda_E,t)$ (Equation 2.24); and $I^R(\lambda_e,t)$ (Equation 2.25).

$$\begin{aligned} I^R(\lambda_E,t) &= E(\lambda_a,t)\otimes G^R(\lambda_E,t)\otimes H(\lambda_E,t) \\ &= E(\lambda_a,t)\otimes G^R(\lambda_e,t)\otimes H(\lambda_E,t) \qquad (2.24) \end{aligned}$$

$$I^R(\lambda_e,t) = E(\lambda_a,t)\otimes G^R(\lambda_e,t)\otimes H(\lambda_e,t) \qquad (2.25)$$

Convolve $I(\lambda_e,t)$ with $I^R(\lambda_E,t)$ to get

$$I(\lambda_e,t)\otimes I^R(\lambda_E,t) = G(\lambda_e,t)\otimes I^R(\lambda_e,t)\otimes P(\lambda_E,t). \qquad (2.26)$$

Since $I^R(\lambda_e,t)$ and $P(\lambda_E,t)$ have been measured, Equation 2.26 is deconvolved to yield the decay of interest, $G(\lambda_e,t)$. Although the proposers of this technique have achieved excellent results, some drawbacks are obvious. For instance the reference compound, R, must have a broad fluorescence wavelength range suitable for each sample of interest. It is therefore likely that a number of reference compounds, all in a high state of purity, will be required. In addition, the decay of the reference fluorescence must be wavelength independent; this property of decay times has been investigated only to a very limited extent. Finally, propagated errors are likely to be very large since four functions rather than two are involved in deconvolution. Nevertheless, the ratio correction technique is at present the best alternative to the zero-time shift outlined in Chapter 6 in experiments affected by a wavelength-dependent PM response.

2.4 Standards

In the following four chapters we describe the individual components of an SPC fluorimeter and the mathematical methods with which SPC data are analysed. During the course of this description we shall emphasize that a large number of instrumental and mathematical factors, some of them quite subtle, may interfere with the success of a fluorescence lifetime measurement. For example, excessive discrimination of the fluorescence PM pulses can bias the data collection towards multiphoton events in such a way that, while measured lifetimes will be in error, simple criteria by which data analysis is judged may indicate success. Consequently it is of the utmost importance to have available a range of standard compounds with definitely established single exponential decay times, which can be used to test the performance of the instrument.

Recently we have, together with our colleagues, reviewed the literature for standard compounds and prepared a list of lifetimes that we have measured ourselves (Lampert *et al.*, 1983). In general, the agreement between lifetimes measured in different laboratories is good, but there are some startling discrepancies. These discrepancies may arise from different states of purity of solvent or solute, from strong temperature dependence in the lifetime of interest or from maladjusted instruments or inadequate mathematical analysis techniques. For one compound, quinine bisulphate in sulphuric acid, the reported lifetimes show a wide variation. We believe that this compound has a double exponential fluorescence decay (Meech *et al.*, 1982) and we strongly caution against its use as a fluorescence lifetime standard. Perhaps the most popular standard is a dilute solution of anthracene in cyclohexane, the decay time of which at 25°C is independent of excitation and emission wavelength. Before use both the anthracene and cyclohexane must be rigorously purified as residual impurities have in the past led some workers to assume a rather low value for the decay time. It should also be noted that the decay time is temperature, viscosity and concentration dependent (Blatt *et al.*, 1982). In

Table 2.1

Compound	Solvent	Wavelength of observation of emission/nm	Lifetime, τ_0/ns
PPO[a]	Cyclohexane	440	1.42
PPO[a]	Cyclohexane (undegassed)	440	1.28
Anthracene	Cyclohexane	405	5.23
Anthracene	Cyclohexane (undegassed)	405	4.10
1-Cyanonaphthalene	Hexane	345	18.23
1-Cyanonaphthalene	Vapour[b]	345	23.77
1-Methylindole	Cyclohexane	330	6.24
3-Methylindole	Cyclohexane	330	4.36
3-Methylindole	Ethanol	330	8.17
1,2-Dimethylindole	Ethanol	330	5.71
DMNA[c]	CH_2Cl_2	375	2.40

[a] PPO is the symbol for 2,5-diphenyloxazole.
[b] At 0.3 Torr vibrationally relaxed with 1 atmosphere cyclohexane; $t = 185°C$.
[c] DMNA is the symbol for *N,N*-dimethyl-1-naphthylamine.

Table 2.1 we list a number of compounds that can be purified by standard techniques and that have single exponential decay times. The lifetimes have been measured at least twice on our SPC instrument, which employs a synchronously pumped dye laser as an excitation source. The method of data analysis was reiterative least-squares fitting (see Chapter 6). Further lists of standard compounds are given by Birks (1970) and Berlman (1965) but many of the results in these publications were obtained with inaccurate, or at least, unsophisticated techniques and it is now advisable to rely on more recent data.

We have not treated the transfer of data to a computer, nor do we consider the type of computer to be used since there will be enormous variation from laboratory to laboratory.

References

Bailey, E. A. and Rollefson, G. K. (1953). *J. Chem. Phys.* **21**, 1315–1322.

Berlman, H. B. (1965). "Handbook of Fluorescence Spectra of Aromatic Molecules", passim. Academic Press, New York and London. (2nd edn, 1971).

Bevington, P R. (1969). "Data Reduction and Error Analysis for the Physical Sciences", p. 36. McGraw-Hill, New York.

Birks, J. B. (1970). "Photophysics of Aromatic Molecules", pp. 120–132. Wiley-Interscience, New York.

Blatt, E., Treloar, E. F., Ghiggino, K. P. and Gilbert, R. G. (1981). *J. Phys. Chem.* **85**, 2810–2813.

Britten, A. and Lockwood, G. (1976). *Mol. Photochem.* **7**, 79–84.

Coates, P. D. (1968). *J. Phys. E. Ser. 2* **1**, 878–879.

Cramer, L. E. and Spears, K. G. (1978). *J. Amer. Chem. Soc.* **100**, 221–227.

Demas, J. N. and Crosby, G. A. (1971). *J. Phys. Chem.* **75**, 991–1024.

Easter, J. H., De Toma, R. P. and Brand, L. (1976). *Biophys. J.* **16**, 571–583.

Eastman, J. W. (1967). *Photochem. and Photobiol.* **6**, 55–72.

Grinwald, A. (1976). *Anal. Biochem.* **75**, 260–280.

Harris, J. M., Gray, L. M., Pelletier, M. J. and Lytle, J. E. (1977). *Mol. Photochem.* **8**, 161–174.

Harris, C. M. and Selinger, B. K. (1979). *Aust. J. Chem.* **32**, 2111–2129.

Haugen, G. R., Wallin, B. W. and Lytle, J. E. (1979). *Rev. Sci. Instrum.* **50**, 64–72.

Haugen, G. R. and Lytle, J. E. (1981). *Anal. Chem.* **53**, 1554–1559.

Hazan, G., Grinwald, A., Maytal, M. and Steinberg, I. Z. (1974). *Rev. Sci. Instrum.* **45**, 1602–1604.

Imasaka, J., Kawabata, Y. and Ishibashi, N. (1981). *Rev. Sci. Instrum.* **52**, 1473–1477.

Knight, A. E. W. and Selinger, B. K. (1973). *Aust. J. Chem.* **26**, 1–27.

Koechlin, Y. (1961). Thesis. University of Paris.

Koester, V. J. and Dowben, R. J. (1978). *Rev. Sci. Instrum.* **49**, 1186–1191.

Lampert, R. L., Chewter, L. A., Phillips, D., O'Connor, D. V., Roberts, A. J. and Meech, S. R. (1983). *Anal. Chem.* **55**, 68–73.

Lewis, C., Ware, W. R., Doemeny, L. J. and Nemzek, T. L. (1973). *Rev. Sci. Instrum.* **44**, 107–114.

Lopez-Delgado, R., Tramer, A. and Munro, I. H. (1974). *Chem. Phys.* **5**, 320–326.

Mandel, L. and Wolf, E. (1965). *Rev. Mod. Phys.* **37**, 231–287.

Meech, S. R., O'Connor, D. V. and Phillips, D. (1982). *Chem. Phys. Lett.* **88**, 22–24.

Millar, D. P., Robbins, R. J. and Zewail, A. H. (1980). *Proc. Natl. Acad. Sci. USA* **77**, 5593–5597.

Morton, G. A. (1968). *Appl. Optics* **7**, 1–10.

Murao, T., Yamazaki, I. and Yoshihara, K. (1982). *Appl. Optics* **21**, 2297–2300.

O'Connor, D. V. (1977). Thesis. University of Western Ontario.

Pfeffer, G., Lami, H., Laustriat, G. and Coche, A. (1962). *Comptes Rendus* **254**, 1035–1037.

Rayner, D. M., McKinnon, A. E., Szabo, A. G. and Hackett, P. A. (1976). *Canad. J. Chem.* **54**, 3246–3259.

Rayner, D. M., McKinnon, A. E. and Szabo, A. G. (1977). *Rev. Sci. Instrum.* **48**, 1050–1054.

Ricka, J. (1981). *Rev. Sci. Instrum.* **52**, 195–199.

Robbins, R. J., Fleming, G. R., Beddard, G. S., Robinson, G. W., Thistlewaite, P. J. and Wolfe, G. J. (1980). *J. Amer. Chem. Soc.* **103**, 6271–6279.

Shave, L. A. and Clive Love, L. J. (1975). *Appl. Spectrosc.* **29**, 485–489.

Swords, M. (1977). Thesis. University of Southampton.

Wahl, Ph., Auchet, J. B. and Donzel, B. (1974). *Rev. Sci. Instrum.* **45**, 28–32.

Wahl, P. R. (1975). *New Tech. Biophys. Cell. Biol.* **2**, 233–241.

Ware, W. R. (1971). In "Creation and Detection of the Excited State", Vol. 1A (Lamola, A. A., ed.) p. 250. Marcel-Dekker, New York.

Weber, G. and Teale, F. W. J. (1957). *Trans. Faraday Soc.* **53**, 646–655.

West, M. A. (1979). In "Photochemistry", Vol. 10 (Bryce-Smith, D., ed.) p. 40. The Chemical Society, London.
Wild, U. P., Holzwarth, A. R. and Good, H. P. (1977). *Rev. Sci. Instrum.* **48**, 1621–1627.
Wong, D. K. and Halpern, A. M. (1976). *Photochem. and Photobiol.* **24**, 609–611.
Yguerabide, J. (1972). *Meth. Enzymol.* **26**, 498–578.
Zimmerman, H. E. and Cutler, J. P. (1975). *Chem. Commun.* 598–599.

3 Light Sources

3.1 Introduction

Among the properties desired in an SPC excitation source are high intensity, repetition rate variable from khz to 100 MHz, complete polarization of radiation, tunability over some 500 to 600 nm, wavelength-independent pulse shape, perfect reproducibility of pulse shape and intensity, and picosecond pulse widths. This ideal source would also be inexpensive and relatively easy to transport from one laboratory to another. Such a source does not, of course, exist. In general, sources that fulfil eminently some of these conditions have severe disadvantages in other respects. For instance, synchrotron storage rings furnish extremely stable sub- nanosecond pulses at repetition rates up to 40 MHz. But the intensity per pulse is quite low and, for the majority of workers, the facility is, to say the least, inconvenient to use. Again, mode-locked lasers furnish high intensity picosecond pulses at MHz repetition rates but are not easily tunable and can be very expensive. The spark gap discharge lamp can be tuned easily but suffers from lack of intensity in a narrow spectral bandwidth, furnishes pulses of relatively long duration, and does not enjoy long-term stability of pulse profile.

In this chapter, we shall discuss in more detail the three sources already referred to. We shall not consider the non-optical sources that have occasionally been used in SPC, such as pulsed X-rays (Phillips and Swank, 1953; Henry and Helman, 1972) and electron bursts (Bennett *et al.*, 1965). Nor shall we describe the ingenious methods of pulse generation based upon a continuous light beam impinging on a rotating mirror system, which have attained little popularity in SPC instruments, presumably because of low intensity per pulse (Garbury *et al.*, 1957) and poor pulse shape reproducibility. For a review of these methods the reader is referred to the paper by Jones (1964). The discussion of synchrotron radiation will be brief, since, although it is in many ways an attractive excitation source, it is almost impossible for most research workers to use it on a regular basis for lifetime measurements.

Prospective purchasers of SPC equipment will clearly have to opt for the quality of items for which they have funds available. In Appendix 3.A1 we list suppliers of excitation sources, from whom prices may be obtained. Other important factors such as wavelength tunability, reliability, ease of service, radio frequency noise generation, etc., will be considered in each section. At this stage, it might be useful to point out that, in general, a sample that is likely to cause complications because it is liable to undergo photodecomposition will do so just as readily under flash lamp as under laser excitation, because of the longer data collection times required for flash lamps.

3.2 Storage Ring Radiation

The classical mechanism by which electrons in an accelerating field lose energy is by the emission of electromagnetic radiation. In a circular accelerator such as a synchrotron or a storage ring, the accelerating field is created by magnetically constraining the electrons to travel round a closed path; the consequent centripetal acceleration leads to the emission of synchrotron radiation (SR). In actual synchrotron machines electron energies are high ($>m_0c^2$, $m_0=$electron mass) and relativistic effects result in SR emission in a well-defined distribution of directions about the tangent to the electron orbit. Most of the intensity from a point at the source is confined within a cone having a small half-angle given by $m_0C^2/E\approx 0.1$ milliradians with respect to the tangent. Therefore, for these electrons, with E in the GeV range, almost all the emitted light lies in the plane of the electron orbit (Lopez-Delgado *et al.*, 1974).

As an excitation source for SPC the storage ring is more suitable than the synchrotron. In the latter electrons are continuously injected and accelerated; during the accelerating cycle the properties of the radiation are varying rapidly with time. In a storage ring, on the other hand, electrons are pre-accelerated, injected into the ring and then maintained in a stable orbit in an ultra-high vacuum chamber (10^{-10} Torr) by a constant magnetic field. Therefore the properties of the radiation important in SPC, such as pulse repetition rate, pulse width, angular distribution, etc., remain constant over many hours of continuous operation.

Storage ring radiation has the following important features:

(1) It is a continuum from the hard X-ray region to 10 μm (see Fig. 3.1).
(2) It provides a small natural aperture with relatively high photocollection efficiencies.
(3) The repetition rate of the pulses (depending on the particular machine) can be as high as 400 MHz.
(4) The pulse width, which depends on the length of the circulating electron bunch, can be as low as 150 ps and is usually $\leqslant 1$ ns.
(5) The radiation is 100% linearly polarized with the electric vector in the orbital plane.

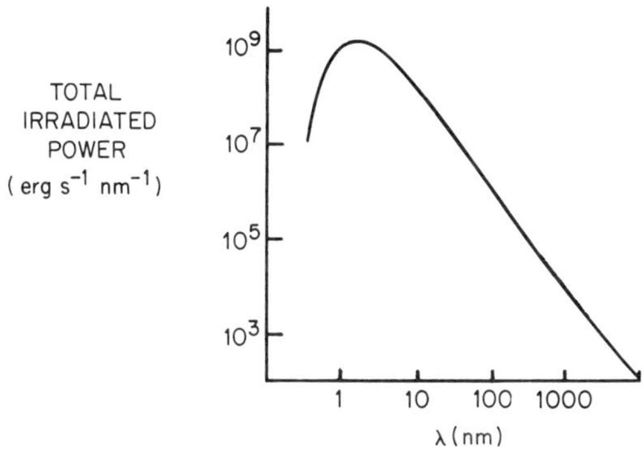

Figure 3.1 Calculated wavelength distribution of radiation in the ACO synchrotron source at 540 MeV electron energy and 100 mA circulating current (adapted from Lopez-Delgado *et al.*, 1974).

(6) The electron distribution in the bunch can be calculated exactly for a real accelerator. It is nearly Gaussian. The photon distribution in the pulse is also nearly Gaussian. Statements that the pulse shape in real machines is exactly calculable are probably over-optimistic.
(7) There are strong theoretical and experimental grounds for believing the pulse shape to be wavelength independent.
(8) Interpulse optical noise is negligible.
(9) Intensities per pulse delivered to the sample have been calculated to be 1.5×10^4 photons with 4 MHz repetition rate at 300 nm (Lopez-Delgado *et al.*, 1974). This intensity (which has not been verified experimentally) is greater than that furnished by flash lamps, but less than pulsed laser intensities.

It will be seen that storage ring radiation has many features that recommend it for the SPC experiment. In particular, the wide and easy wavelength tunability, polarization, narrow pulse width, and invariance of pulse shape with wavelength are highly attractive properties. The greatest disadvantage of the source is its inconvenience. The machine may be quite far from the user's place of work with consequent difficulties in experimental programme planning and sample transport. Furthermore, in many storage ring facilities, organization of user time leaves something to be desired, with the result that measurements have to be carried out hurriedly. It is probably preferable, on the whole, to employ a less inconvenient source for regular lifetime determinations. Nevertheless, in view of the merits of SR radiation, one would be well advised, before purchasing or constructing one's own SPC instrument, to ascertain the user facilities at the nearest storage ring machine and to consider whether one's needs could be met by one or two visits a year to the machine. Storage rings in operation at present as well as those under construction are listed in Table 3.1.

3.3 Flash Lamps

3.3.1 Introduction

The first flash lamp for use in single photon counting was constructed by Malmberg (1957), and its application was described by Brody (1957). Almost all flash lamps designed since then consist, essentially, of two metal electrodes enclosed in a gas, which, when a sufficiently high voltage is applied through a charging resistor, undergoes periodic dielectric breakdown providing flashes of light. Flash lamps may be classified into four basic types according to the pressure of the filling gas and the method by which the low voltage electrode is connected to earth. Details peculiar to each type will be given in Section 3.3.2. Some practical considerations, common to all types, are discussed first.

First, in most lamps the gas is enclosed in a sealed chamber, which should be fitted with one or two optical windows. The windows must be sealed tightly in order to prevent leakage but should also be easily demountable so that they may be cleaned conveniently. In glass-bodied lamps black wax is preferable to epoxy resins for sealing purposes since it requires only moderate heat to soften and can be removed easily with trichloroethane. Less messy than either is a threaded glass tube onto which a plastic ring is screwed bearing down on the optical flat. The seal is made by means of an O-ring between the flat and the rim of the threaded tube. If a photomultiplier provides the trigger pulse for the electronics the lamp should have two optical flats. Suprasil is the usual window material. For excitation in the vacuum ultra-violet, magnesium fluoride or lithium fluoride discs give transmission down to about 110 nm and are not particularly difficult to handle. Lithium fluoride, however, may form colour centres. In high pressure lamps the attachment of windows is more difficult and the method chosen should first be thoroughly tested at the appropriate pressures. If the spark is to be struck exclusively in air at 1 atm., windows are not necessary.

Electrode sputtering is the main cause of loss of window transmittance, for which reason the window should not be placed too close to the electrode gap. A distance of 2 cm between window and gap has been found satisfactory (Lewis *et al.*, 1973). Platinum as an electrode material tends to deposit relatively heavily on the windows; hence thoriated tungsten is now more common. Although preferable to platinum, tungsten electrodes usually develop, over many hours of operation, a tip of deposited material, which reduces the length of the spark gap and hence the intensity. Flashing stability

Table 3.1 Selected parameters of storage rings which are used as synchrotron radiation sources. E_0: electron energy, R: bending radius, E_c: critical energy, T_0: circulation time, N: harmonic number, $T_{r.f.}$: period of r.f. field, t: bunch length.

	E_0/GeV	R/m	E_c/keV	T_0/ns	N	$T_{r.f.}$/ns	t/ns
SURF (Washington, USA)	0.25	0.83	0.041	17.5	2	8.8	
TANTALUS (Stoughton, USA)	0.24	0.64	0.048	31.4	1	31.4	1–4
INS-SOR (Tokyo, Japan)	0.3	1.1	0.054	57.8	7	8.3	
ACO (Orsay, France)	0.54	1.11	0.31	73.5	2	36.7	0.8–1.4
BROOKHAVEN (USA) VUV Ring	0.7	1.9	0.40	170.3	9	18.9	
BESSY (Berlin, West Germany)	0.75	1.8	0.52	208.3	26	8.0	0.01–0.13
					104	2.0	
VEPP 2M (Novosibirsk, USSR)	0.67	1.22	0.54	59.9	20	3.3	
ALADDIN (Stoughton, USA)	1.0	2.08	1.1	295.0	15	19.7	
ADONE (Frascati, Italy)	1.5	5.0	1.5	349.7	3	111.7	
DCI (Orsay, France)	1.8	3.82	3.4	316.4	8	39.5	
SRS (Daresbury, UK)	2.0	5.55	3.9	320.5	160	2.0	0.2
PHOTON FACTORY (Tsukuba, Japan)	2.5	8.33	4.1	598.8	285	2.1	
BROOKHAVEN (USA) X-ray ring	2.5	6.8	5.1	568.2	30	18.9	
DORIS (Hamburg, West Germany)	3.5	12.1	7.8	961.5	481	2.0	0.13
SPEAR (Stanford, USA)	4.0	12.7	11.1	781.2	279	2.8	0.08–0.4
CHESS (Cornell, USA)	8.0	88	12.8	2222.0	1111	2.0	0.1
VEPP-4 (Novosibirsk, USSR)	7.0	29	26.0	1250.0	227	5.5	

and pulse profile are also degraded. The deposit, thought to arise from impurities extracted from the tungsten by the strong electric field (Berlman *et al.*, 1968), must be removed. Application of the high voltage to the electrodes when the lamp is completely evacuated may reduce the deposit but it will usually be necessary, at some stage, to remove the electrodes and clean them with emery paper. Removable electrodes in glass-bodied lamps may be force fit through Teflon stop-cocks. For machineable materials swagelock or similar fittings are satisfactory. In either case, provision should also be made for adjustment of the interelectrode spacing.

The electrode to which the high voltage is applied is generally machined to a point while the other electrode is either rounded or flat ended. It would appear that the reason is purely empirical, these being the shapes that give best performance. If both electrodes are pointed, the arc wanders and the flash does not stabilize, whereas if both are rounded a fuzzy arc is generated, again yielding a flash that cannot be timed precisely. When the electrodes have been cleaned it is necessary to apply the high voltage to the lamp for at least one hour before commencing measurements. During this time the arc will be seen to wander round the earthed electrode until finally "burning a hole" in it, at which stage the discharge arcs between two fixed points and the flash profile is stable. It will frequently be found that the arc does not stabilize; in these cases a slight adjustment to the arc gap should be made. When the high voltage is first applied to a newly-filled lamp, the electrodes of which have not been cleaned, the lamp should be allowed to run for at least fifteen minutes so that the plasma can reach equilibrium conditions. It is also worthy of note that unstable performance has been reported in misaligned electrodes, that is when the electrodes are not collinear (Lyke and Ware, 1977).

For most designs of lamp, it is advisable to have attached a connection for evacuation and gas filling. Operations can then be performed without removing the lamp or if possible, the back plate from the lamp housing. It is not necessary to evacuate with a diffusion pump but it is vitally important to incorporate an oil trap between the rotary pump and the lamp. Liquid nitrogen is of course most effective, but even a molecular sieve of silica gel granules will suffice. Oil vapour in the plasma region causes grossly unstable operation. Unstable operation may also result if the filling gas is not of high purity (Lyke and Ware, 1977).

For ease of alignment the lamp should be mounted in such a way that the position of the arc can be adjusted, preferably without removing the access plate from the housing. A simple X–Y stage should be satisfactory for this purpose. It should of course be borne in mind that wires carrying extremely high voltages are attached to the electrodes and must not be allowed to come into contact with earthed components. If the electrodes are carried in teflon stop-cocks there will always be the temptation to adjust the arc gap while the high voltage is applied. This temptation should be resisted. Furthermore, when the high voltage has been switched off the stray capacitance in the circuit should be discharged before current-carrying components are touched. Discharge is most conveniently achieved by connecting the high voltage to earth through a screw driver while holding the insulated (plastic) handle. A further hazard associated with hydrogen- or deuterium-filled lamps is the possibility of an explosion if air has leaked into the electrode chamber. Glass-bodied lamps, in particular, may shatter under the stress of such an explosion with the consequent possibility of injury to the user. Therefore, when the high voltage is first applied to a lamp that has not been filled for some time, the access plate of the housing should be securely in position.

After construction or purchase the lamp should be tested for leaks and if any are found they must be eliminated. With hydrogen or deuterium as filling gases in low pressure lamps leakage may be more severe. Small leaks can be detected by monitoring the pressure in the spark gap chamber during operation. The initial rise in pressure as the plasma heats is usually followed by a slight drop to a steady value. If the pressure continues to drop gas is probably leaking from the chamber. Slightly larger leaks, through which air diffuses into the chamber, will lead to a change in the colour of the arc from an initial blue to a purple-violet.

In all types of flash lamp stray or added capacitance, C, is charged through a resistor R and discharged through the arc gap, causing the flash (Fig. 3.2). In general the duration of the flash is proportional to the product RC. Consequently for a narrow pump pulse profile the capacitance must be as

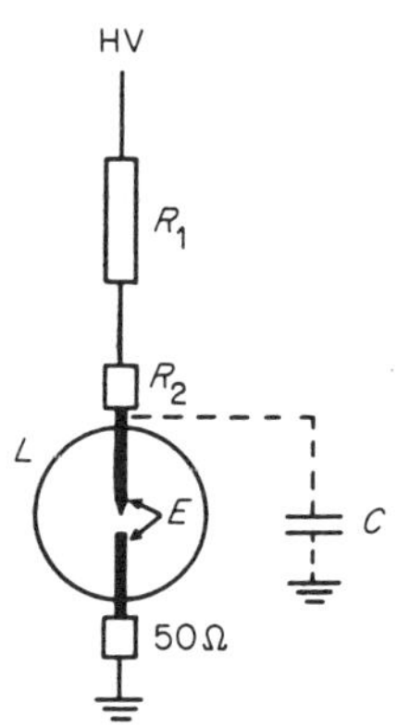

Figure 3.2 Circuit representation of free-running flash lamp. HV, high voltage; R_1, resistor or resistor string; R_2, resistor; C, capacitor (stray or added); L, lamp envelope; E, electrodes.

small as possible. It can be reduced by replacing a single bulky resistor R with two resistors R_1 and R_2 one of which is physically small (Yguerabide, 1965). By attaching R_2 as close as possible to the electrode, C and consequently the flash duration is reduced. By increasing C, usually by adding a high voltage capacitor between the tip of the high voltage electrode and earth, increased intensity per flash can be attained at the expense of short pulse width.

Low pressure lamps, which are usually operated under partial vacuum, tend to collect foreign vapours in the flash chamber. These vapours may be absorbed into the electrodes, survive the initial pumping and flushing and be outgassed at the high plasma temperatures. We have already referred to pulse instability resulting from the rotary pump oil vapours. Another source of instability is the plasticizer that may be present in polythene tubes connecting the lamp to the pump or even in the material of the lamp itself.

Lamps made from a machineable plastic material, Rexolite*, give stable operation over quite long periods, but recently a ceramic material has been reported to give better performance (Lyke and Ware, 1977). This high vacuum dielectric material, although easily machineable, requires careful handling because it tends to crack if screws or tube fittings are tightened too much. Like Rexolite and metal it is less fragile than glass, and does not require epoxy or black wax sealant for the windows. In lamps constructed from machineable materials the electrodes can be secured by means of nylon straight thread fittings with O-ring seals (see Fig. 3.7). Nylon, teflon, flexible metal, or glass tubing should provide the connection between pump and lamp on the lamp side of the trap. Yet another source of contamination is vacuum grease that may be used on joints in the vacuum system or on O-rings that seal on the windows. Wherever possible grease vapour should be trapped; there is generally no need for grease on O-rings. Since apiezon grease has higher fluorescence than silicone grease the latter is recommended for the vacuum joints. Greaseless joints, however, are preferable.

Flash lamp housings will be discussed in Section 3.3.6 as will r.f. suppression. Pulse intensities will be discussed in Section 3.3.4. It should be mentioned here that undistorted pump pulse profiles have seldom been published owing to the difficulty of deconvolving the detector response from the measured instrument response function. Almost all the available data suggest that flash lamp pulses have a low intensity, relatively long lived tail, although the secondary peak observed in most instrument response functions is almost entirely a result of photomultiplier effects (Lewis *et al.*, 1973).

*Rexolite is manufactured by Atlantic Laminates, Franklin, New Hampshire, USA.

3.3.2 Four basic types of lamp

3.3.2(a) High pressure, ungated lamps

In an ungated, or free-running, lamp one of the electrodes is connected directly to earth and, as Fig. 3.2 illustrates, the lamp is essentially a relaxation oscillator. The discharge takes place at or near the breakdown voltage of the gas. Pulse width and repetition rate are determined by the time constant, $R \times C (R = R_1 + R_2)$.

Filling gas pressures in excess of 1 atm. are referred to as high; those reported in high pressure lamps vary between 1.5 atm. and 30 atm. Usually the lamp remains attached to the pressurizing system during operation but it may (with some risk to the handler) be sealed off from the gas filling line by means of, for example, a special crimping tool (Yguerabide, 1965). In the earlier designs of this type of lamp the spark was struck between mercury wetted electrodes (D'Alessio *et al.*, 1964; Yguerabide, 1965) but more recently mercury has been dispensed with. In the design of Hundley *et al.* (1967), illustrated in Fig. 3.3, thoriated tungsten electrodes are sealed into a quartz cylinder which is surrounded by a copper cylinder. It will be noticed that the charging resistors are placed inside the pressurized region in order to reduce surface breakdown. It is more usual to cool the resistor, placed outside the lamp body, by for example, immersing it in oil cooled by a water jacket (Rayner *et al.*, 1976).

In order to minimize the duration of the flash, the stray capacitance, C, should be as small as possible. As was mentioned already, a short high voltage electrode appears to yield narrower pulses, as does an increase in the width of the metal container in the region of this electrode. Coaxial electrode enclosures are also reported to yield slightly narrower pulses owing to reduced capacitance between the high voltage electrode and earth. In all types of flash lamp it is a general rule of thumb (admitting of some exceptions) that the optical pulse width is proportional to the width of the electrode spacing. Spacings in high pressure lamps are usually quite small (<1 mm) since a narrow arc gap induces discharge at reasonably low voltages. However, the pulse widths in low pressure and high pressure lamps seem to be comparable. Some optical pulse width measurements have been made with very fast detectors such as photodiodes or image converter tubes. As a guide to the duration of pulses from high pressure discharge the results of D'Alessio *et al.* (1964) (FWHM = 0.5 ns for hydrogen at 18 atm., 1–10 kV, 0.7 mm arc gap), Yguerabide (1965) (FWHM = 0.5 ns for hydrogen at 20 atm.; 0.5 ns for air; 1 ns for xenon at 20 atm., all with 0.2 mm arc gap), Hundley *et al.* (1967) (FWHM = 1.25 ns for oxygen at 6 atm., 12 kV, 0.762 mm arc gap) and Wahl *et al.* (1974) (FWHM = 0.97 ns for deuterium at 15 atm.) may be quoted. Pulse profiles from this type of lamp are shown in Fig. 3.4.

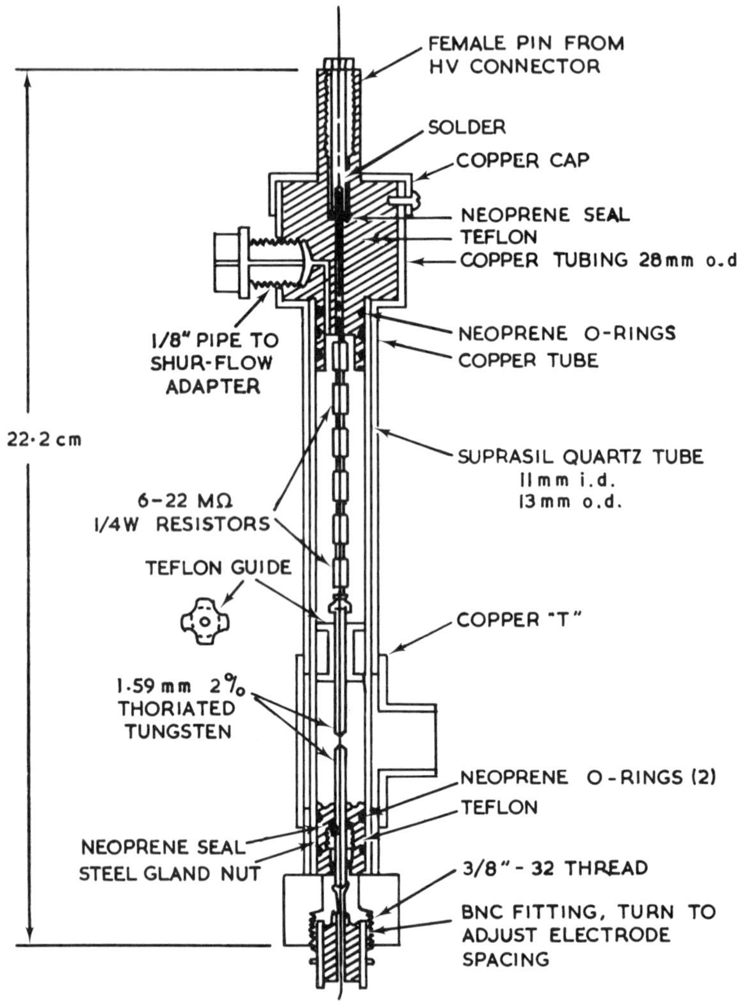

Figure 3.3 Schematic diagram of a high pressure free-running flash lamp (after Hundley *et al.*, 1967).

In free-running lamps the pulse repetition rate cannot be varied at will. It is determined by the applied voltage, the breakdown voltage (electrode spacing, gas and pressure) and the *RC* time constant. Firing as fast as 50 000 pulses a second may be attained, although at the higher frequencies the power per pulse is drastically reduced. For this reason high pressure lamps are, in general, operated at lower repetition frequencies than gated lamps. Typical operating frequencies lie in the range 1 kHz to 15 kHz. Intensity per pulse will

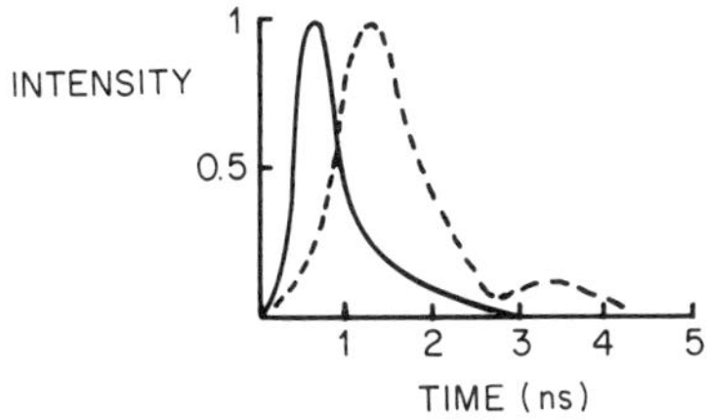

Figure 3.4 Time profiles of pulses generated in a high pressure discharge (with a small amount of detector convolution effects). —, H_2 at 20 atm. (adapted from Yguerabide, 1965); ..., O_2 at 6 atm. (adapted from Hundley *et al.*, 1967).

be discussed in Section 3.3.4. The voltage required to break down the gas depends on pressure and electrode spacing. Voltages ranging between 2 kV and 25 kV have been reported.

A number of gases are suitable for high pressure operation, the most common being D_2, H_2, N_2, O_2 and air. It is surprising that oxygen, which does not produce stable pulses in a discharge at low pressures, seems to perform well when the pressure exceeds 5 atm. The spectral output of the O_2 high pressure lamp (Fig. 3.5(a)) is more or less flat from 250 nm to 600 nm. Hydrogen has a similar spectrum (Fig. 3.5(b)). Nitrogen, which shows its characteristic line spectrum at low pressures (see Fig. 3.8), produces at high pressures a continuum from 200 to 400 nm (Thomaz *et al.*, 1975; Yagi *et al.*, 1978). Air has essentially the same spectral output as nitrogen but the time characteristics of the light pulses differ somewhat. It is likely that, in high pressure lamps, the pulse shape is independent of wavelength, although experimental evidence for this statement is slight.

High pressure lamps are probably less common now than formerly, presumably because of the inconvenience associated with pressures in excess of atmospheric. In addition, gated lamps offer more flexibility of operation, the repetition rate being adjustable by the simple means of a frequency control on the thyratron grid pulser. One commercial manufacturer of SPC equipment (Applied Photophysics Ltd.) has discontinued its high pressure free-running lamp in favour of the low pressure gated variety. The reason is not clear but may perhaps be connected with instabilities encountered in the high pressure discharge. Recently a high pressure lamp has been marketed by Horiba Ltd. The performance of the lamp has not yet been tested "in the field", but is reported to be very stable.

3.3.2(b) Low pressure, ungated lamps

For this type of lamp, the charging circuit is again as illustrated in Fig. 3.2, but the discharge chamber is filled to 1 atm. or less with the flashing gas. Low

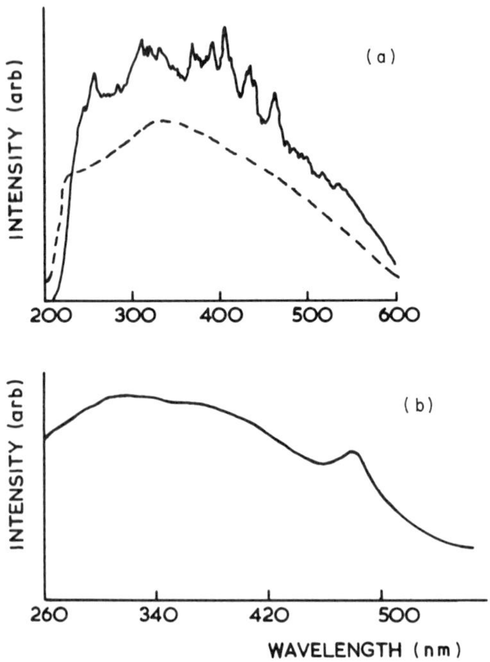

Figure 3.5 Measured spectra of light emitted from high pressure flash lamps. (a) O_2 at 6 atm. (—) (adapted from Hundley, 1967); (b) H_2 at 20 atm. (adapted from Yguerabide, 1972). Both spectra are uncorrected for the spectral response of the detecting system (monochromator and 1P28 PM). The relative sensitivity of the 1P28 PM is plotted as the dotted line in (a).

pressure free-running lamps are, therefore, simple and relatively inexpensive to construct and operate. The charging resistor $R(=R_1+R_2$ in Fig. 3.2) is again fairly large (20–30 MΩ) but because of the low pressures applied, high voltages are usually less than 10 kV. Consequently, r.f. noise generation is likely to be less severe. A simple design for this type of lamp that can be made quite easily from pyrex glass is shown in Fig. 3.6 while a design for which some machineable insulator (such as Rexolite or ceramic) would be suitable is shown in Fig. 3.7. In the first design the tungsten electrodes are force fit through the teflon; if the seal is not complete a bead of epoxy or Torr-seal can be applied at the point of exit of the metal from the teflon. In this design the arc gap is adjusted merely by turning either one of the stop cocks. The window can be attached with black wax or epoxy. In the second design a disc of the insulating material is made to bear down on the window which forms a seal with an O-ring slotted into the lamp body. The swagelock fittings are of the compression type and can be of brass or, preferably, nylon.

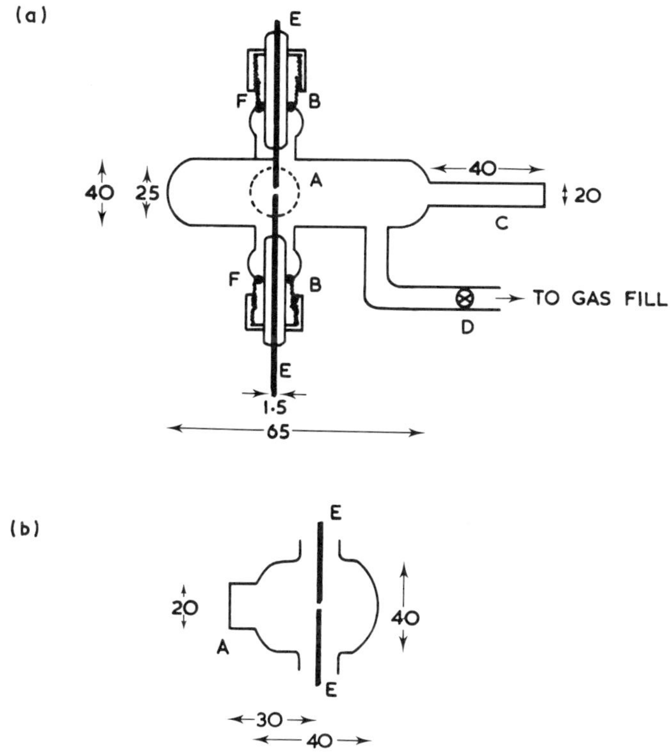

Figure 3.6 Design of low pressure Pyrex flash lamp (not drawn to scale: all dimensions in mm). A, Suprasil optical flat; B, Youngs PTT/10 stop cocks; C, arm for mounting; D, stop cock; E, tungsten electrodes; F, O-rings (included in B). (a) Front view; (b) side view.

Typical operating conditions for the lamp illustrated in Fig. 3.6 are: R, 20 MΩ; applied high voltage; 6 kV; arc gap, 2 mm; gas pressure, 0.5 atm. With deuterium as the filling gas these conditions produce a repetition rate of about 90 kHz and a pump pulse profile having a FWHM of 3 ns. (This is crudely estimated by subtracting 1 ns from the FWHM of the measured instrument response function.) Narrower pulses and higher repetition rates can be attained by decreasing the arc gap but rates in excess of 100 kHz seem to yield reduced overall intensity and unstable operation. We would stress that these conditions are suitable for the specified design and a 20 MΩ charging resistor. Instrument response functions of FWHM as low as 800 ps have been obtained in a commercial flash lamp, Ortec model 9352 (no longer on the market), with air as the filling gas (Leskovar *et al.*, 1976).

There need be no difference in construction between low pressure lamps

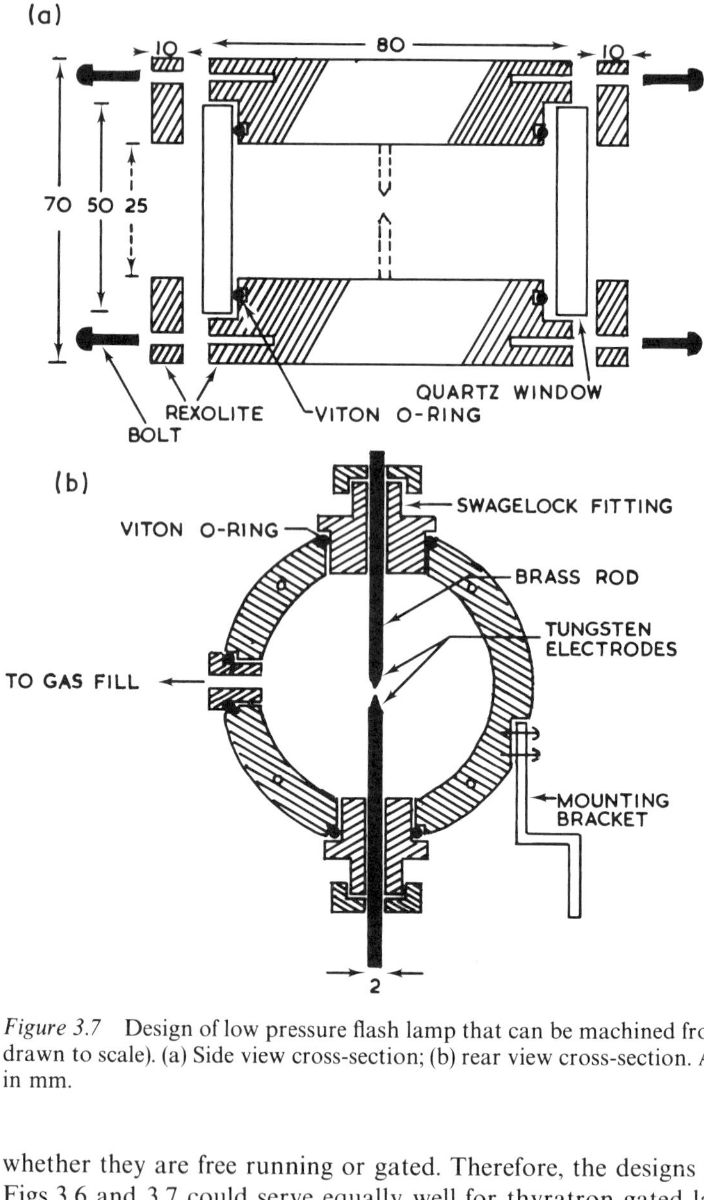

Figure 3.7 Design of low pressure flash lamp that can be machined from plastic (not drawn to scale). (a) Side view cross-section; (b) rear view cross-section. All dimensions in mm.

whether they are free running or gated. Therefore, the designs illustrated in Figs 3.6 and 3.7 could serve equally well for thyratron-gated lamps. Filling gases that are suitable for low pressure operation are also the same. As was mentioned, oxygen does not perform well at low pressures. On the other hand, air yields an optimum performance at pressures of 1 atm. or less. The low pressure spectrum of air, which consists mostly of emission from the $C^3\pi_u$

$\rightarrow B^3\pi_g$ transition of molecular nitrogen, is illustrated in Fig. 3.8(a). Emission is mainly between 300 and 400 nm. Air, even at low pressures, is frequently found to be troublesome to work with since it causes electrode sputtering. As a consequence the flash repetition rate becomes less stable, and a deposit, formed on the window and electrodes, leads to drastically reduced intensity. Window blackening may not be severe if the distance between electrode gap

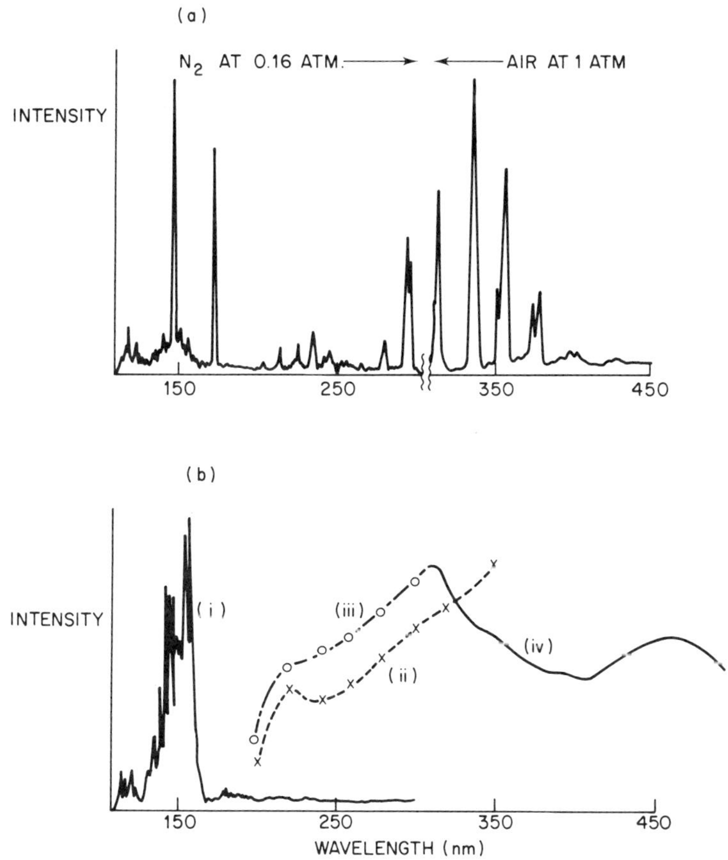

Figure 3.8 Measured spectra of light emitted from low pressure flash lamps. (a) N_2. Measured in different laboratories and uncorrected. Spectra in either side of the wavelength indicated by the arrow are not normalized relative to each other. From 110 to 310 nm correction factor not known. From 310 to 450 nm rough correction can be achieved with 1P28 sensitivity curve (Fig. 3.5(a)).

(b)(i) H_2 at 0.1 atm. (adapted from Lyke and Ware, 1977); (ii) H_2 at 0.16 atm. (adapted from Berlman *et al.*, 1968); (iii) D_2 at 0.16 atm. (adapted from Berlman *et al.*, 1968); (iv) D_2 at 1 atm. (adapted from Badea and Brand, 1979).

and window is reasonably large. Leskovar *et al.* (1976) obtained a stable repetition rate by flushing a steady flow of dry air through the discharge chamber during operation and by cleaning the electrode tips after every ten hours of operation. Essentially the same spectral output and increased intensity can be obtained from nitrogen but at the cost of a broader pulse profile. In addition, the exact conditions of arc gap, high voltage and pressure necessary to obtain stable operation with nitrogen are difficult to determine. Carbon monoxide, which has a highly structured emission between 150 and 250 nm, is reported to give rise to particularly loathsome reaction products (Lyke and Ware, 1977) and cannot be recommended.

Hydrogen and deuterium are probably the most common discharge gases in low pressure lamps. Although the intensity from these gases is less than that available from air or nitrogen, the pulse width is narrower and stable operation can be attained over a wide range of operating conditions. The approximate relative spectral output of these gases is illustrated in Fig. 3.8(b). The intensity distribution in H_2 and D_2 is thought to result from transitions from lower vibrational levels of the $1s\sigma 2s\sigma {}^3\Sigma g$ state to a repulsive $1s\sigma 2p\sigma {}^3\Sigma u$ state (Coolidge and James, 1939). Although spectrum (i) is not corrected for the wavelength response of the detection system, the intensity of the many line spectrum below 170 nm is known to be in excess of the continuum extending from 200 to 600 nm. The relative intensity in this continuum is flatter than is indicated by the uncorrected spectrum. Published spectra with peaks in the wavelength range from 300 to 400 nm (e.g., Brody, 1957) probably result from air contaminated hydrogen plasmas rather than from, as has been suggested by Knight and Selinger (1973), impurities in the tungsten. As spectrum (iii) illustrates, the emission from D_2 is more intense than that from H_2; this enhancement pertains at all wavelengths. Optical pulse widths from both gases are the same so that D_2, were it not so expensive, would obviously be recommended over H_2.

3.3.2(c) Low pressure, gated lamps

Low pressure lamps are usually gated, in accordance with the charging circuit illustrated in Fig. 3.9. With discharge triggered by the thyraton, voltages in excess of the gas breakdown voltage can be applied with a consequent increase in the pulse intensity. In addition, the pulse repetition rate can be independently adjusted by means of an oscillator driving the thyratron pulser. A lamp of this type was described by Malmberg as early as 1957. Gated lamps are probably still the most common type of flash lamp and are the type offered by three of the commercial companies marketing SPC instruments (see Appendix 3.A1). One design of lamp body is illustrated in Fig. 3.10. This lamp, together with the thyratron, will be housed in a larger

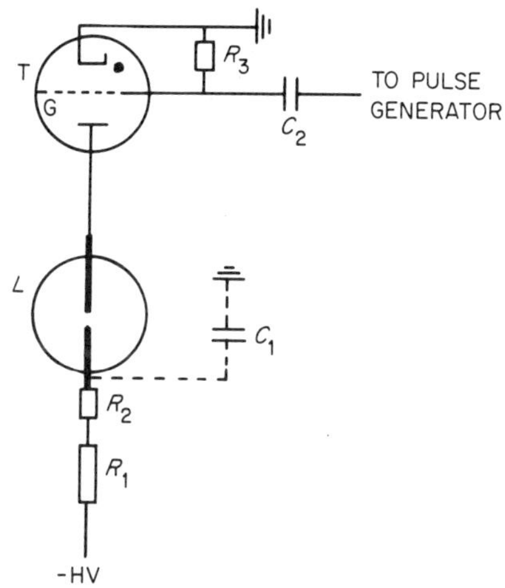

Figure 3.9 Low pressure flash lamp gating circuit L, lamp; T, hydrogen thyratron; G, grid of thyratron; C_1, stray capacitance; C_2, 0.01 μF; R_1, *c.* 1 MΩ; R_2, *c.* 0.5 MΩ; R_3, 22 kΩ; HV, 4–10 kV.

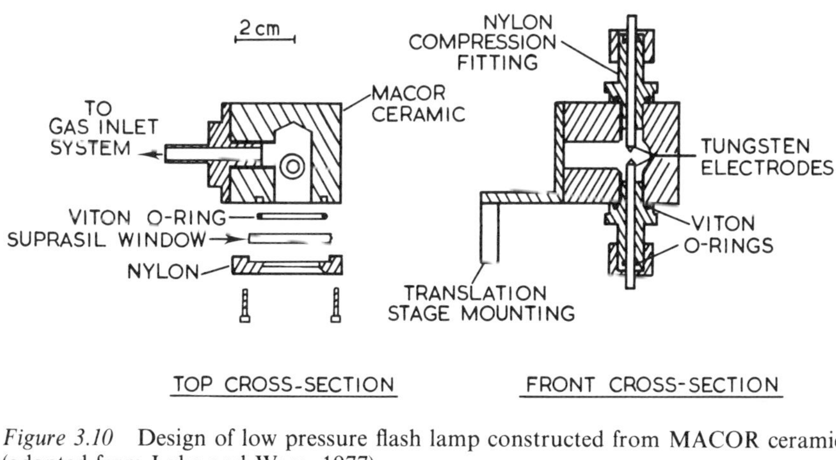

Figure 3.10 Design of low pressure flash lamp constructed from MACOR ceramic (adapted from Lyke and Ware, 1977).

metal box. In a more compact design, illustrated in Fig. 3.11, the thyratron is coaxial with, and physically close to, the electrode. The lamp body is of metal, removing the need for a separate metal housing.

Among the thyratrons in common use are the E, G and G Models HY2 (rated at 8 kV) and HY6 (16 kV), ITT Model 7621 and English Electric Model FX 2530/6777. The grid pulse for both the HY2 and HY6 should be

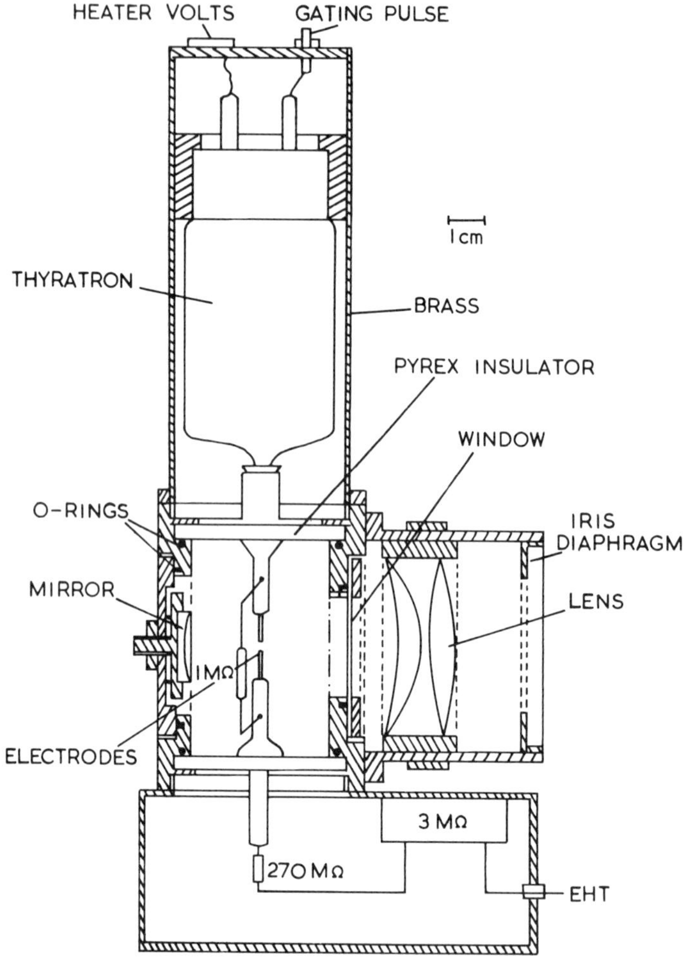

Figure 3.11 Design of low pressure triggered flash lamp constructed from brass. Connection to gas filling system and window for trigger photomultiplier are not shown. The resistor across the electrodes is reported to improve pulse stability (adapted from Birch and Imhof, 1981).

from 100 V to 300 V amplitude with a rise time of about 20 ns and a FWHM of about 50 ns. A circuit for producing the required pulses has been described by Halpern (1974) and includes 24 p–n diodes (thyristors) in series, each having a forward breakdown potential of about 15 V. A circuit diagram has been given by Knight and Selinger (1973) and is illustrated in Fig. 3.12. For all thyratrons the plate breakdown factor (repetition rate × peak anode

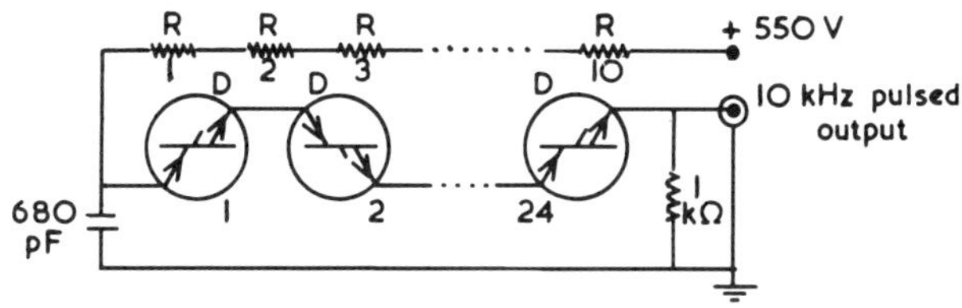

Figure 3.12 Design of pulse circuit for gating thyratron. R, 5 kΩ, 2 W; D, Western Electric 443A thyristor diode (after Knight and Selinger, 1973).

voltage × peak anode current) should not be exceeded. Even if the circuit can deliver pulses to the grid at a faster rate the breakdown factor may place an upper limit on the pulse repetition rate. In two commercial SPC instruments (Applied Photophysics Ltd., Edinburgh Instruments Ltd.) the thyratron can be operated at up to 200 kHz but pulse widths at these high repetition rates are extremely large (6–10 ns). The critical factor causing this effect seems to be the limited currents available from the flash lamp high voltage supply (West, 1981). In any case, the best all round performance for gated lamps is achieved with the repetition rate in the range 20–50 kHz. It is possible to derive the trigger pulse for the counting electronics from the thyratron pulser but, because of instabilities in the firing circuit, it is preferable to time the actual flash (see Section 3.3.5). It should be possible to limit the flashing rate instability to less than 10 Hz in 50 kHz.

In hydrogen thyratrons of the type employed in flash lamp circuits the gas reservoir must be maintained at a certain temperature by means of an externally supplied heater voltage. This voltage is 6.2 V a.c. for the ceramic–metal thyratrons HY2 and HY6. It can be derived from the mains through a simple step-down transformer. However, if the mains voltage is not stablized it is convenient to route it through a Variac transformer before input to the 6.3 V transformer. In this way a slightly higher heater voltage can be supplied to the reservoir and observed malfunctioning resulting from insufficient heating avoided. It is also advisable to route the heater voltage through a passive r.f. filter (see Section 3.3.6). Heat generation by the thyratron may result in high temperatures inside the lamp housing. Therefore a small fan is sometimes attached to the back or side wall. It should be noted, however, that the ventilation holes constitute an outlet for r.f. leakage. In addition, care must be taken to avoid overcooling the thyratron. Unless resistors, capacitors or the lamp body are being damaged by overheating, therefore, fan cooling is best avoided. While no evidence has been reported connecting optical pulse width with the lead length from earth electrode to thyratron, on general principles of optimum conditions for pulse transmission the lead should be kept as short as is compatible with the geometry of the housing. Thyratrons are quite robust and should yield years of operation before failure. Nevertheless, whenever lamp malfunctioning is observed the thyratron should be rigorously tested for internal mechanical failure.

There has been some discussion concerning the factors that determine the pulse shape in triggered lamps. Berlman *et al.* (1968) and Birch and Imhof (1981), on observing a narrowing of the time profile at increased pressures, concluded that the sub-nanosecond decay times resulted from collisions between excited and unexcited molecules, higher pressures causing greater frequency of collisions. Collisional quenching would also account for the decrease in intensity sometimes observed at higher pressures. Ware and co-workers on the other hand, on observing an almost negligible shape change in pulses from the many-line H_2 spectrum at 170 nm and the continuum at 220 nm postulated that the emitting states are quenched so far that the observed time profile is determined by complex plasma processes (Andre *et al.*, 1979). No definite conclusion can be reached until a reliable method for determining the actual pulse shape (as opposed to the instrument response function) has been devised. At least part, if not all, of the long tail sometimes observed in instrument response functions, which, it has been suggested, results from recombination of electrons and positively charged molecules (Malmberg, 1957), is generated in the detection photomultiplier. It is likely however, that with N_2 and air, for which this tail is much more pronounced, recombination effects are important. In addition, since the air pump pulse is narrower than the nitrogen pulse and the intensity is also considerably reduced, oxygen is likely to be quenching at least some of the nitrogen emission processes. It is interesting to note, however, that with the lamp illustrated in Fig. 3.11 an instrument response function with FWHM as short as 1.2 ns has been measured for nitrogen.

Of the four gases that are most popular for low pressure lamps, H_2, D_2, N_2 and air, H_2 and D_2 probably yield the narrowest pulse profiles. However, it is difficult to reconcile all the reported values of FWHM. We have already mentioned the FWHM of 0.8 ns reported for the instrument response function of an air lamp (Leskovar *et al.*, 1976) which is shorter than the FWHM of the actual pulse profile calculated for H_2 by Andre *et al.* (1976). Birch and Imhof (1981) obtained similar FWHM values in instrument response functions from H_2 and N_2 (0.95 and 1.2 ns, respectively) whereas a difference of almost a factor of 4 was reported by Birks *et al.* (1962). Again one laboratory has reported a broader FWHM for D_2 pulses compared with H_2 pulses (Berlman *et al.*, 1968) whereas another has reported no difference (Lewis *et al.*, 1973; Lyke and Ware, 1977). Until a complete set of results has been obtained with the newest generations of flash lamps and detection components, perhaps the most reasonable conclusion should be based on the results from a single lamp operated under accurately controlled conditions. If such a basis is accepted the widths of pulse profiles from H_2, D_2 air and nitrogen are in the ratio 1:1:1.25:1.5 (Lewis *et al.*, 1973).

As the discussion in Section 2.3 makes clear a variation in pump pulse shape $E(\lambda,t)$ with wavelength is a potentially serious problem, particularly if the photomultiplier response is also wavelength dependent. Pulse shapes from air and nitrogen lamps are indeed wavelength dependent. Fortunately, the wavelength dependence of the response in many PM tubes can be approximated by a simple time shift which can be incorporated in the deconvolution routine. Since shift routines are not without their own problems it is advisable, whenever the nature of the experiment makes it possible, to observe the fluorescence at a wavelength as close as possible to the excitation wavelength. Alternatively, the experimental correction of Szabo and co-workers, described in Section 2.3, can be applied. It is generally believed that the pulse shape of H_2 or D_2 is independent of wavelength in the range 200 to 400 nm. Direct evidence for this belief is however lacking. Methods of deducing pulse profiles, $E(\lambda,t)$, from measured instrument response functions, $P(\lambda,t)$, are discussed in Section 4.4. The result of application of one of these methods to the H_2 pump pulse profile between 170 nm and 220 nm is shown in Fig. 3.13. Owing to the limited resolution of the method, it is not possible definitely to attribute the observed difference in the second decade of intensity to a physical cause. It is clear, however, that over almost a decade and a half of intensity the pulse profiles are identical.

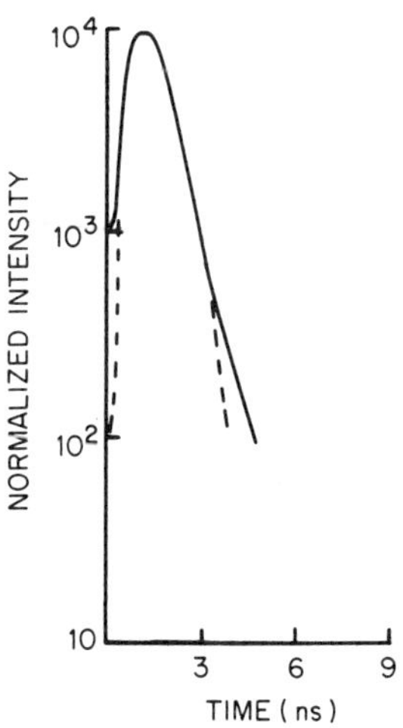

Figure 3.13 Pulse profile for H_2 flash lamp ....., 170 nm; ——, 220 nm (adapted from Andre *et al.*, 1979).

3.3.2(d) High pressure, gated lamps

Although high pressure lamps are thought to yield a more intense flash than low pressure lamps, their flash repetition rates, which are not independently variable, are generally lower. An obvious modification in these lamps is to gate the gas breakdown with a thyratron. The plate breakdown factor, however, discussed in the previous section, imposes a limit on the repetition rate that can be achieved for a given pressure. Successful gated operation of N_2 at 2 atm. has been reported (Birch and Imhof, 1981) but intolerable instabilities have been observed when gating is attempted at higher pressures (O'Connor, 1977). At time of writing it appears that gating is most successful for pressures less than or in the region of 1 atm.

3.3.3 Operating conditions and performance

The pump pulse reproducibility is of critical importance in the SPC experiment. It is essential, therefore, to determine the operating conditions (applied high voltage, pressure, arc gap, charging resistor and capacitance, and flash repetition rate) that yield optimum performance for a given lamp body, electrode material and filling gas. Since optimum conditions change not only from one design of lamp to the next but even among different lamps of the same design, the choice of these conditions is largely an empirical process. Each operating variable is adjusted independently (as far as possible; increased pressures will usually require increased voltages for gas breakdown) and the pulse shape, as observed on an oscilloscope and as measured in the instrument response function, is noted. In this section we mention briefly some points that might be overlooked by a first time user. Some of these points have been discussed in more detail previously (e.g., Berlman *et al.*, 1968; Birch and Imhof, 1981).

Since pulse characteristics are strongly dependent on applied high voltage, the power supply should incorporate a voltage regulator. If it does not, the mains voltage to the power supply should be regulated. In home-built supplies, any a.c. ripple superimposed on the applied d.c. voltage should be eliminated. In gated lamps at repetition rates of about 30 kHz the current drawn by the arc is about 2 mA, well within the capability of most power supplies. Instability of pulse shape at higher repetition rates may result in part from inability of the power supply to deliver the requisite current. The optical power is much more dependent on applied voltage in gated lamps than in free running lamps. In all low pressure lamps optimum stability is usually found at an applied voltage less than that required for maximum intensity. A high charging voltage may also produce increased r.f. noise.

Increased pressure is observed to lead to a narrowing of the pulse profile in low pressure lamps. As with intensity, however, the narrowest pulse is not necessarily the most stable. In addition, as pressure is increased above a certain value, intensity decreases. These effects can be attributed to quenching of the emitting state although it is worthy of notice that pulse narrowing does not persist when pressure is raised above about 1 atm. In general, optimum pressures for H_2 or D_2 lie in the range 1/3 to 2/3 atm., whereas for air and nitrogen the optimum pressure seems to depend strongly on the particular lamp.

As might be expected, at smaller arc gaps intensities and pulse widths are reduced, and r.f. noise generation increases, especially if the applied voltage is raised in an effort to recover intensity. In low pressure lamps arc widths of from 2 to 3 mm are standard, whereas for high pressure lamps widths range from about 1.5 mm to less than 1 mm. As was stated previously, the electrode tips should be kept clean. It is sufficient to remove them from the lamp and rub them with emery paper when a whisker appears. When erosion has become severe, the high voltage electrode should be re-ground to a point.

The flash repetition rate to be adopted will depend to some extent on the capabilities of the thyratron. It should also be noted that at high repetition rates intensity per pulse drops drastically, with the result that overall intensity per second reaches a maximum at some intermediate value of the repetition rate. It has already been noted that pulse widths increase as repetition rate increases. In most gated lamps repetition rates in the range 20 kHz to 60 kHz appear to yield optimum performance.

In general, flash lamps are run on stray capacitance. Measurements of the capacitance range from 6 pF (Lewis *et al.*, 1973) to 20 pF (Schuyler and Isenberg, 1971). If it is desired to increase intensity, a capacitor, rated at the voltage applied to the electrode, is connected between the high voltage electrode and earth. Intensity should be roughly proportional to capacitance (C) according to the formula

$$E = \tfrac{1}{2}CV^2, \tag{3.1}$$

where E is the electrical energy dissipated and V is the applied voltage. It is not known whether the optical energy is linearly dependent on capacitance, but it has been reported that a considerable increase in intensity resulted when a 30 pF capacitor was added in a gated lamp (Ware and O'Connor, 1977). The stability of the pulse profile was maintained while the FWHM of the instrument response function increased from 2 ns to 10 ns. Although the type of charging resistor is not critical, its wattage rating must be chosen so that it does not overheat at the applied high voltage. If carbon film resistors are used they can be grouped in parallel pairs and added in series until the desired resistance is attained. A close check should always be kept on resistors to confirm that breakdown due to overheating has not occurred.

A useful diagnostic of flash lamp performance is the count rate as measured on a counter timer. A stable repetition rate is not, of course, a requisite in the SPC experiment; however, instabilities are usually accompanied by pulse-to-pulse profile changes, which can seriously distort the collected data. If the repetition frequency shows large fluctuations the trigger pulse should be examined for jitter on an oscilloscope.

3.3.4 Power

There have been relatively few measurements of the absolute intensity of flash lamp pulses. It is well established however that intensity in nitrogen is greater than in air, which is in turn over an order of magnitude more intense than hydrogen. Deuterium delivers about twice the power of hydrogen. It is worthy of note that optical power does not increase as the square of the applied voltage (see Equation 3.1) as has been reported (Malmberg, 1957); the dependence appears to be more or less linear (Birch and Imhof, 1981).

Table 3.2 Experimentally determined intensities for various types of flash lamp.

Gas	Type of lamp[a]	Photons per flash	Flash repetition rate/kHz	Reference[b]
H_2	LP, G	4×10^8	3	1
	HP	4×10^8	20	2
	LP, G	8×10^8	25	3
	LP, G	4×10^8	25	4
D_2	LP, G	1.3×10^9	3	1
	LP, G	1.6×10^9	25	3
Air	LP, G	2×10^{10}	25	3
	LP, G	1.4×10^{10}	25	5
	LP, F	3×10^6	15	6
	LP, F	$<5 \times 10^6$	90	5
N_2	LP, G	8×10^{10}	25	3
	LP, G	1.2×10^{11}	15	7
	LP, G	2×10^{10}	25	4
O_2	HP	4×10^9	20	2
	HP	6×10^{11}	2	8

[a] LP, G = low pressure, gated.
LP, F = low pressure, free-running.
HP = high pressure.
[b] References: 1. Berlman *et al.*, 1968; 2. Yguerabide, 1972; 3. Lewis *et al.*, 1973; 4. Birch and Imhof, 1981; 5. Phillips and O'Connor, 1981; 6. Leskovar *et al.*, 1976; 7. Stuart and Kirk, 1977; 8. Hundley *et al.*, 1967.

Optical output is usually determined by ferrioxalate actinometry (Hatchard and Parker, 1956). A less troublesome, though less reliable, procedure is measurement with a calibrated radiometer. In Table 3.2 we have assembled some representative data on flash lamp intensities. We have attempted to correct published figures so that the numbers in column 3 represent intensities per pulse integrated over all wavelengths and emitted into 4π steradians. Since data collection times in the SPC experiment depend on the total number of photons incident on the sample we have also given the flash repetition rate at which the measurements were made.

Give the errors inevitably accompanying chemical actinometry and those associated with our corrections based on the published geometry details (sometimes incomplete) there is in general rather good agreement among the results of different laboratories. It is interesting to notice that high pressure lamps operated at high repetition rates do not offer enhanced intensity over low pressure lamps. On the other hand as the results for oxygen indicate it is more efficient to operate a high pressure lamp at low repetition rates. As can be seen from the data for air, gating produces orders of magnitude more intensity in low pressure lamps.

It would appear from these data and the considerations discussed in previous sections that gated lamps provide the most satisfactory all round performance among flash lamps. High pressure free-running operation does not result in increased intensity and is less flexible as far as varying the operating parameters is concerned. Moreover, high pressure lamps are less simple to construct and operate safely.

3.3.5 The trigger pulse

With flash lamp excitation the trigger pulse in the SPC experiment is usually derived from an antenna, which detects an r.f. signal every time the lamp flashes, or from a photomultiplier which views the lamp directly, or through a fibre optic cable. If the lamp is thyratron-gated a synchronous pulse may be available from the thyratron grid pulser electronics. However, because the lamp may occasionally fire without a gating pulse or fail to fire when the gating pulse is delivered it is no longer common practice to trigger with the pulser sync. pulse.

Antennae are particularly convenient to operate since they require only a BNC bulk head connector on the lamp housing, a 50 Ω load resistor and a short length of coaxial cable. The centre conductor on the cable is connected to the bulkhead connector. The stiffness of the cable is then sufficient to support it so that the other end is about 3 to 6 cm from the lamp body. The 50 Ω resistor is connected between a convenient earthing point on the housing and the central conductor of the coaxial cable at the lamp end.

Photomultipliers such as the RCA 1P28 are ideal for trigger pulse purposes since they are easy to accommodate together with their voltage divider network in a small box attached to the side or back of the flash lamp housing. The interstage resistance can be constant (e.g., 22 kΩ) and the anode is terminated in 50 Ω. In order to avoid saturation effects neutral density filters (or wire mesh) may have to be inserted between the lamp and the photomultiplier window.

3.3.6 Lamp housings and r.f. suppression

A radio frequency electrical signal is always generated at the time of arc gap discharge. Because this is picked up in the fluorescence channel electronics, it can distort the observed instrument response functions and decay curve (see for example, Grinvald, 1976). If the interference is severe the instrument response function will appear to narrow and a ripple will be observed in the decay curve. Less severe interference may not be obvious from a visual inspection of the recorded data but will lead to erroneous data analysis since a mathematical correction for this distortion is not possible. The fluorescence PM and electronics must therefore be shielded from r.f. noise. Somewhat reduced r.f. interference is expected for flash lamps of the type illustrated in Fig. 3.11 which are of coaxial design and are not constructed from an insulating material (Birch and Imhof, 1981). Since it is more convenient to construct the lamp from glass, ceramic, or plastic we give here some details of r.f. suppression that have been found helpful in the past. For a further treatment of this topic the reader is referred to Morrison (1967).

The flash lamp housing should be designed in such a way that the number of apertures, through which r.f. can leak, is as small as possible. In addition, unless the apertures can be fitted with a waveguide the length of which is about twice as long as the diameter, they should be as small as possible. Solder gasket seals should be inserted between bulkhead electrical connectors and the metal of the lamp housing. Paint should be cleared from areas where there is metal-to-metal contact. It may be impossible to follow these directions in all respects since the lamp housing must accommodate not only an aperture for light exit but also a mounting stage for the lamp, an input for high voltage and either an antenna bulkhead connector or an aperture for viewing by the trigger photomultiplier. In addition, it may also have a thyratron mounting stage and bulkhead connectors for input of the thyratron grid pulse and heater voltage, a gas inlet, fan, shutter, etc. However, if screw holes are kept as small as possible and well-contacted wire mesh covers the photomultiplier viewing point or fan vent, r.f. leakage can be reduced to negligible levels.

Welding is recommended for all joints in the housing except at one side, usually the back, where a removable plate is necessary to allow access to the inside of the housing. This plate should be stepped and have a lip wide enough to allow insertion of a solder or indium gasket inside the screw holes. Since the aim is to maintain continuous metal-to-metal contact the screw holes should be closely spaced. Each time the back plate is tightened down and subsequently removed, the solder gasket should be replaced. A design for the back plate is shown in Fig. 3.14. A waveguide for the aperture by which

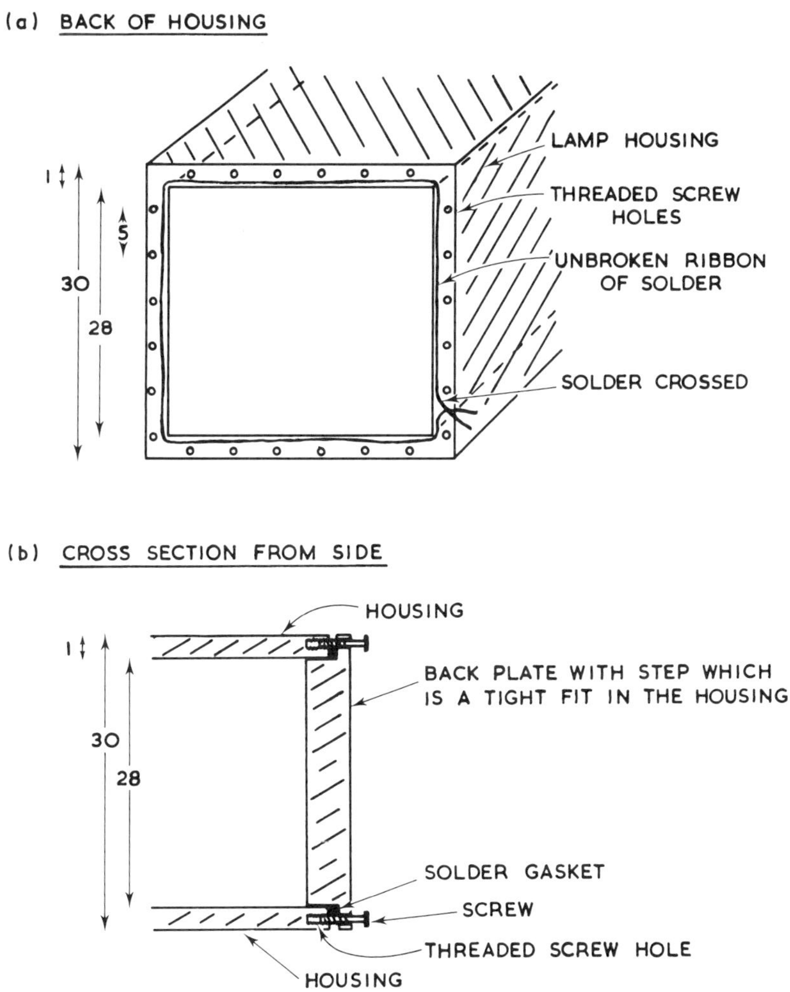

Figure 3.14 Stepped backplate for lamp housing (not drawn to scale). Dimensions (in cm) are intended only as rough guide.

light exits to the sample is relatively easy to fit and can be threaded to accommodate a lens holder. Again a lip on the tube makes insertion of a solder gasket easy, as illustrated in Fig. 3.15. A shutter or iris diaphragm can also be accommodated in the waveguide but the mechanism should be so designed that only a small hole is necessary for the control lever. Some high voltage and BNC bulkhead connectors are of design similar to that of the waveguide and can be used with a solder gasket.

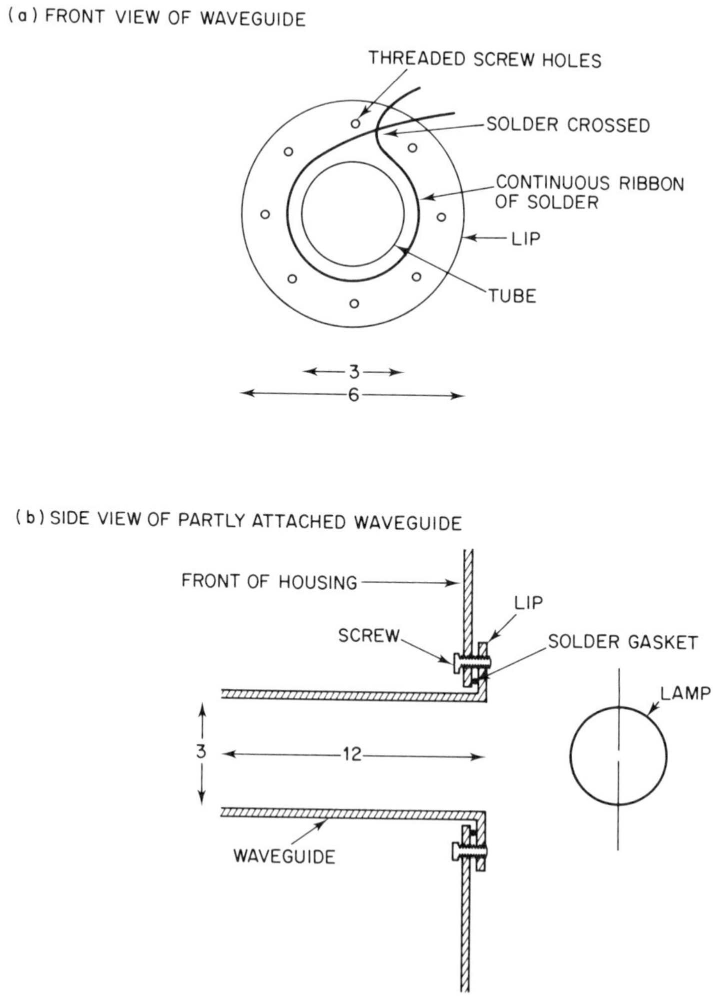

Figure 3.15 Illustration of waveguide for r.f. filtering on housing light aperture (not drawn to scale). Dimensions (in cm) are intended only as a rough guide.

Since r.f. may also be generated in the high voltage power supply and be transferred along the high voltage cable into the housing, it may also be necessary to filter this voltage. A four stage π-type RC filter suitable for this purpose has been suggested by Lyke and Ware (1977). This filter should, of course, be accommodated in a shielded box which is attached to the lamp housing in such a way as to introduce no new ports for r.f. leakage. If the lamp is gated the thyratron heater voltage can be filtered by an LC filter of the type illustrated in Fig. 3.16. The small inductance, created by a number of coils of conductor wound round a reasonably high resistor, blocks r.f. signals but not the 50 or 60 cycles of the a.c. signal. The capacitors in the filter will pass r.f. but not the 60 cycle signal. They should be physically small so that they do not act as pick-up points.

For the transmission of signals from the trigger source and fluorescence PM to the electronics doubly-screened coaxial cable is recommended. If it is not available singly-screened cable may be further screened by drawing the braid from a larger diameter cable over the coaxial cable and clamping it tightly at the connectors. Lyke and Ware (1977) recommend passage of the cable carrying the trigger signal to a pulse transformer in a screened box in order to eliminate ground loops and r.f. carried along the cable skin. The transformer consists of two turns of No. 50 wire for both input and output in a 3.18 mm ferrite core.

It should be emphasized that with many types of lamp r.f. interference will not be severe and the foregoing precautions will not be necessary. If r.f. generation is suspected an antenna connected to an oscilloscope may help to trace the source. Occasionally interference originating in some equipment unconnected with the SPC instrument may be present. Because it is a regular

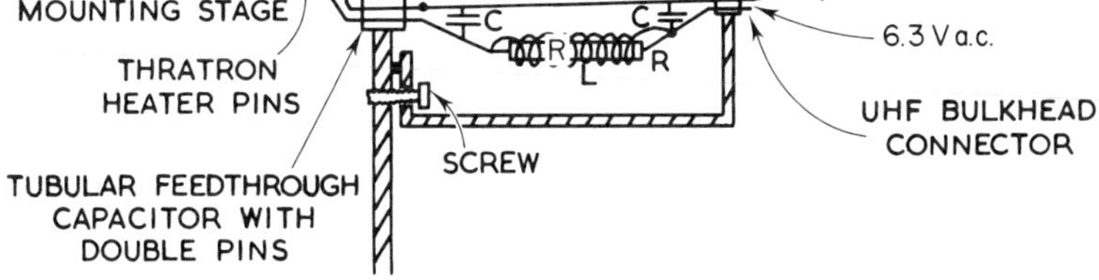

Figure 3.16 Schematic diagram of π-type r.f. filter for thyratron heater voltage (not drawn to scale). L, inductance; R, resistor (47 kΩ); C, capacitors (0.0015 μF).

signal with an interpulse spacing in the nanosecond time range it will also distort the data if picked up in the electronics. It may not be possible to filter this signal at source. In this case more elaborate screening procedures may be necessary. If adoption of doubly-screened coaxial cable for all intercomponent connections and slight raising of discriminator levels (see Section 5.3) does not eliminate the interference, specialized advice should be sought.

3.4 Pulsed Lasers

3.4.1 Introduction

Since the early 1970s, lasers, and in particular synchronously pumped dye lasers, have found increasing application in SPC experiments. The ways in which lasers can provide excitation pulses are numerous; however, in the following sections we shall deal only with the most commonly used laser sources. We have omitted discussion of nitrogen laser pumped dye lasers (Imasaka *et al.*, 1981) since at the present stage of development the nitrogen laser does not provide a high repetition rate, is relatively unstable, and frequently has extreme r.f. noise generation associated with its operation. Solid state and excimer lasers are again of low repetition rate and have not found wide applicability in SPC measurements and will also be excluded from the discussion. We shall concentrate almost exclusively on Ar^+ ion laser pumped dye lasers. Recently the Kr^+ ion laser has been developed to replace the Ar^+ ion laser in certain wavelength regions. At the time of writing it is too early to comment on its performance in the SPC instrument, but we expect that there will be very little difference between Ar^+ and Kr^+ ion lasers as regards general effectiveness of operation. The cavity length in reasonably sized ion lasers is such that the mode-locked pulse train has an interpulse separation of only a few nanoseconds, rendering it rather unsuitable for nanosecond lifetime measurements (but see Wild *et al.*, 1977; Spears and Hoffland, 1977). Similar remarks apply to the mode-locked train from a synchronously pumped dye laser, although such a laser can also serve for excitation of very short-lived samples (Kinoshita *et al.*, 1981). Consequently we confine our attention to laser systems in which the rate at which pulses are obtained from the laser is determined either by intra-cavity cavity-dumping or extra-cavity pulse selection with a Pockels cell.

3.4.2 Cavity-dumping in CW lasers

Pulses, at repetition frequencies suitable for SPC measurements, can be obtained from lasers operating CW if an acousto-optic modulator (cavity-dumper) is incorporated in the cavity (Roberts *et al.*, 1980; Rumbles *et al.*, 1982). In cavity-dumping the partially transmissive output mirror of the laser is replaced by a quartz plate placed at the focus of a high reflection antistigmatically compensated folded cavity, as illustrated in Fig. 3.17. To a non-optical face of the plate is attached a piezoelectric transducer which converts a pulsed r.f. signal into an acoustic wave propagating across the plate. This quartz plate now functions as a Bragg diffraction cell. The propagating acoustic wave causes regions of compression and rarefaction and, hence, periodic fluctuations in the index of refraction. In this way intra-cavity laser light is periodically deflected from its normal path and can be switched as pulses out of the cavity at repetition rates of from single shot to 5 MHz or greater.

The width of the output pulses is determined by the duration of the acoustic pulse and the fractional diffraction efficiency of the Bragg cell (Johnson, 1973). Pulse widths of 10 ns in SPC instruments have been reported (Roberts *et al.*, 1980). Peak power per pulse also depends on the diffraction efficiency and may in theory be as high as 50 times the average output power of the laser (Maydan, 1970). As a rough guide to the intensities obtainable in practice, a peak power of 20 W is reported for the cavity-dumped pulses from a 4 W Ar^+ ion laser (Rumbles *et al.*, 1982) while peak powers of 10 W have been obtained from the cavity-dumped output of a Rhodamine 6G dye laser pumped CW by this ion laser (Roberts, 1980). Cavity-dumped pulses have in general a highly reproducible interpulse separation (better than 1 part in 10^5) and a repetition rate that is easily varied over a wide range.

In older generation commercial cavity-dumpers reflection of the acoustic wave from a second face of the Bragg cell led to a secondary peak in the tail of the primary pulse. However, this peak, though sometimes fairly large (*c.* 5%

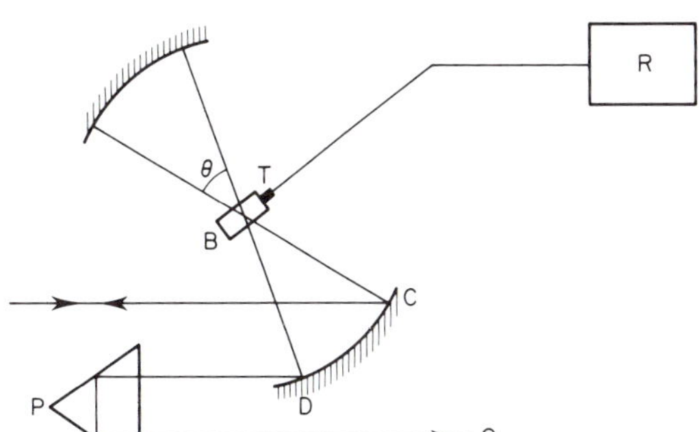

Figure 3.17 Schematic outline of accousto-optic cavity-dumping. A, input and returned beams; B, Bragg cell; O, output beam; P, prism; R, r.f. driver; T, piezoelectric transducer.

of the primary), is a true optical pump and is found not to interfere with deconvolution. In recent models it has been eliminated and interpulse optical noise is negligible. A potential source of distortion in SPC data is the modulation in the carrier wave for the r.f. pulse in the cavity-dumper. If the acoustic frequency is large a smooth optical pulse is obtained (Maydan, 1970), but if it is low enough to be resolvable by the SPC detection system the optical pulse observed will be modulated by a frequency twice that of the acoustic carrier wave. This modulation has been observed in SPC instrument response functions (Spears *et al.*, 1978) and decay curves (Roberts, 1980) and can distort data reduction. However, it can be eliminated by slight adjustments in the Bragg cell orientation.

The cavity-dumped CW laser is easy to operate and furnishes stable high intensity pulses at MHz repetition rates. Since mode-locking is not involved mechanical stability in the laser table need not be perfect. With commercial ion lasers a reasonably sized cavity-dumper can fit in the existing laser cavity; in dye lasers however the cavity may have to be extended. A cavity-dumper that has found wide acceptance in SPC instruments is manufactured by Spectra Physics Inc*. A similar device is now marketed by Coherent Radiation Ltd*. Details regarding the construction, pulsing and operation of a home built cavity-dumper have been given by Spears and Larsen (1977).

Cavity-dumping from CW lasers suffers from the crucial drawback of producing relatively broad excitation pulses. Pulses generated in this way, even after frequency doubling, do not have FWHMs much shorter than 7 ns. Consequently, in spite of high stability, this type of pulse generation is not ideal for the measurement of picosecond decay times.

3.4.3 Cavity-dumping from mode-locked lasers

Laser mode-locking has long been employed as a means of producing narrow pulses of high peak power. In SPC applications a mode-locked ion laser with a cavity-dumper selecting a fraction of the mode-locked pulses can furnish an effective excitation source (Roberts *et al.*, 1980). A dye laser pumped CW by an ion laser may also be mode-locked (Spears and Cramer, 1978). In either method good pulse amplitude stability may be achieved by synchronizing the acoustic pulses in cavity-dumper and mode-locker (Spears and Cramer, 1978).

Among the methods by which mode-locking may be achieved (Smith, 1970), the most common in SPC sources is loss modulation through acousto-optic Bragg diffraction. In this method the loss is modulated at a frequency of $c/2l$ where c is the velocity of light and l is the length of the cavity resonator.

*See Appendix 3.A1.

Stable mode-locking therefore depends on an exact match between this frequency and the cavity length. Variations in cavity length arising from vibrations and temperature fluctuations must be avoided, preferably through operation of the laser in a constant temperature environment on a pneumatically stabilized vibration-free table.

Mode-locked lasers usually have average output powers 25% to 35% of the average power that results from CW operation. Pulse widths are in the range 100–200 ps. For the dyes Rhodamine B, 6G and 110 at a dump rate of 5 MHz Spears and Cramer achieved average power in the mode-locked pulses of between 25 and 100 mW. These powers are much greater than output powers from flash lamps (see Section 3.3.4) and combined with the narrow pulse width render lasers operated in this way a very attractive excitation source. Pulse widths narrower by more than an order of magnitude may be achieved, however, in synchronously pumped dye lasers.

3.4.4 Synchronously pumped mode-locked dye lasers

If the dye laser cavity is extended so that its length matches to within 2×10^{-6} m the cavity length of the pumping mode-locked ion laser, the dye laser is said to be synchronously pumped (Harris *et al.*, 1977). A diagram of a synchronously pumped dye laser functioning as an SPC excitation source is given in Fig. 3.18. In this configuration the dye laser intermode spacing is an integral multiple of the mode-locker frequency and the dye jet is pumped synchronously by an ion laser pulse and a circulating dye laser pulse. Synchronous pumping yields output dye laser pulses of a few picoseconds duration (Chan and Sari, 1974) as illustrated by the autocorrelation trace in Fig. 3.19.

As in mode-locked ion lasers, the pulse separation in synchronously pumped dye lasers is determined by the length of the ion laser cavity and is usually about 10–15 ns. In one of the commercial synchronously pumped laser systems (Spectra Physics Model 171) the Ar^+ ion laser is mode-locked on the 514.5 nm line at an acoustic modulation frequency of about 41 MHz yielding output pulses at 82 MHz, pulse separation 12.2 ns. The average power of these pulses, the FWHM of which is about 150 ps, is 800 mW compared to 1.6 W for CW operation. This high power is desirable if the Ar^+ ion u.v. lines are to be mode-locked. It should be mentioned that it is difficult to introduce enough loss into the cavity to mode-lock the strong 488 nm Ar^+ ion line, although mode-locking of all the visible Ar^+ ion lines in an SPC excitation source has been achieved (Kinoshita *et al.*, 1981). In the commercial system referred to, dye laser pulses are output with a cavity-dumper usually operating at 4 MHz. Typical average powers at the wavelength of maximum intensity, 596 nm, are 60 mW. Although powers as high as 250 mW

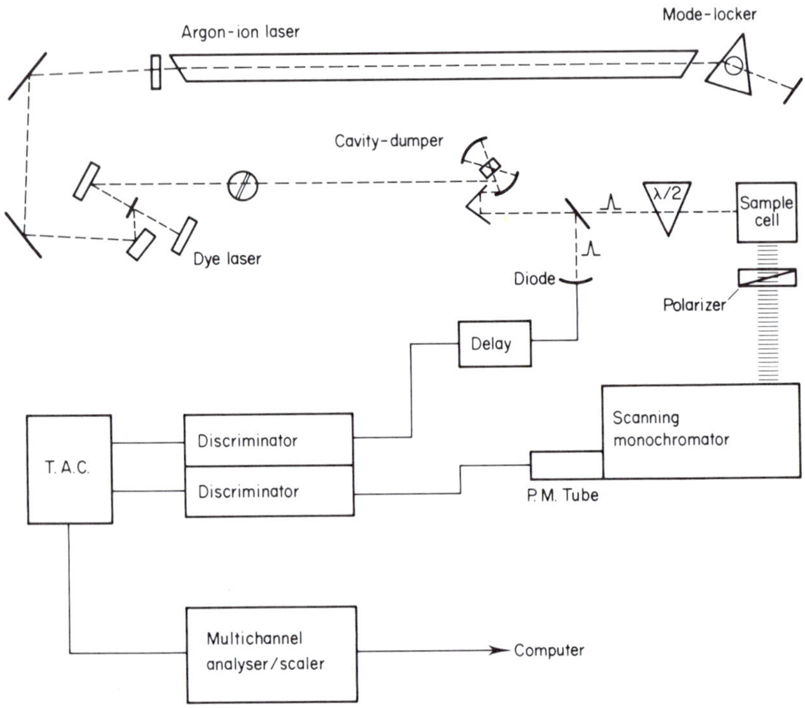

Figure 3.18 Synchronously pumped system configuration.

in cavity-dumped pulses have been reported (Bushaw *et al.*, 1978), in routine operation the power is more likely to be in the range 50–100 mW. Typical average powers for non-cavity-dumped operation are about 250 mW (Robbins *et al.*, 1980). It should be noted, however, that extra-cavity pulse selection, in which the unused laser power is lost, is a less efficient process than cavity-dumping.

In commercial synchronously pumped dye laser systems pulse stability is excellent over a few hours operation, although mechanical drift in the cavity-dumper optics and bubbles in the dye jet may cause some pulse jitter with consequent rising edge deconvolution problems. This jitter can be monitored by deflecting a small amount of the output beam into a photodiode and observing the pulse shape on an oscilloscope. For this purpose, since it is changes in pulse shape that are important, the diode and oscilliscope need not have exceptionally fast rise times. Jitter that cannot be eliminated by adjustment of the optics may have its source in fluctuating non-uniformity in the dye jet. Crystallized dye particles can accummulate on the nozzle and should be removed. Air bubbles in the jet can cause intermittent loss of mode-

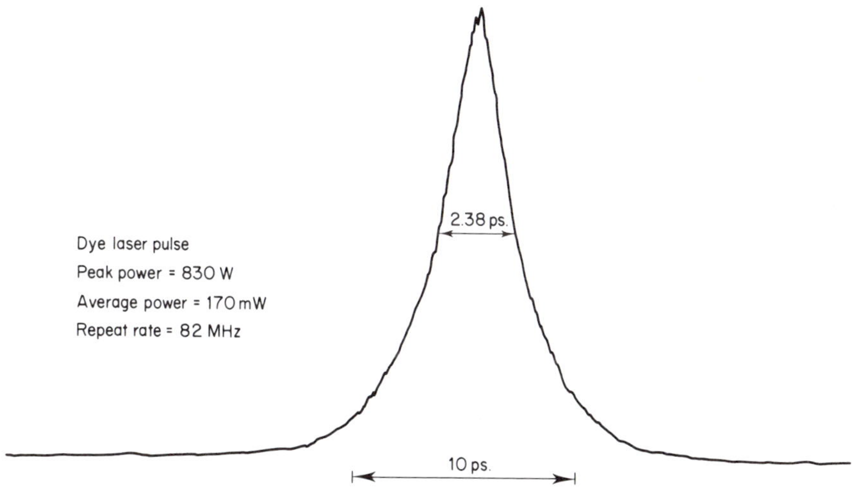

Figure 3.19 The autocorrelation function of pulses obtained from a cavity-dumped synchronously pumped mode-locked dye laser. From this function the pulse width is deduced to be 6 ps. The synchronously pumped system was Spectra Physics Model 171/375/342.

locking and thus affect the pulse stability. If careful attention to the dye pressure fails to remove them a special filter may have to be employed. Vibrations, transmitted along the connecting tubing from the dye pump to the jet, can also cause instability. They can be effectively removed by passage of the dye solution through a simple glass trap of the type illustrated in Fig. 3.20. It has also been reported that an electronic feedback mechanism from a photodiode in the cavity-dumper to the ion laser mode locker driver enhances pulse stabilities considerably (Haugen and Lytle, 1979; Yamazaki, 1982).

Temperature control of the laser environment is important, not only as an influence on mechanical stability, but also as a guard against mode-locker malfunction. If the dissipation of heat by the laser raises the room temperature too much the temperature stabilizer on the mode-locker oven will be unable to maintain control and mode-locking will be interrupted. The laser optics should be screened from airborne dust particles since the beam can burn such particles permanently on to optical surfaces. In many laboratories shoes for use only in the laser room are standard. If a dust remover is not fitted a gentle flow of nitrogen over the mode-locker (and cavity-dumper) crystal should prevent formation of inhomogeneities. Care must be taken to keep the nitrogen flow very low; otherwise instabilities in the ion laser pulses will result.

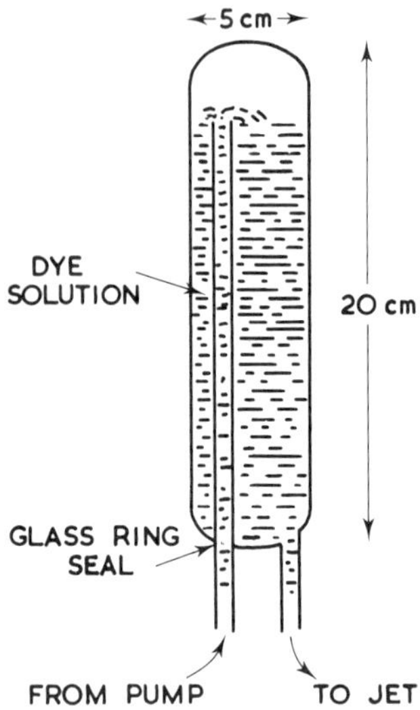

Figure 3.20 Glass trap for damping vibrations transmitted from dye pump.

3.4.5 Pulse selection with a Pockels cell

In the preceding section we have assumed that pulses leave the laser cavity at a repetition rate suitable for nanosecond lifetime measurements. The main disadvantage of selecting pulses in this way is the cost of commercial cavity-dumpers. Costs can be considerably reduced by selecting pulses from the mode-locked train with an extra-cavity Pockels cell shutter.

A Pockels cell is a device which, when subjected to a suitable applied voltage, behaves like a birefringent crystal. The birefringence is directly proportional to the applied voltage for a Pockels cell. When the voltage is zero the cell is not birefringent. The shutter consists of crossed polarizers and a Pockels cell such that upon suitable application of a voltage, of the order of 1–5 kV, to the Pockels cell, the laser pulse is transmitted. Pockels cells can be triggered simply by use of avalanche diodes, at some cost in terms of disturbance to detection electronics by resulting bursts of r.f. (Ware *et al.*, 1983).

SPC experiments with a Nd:YAG laser have also been performed by means of a Pockels cell (Sadkowski, 1981). Since the Pockels cell driver delivers relatively high r.f. powers very much more careful attention must be paid to electrical screening than is necessary with cavity-dumpers. The other main disadvantage of the technique in comparison with cavity-dumping is the relatively low repetition rate that can in practice be attained ($\leqslant 100$ kHz).

3.4.6 Wavelength selection

Lack of easy tunability over a wide wavelength range is the major disadvantage, apart from cost, associated with laser based excitation sources compared to flash lamps. Nevertheless, as is illustrated in Fig. 3.21 a wavelength range from 415 nm to 710 nm is available with a single Ar^+ ion laser–dye laser system. With frequency doubling (*vide infra*) a wide range of wavelengths can be produced.

As can be seen in Fig. 3.21 the dye Rhodamine 6G alone, gives access to the wavelength range from 560 to 650 nm, which is ideally suited for lifetime studies of compounds of interest in solar energy research such as porphyrins. When light at these wavelengths is frequency doubled naphthalenic compounds can be readily excited. In addition, owing to the high intensities available, anthracene and its derivatives can also be studied. In Fig. 3.22 is shown the decay curve of anthracene at 2×10^{-5} M in cyclohexane excited with frequency-doubled light at 300 nm from a Rhodamine 6G synchronously pumped dye laser. In spite of the window in the anthracene absorption at this wavelength the decay curve was collected in a matter of seconds and was deconvolved very successfully. Tryptophan, indole and many biological fluorescence probes are also amenable to study in the wavelength region from 280 to 325 nm, while benzene and its derivatives can be excited by the frequency-doubled Ar^+ ion lines.

With reasonable foresight in planning an experimental programme a change in the dye laser solution should not interrupt research, although a day or two will probably be given to attaining good mode-locked lasing with a new dye. It should be noted that in Fig. 3.21 the y-axis is logarithmic. Therefore considerably reduced output powers should be expected for dyes other than Sodium Fluorescein and the Rhodamines. Nevertheless, provided that the necessary optics and frequency-doubling crystals are available,

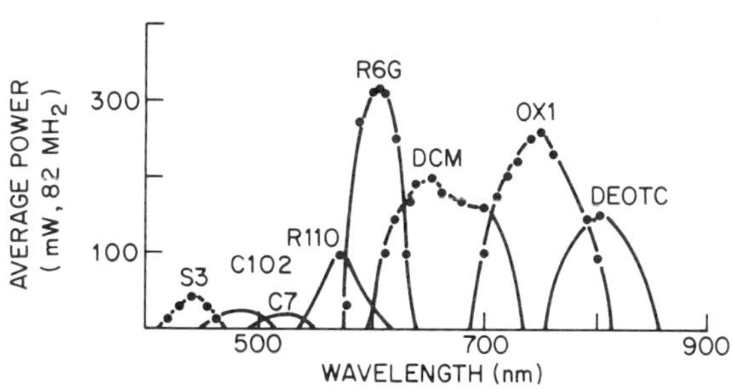

Figure 3.21 Wavelength range available with an Ar^+ ion laser–dye laser combination (from Spectra Physics).

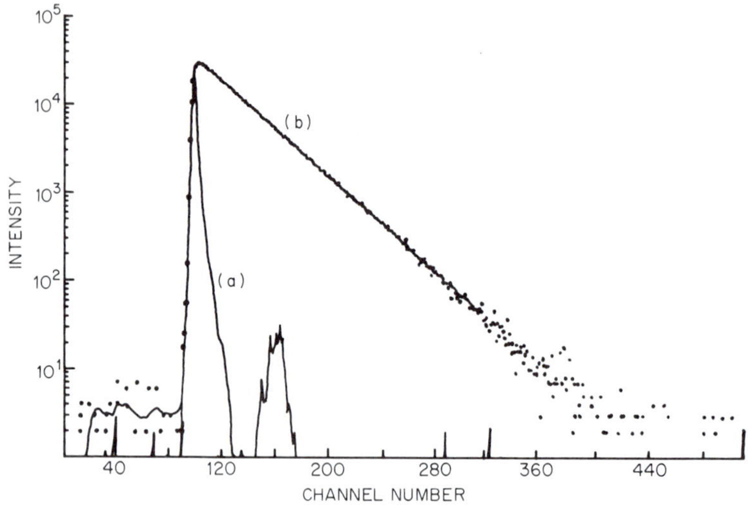

Figure 3.22 Decay of anthracene (2×10^{-5} M) in cyclohexane following laser excitation at 300 nm. (a) Instrument response function; (b) points, observed decay; line, fitted curve. 0.164 ns per channel, $\tau = 5.23$ ns $\chi_\nu^2 = 1.04$ DW $= 1.84$ (see Section 6.7).

intensities should be adequate for most experiments where excitation is in the range 225 to 700 nm.

(We should point out that the element for wavelength tuning in commercial dye lasers cannot be easily fitted with a scanning mechanism. It is to be hoped that the manufacturers will re-design this element in the near future.)

3.4.7 Second harmonic generation

The efficiency of frequency doubling (SHG) in a non-linear crystal is proportional to the square of the peak power of the undoubled pulses. This dependency (Franken and Ward, 1963) is given by the simplified equation:

$$P_S = \frac{J P_I^2 l^2 d^2 \sin^2\theta}{w_0^2} \frac{\sin\left(\frac{\Delta kl}{2}\right)^2}{(\Delta kl/2)} \tag{3.2}$$

in which J is a constant, P_I is the peak power of the incident pulse assumed to be a single mode Gaussian, l is the crystal length, d the SHG coefficient, w_0 the minimum beam waist, θ the angle between the crystal optic axis and the direction of the incident beam under phase matching conditions and Δkl is the phase mismatch between the fundamental and second harmonic frequencies. One consequence of this property of non-linear crystals is that the pulse width is shortened upon frequency doubling. To take an example already referred to, the cavity-dumped (Spectra Physics Model 166) output pulses at 596 nm of a CW dye laser have a width of 10 ns at half maximum compared to 7 ns for the frequency-doubled pulses at 298 nm.

Because of the power squared dependence, frequency doubling is an inefficient process for pulses of the powers encountered in mode-locked lasers operated at high repetition rates. As a typical example, when the cavity-dumped output beam of a synchronously pumped dye laser at 608 nm, repetition rate 4 MHz and average power 20 mW (1.5×10^{10} photons per pulse) was frequency doubled in an angle tuned ADP crystal, temperature stabilized at 35°C, the intensity of the u.v. pulses, as determined by ferrioxalate actinometry, was 5×10^6 photons per pulse (Phillips and O'Connor, 1981). This figure represents an efficiency of less than 0.1%. With some extra effort higher efficiencies may certainly be attained (Robbins *et al.*, 1980; Bushaw *et al.*, 1978) but for routine SPC operation powers at the sample in excess of 1 mW should not be expected. Even with much lower frequency conversion efficiencies u.v. operation of pulsed lasers offer a considerable improvement on conventional flash lamps (see following section).

In Table 3.3 we have listed a number of SHG crystals together with the

Table 3.3 Some frequency doubling crystals.

Symbol[a]	Tuning range of fundamental/nm	Comments
ADP	510–550	temperature tuned
AD*P	515–555	temperature tuned
ADA	565–520	temperature tuned
RDP	615–630	temperature tuned
RD*P	632–650	temperature tuned
RDA	690–700	temperature tuned
CD*A	1040–1060	temperature tuned
CDA	1060–1070	temperature tuned
KPB	434–510	angle tuned
KDP	(518–940)	angle tuned range in region 518–940 nm depends on cut
ADP	600–720	angle tuned

[a]ADP : ammonium dihydrogen phosphate;
AD*P : ammonium dideuterium phosphate;
KDP : potassium dihydrogen phosphate;
ADA : ammonium dihydrogen arsenate;
RDP : rubidium dihydrogen phosphate;
CDA : caesium dihydrogen arsenate;
KPB : potassium pentaborate;
etc.

wavelength range of the fundamental frequency. In general a crystal with a 45°Z cut (90° phase matching) is suitable for temperature tuning. More convenient to use however are angle tuned crystals for which a different cut yields a different tuning range.

3.4.8 Trigger pulse

For lasers, as for gated flash lamps, it is preferable to derive the TAC trigger pulse from a photomultiplier or diode viewing the pulses directly rather than from the electronics in the mode-locker, cavity-dumper, or Pockels cell driver. The diode may be placed behind the sample cell or in a position viewing a small deflection from the exciting laser beam.

3.5 Summary of Flash Lamps and Pulsed Lasers

In this section the relative advantages and disadvantages of pulsed laser and flash lamp excitation sources in SPC experiments will be briefly summarized under a few important headings. While storage ring radiation must be regarded as a somewhat special case, outside the convenient availability of most research workers, its properties will be mentioned in the appropriate sections.

3.5.1 Intensity

In general the intensity of the fundamental frequency from all the laser systems discussed in this chapter is orders of magnitude higher than radiation generated in flash lamps. Even an average output power from a dye laser as low as 20 mW at 600 nm represents 6×10^{16} photons per second, vastly in excess of the intensities available from flash lamps. In the u.v., where most of the output power of flash lamps is concentrated and where inefficient frequency doubling reduces laser powers by two orders of magnitude or more, closer scrutiny is necessary. From Table 3.2 it can be seen that deuterium- and air-filled flash lamps deliver typically, 3.3×10^{13} and 3.5×10^{14} photons per second, respectively, into a wide wavelength range and a solid angle of 4π steradians. If it is assumed that a lens can collect radiation emitted into 0.3 steradians and that losses in a monochromator amount to only 90%, the number of photons incident on the sample per second will be about 7×10^{10} and 8×10^{11} for deuterium and air, respectively. These intensities are still much lower than the 2×10^{13} photons per second available from the synchronously pumped dye laser referred to in Section 3.4.7. Even if wavelength is selected with an interference filter, air lamp intensities in a fairly narrow spectral band do not match those available from lasers. For the ACO storage ring facility at Orsay the intensity at the sample is calculated to be 6 $\times 10^{11}$ photons s^{-1} nm^{-1} at 300 nm (Lopez-Delgado *et al.*, 1974). If the experimental intensity is similar, the storage ring source compares favourably with air flash lamps but lacks the power of the laser.

3.5.2 Data collection time

The time required to collect a decay curve to a given precision is closely related, of course, to the intensity of the excitation source. An additional factor is the flash repetition rate, higher rates leading to more efficient TAC operation and shorter collection times. It is this factor, as much as their higher intensities, that makes measurement times with cavity-dumped lasers so short. The advantage is lost when pulses are selected with a Pockels cell, which is usually operated at <100 kHz. Typical collection times with 4 MHz laser excitation are 0.5 to 1 minute for a decay curve with 30 000 counts in the channel of maximum counts. Storage ring sources also operate at MHz repetition rates.

It should be noted that as the decay time of the samples under investigation gets longer the repetition rate of the cavity-dumper will have to be lowered in order to allow the fluorescence resulting from one flash to decay before the next excitation event, and so prevent miscorrelation in the TAC. Some loss in overall intensity usually accompanies lower dumping rates. However, for most nanosecond lifetime measurements a repetition rate of 800 kHz (1250 ns interpulse separation) will be sufficiently low. Flash lamps are capable of operating at a maximum of only 150 kHz and in routine operation are rarely driven above 100 kHz. Storage ring radiation and cavity-dumped lasers are therefore more efficient than flash lamps and relatively low repetition rate lasers (Pockels cell pulse selection) in time-correlated nanosecond decay time measurements.

3.5.3 Wavelength selection

Wide spectral continua are generated in high pressure lamps filled with nitrogen, deuterium and hydrogen. In the more common low pressure lamps the pressure broadened nitrogen continuum is lost and only a limited number of wavelengths is available. Deuterium- and hydrogen-filled low pressure lamps do furnish broad continua but at reduced intensity compared to nitrogen or air (Table 3.2). Dye lasers, while not yet developed to the stage where wavelength tuning over wide ranges can be performed automatically, are tunable with a single dye over 50 nm and concentrate all the intensity in an extremely narrow wavelength band. For this reason a laser source is probably the only reliable means of excitation in single vibronic level studies. With a certain amount of planning, experiments requiring excitation over a range of a few hundred nanometres can be performed with an ion laser–dye laser combination. It is still true to say, however, that for excitation wavelength scanning over a wide range a flash lamp is necessary. If this type of excitation is required the relatively low intensities provided by deuterium lamps (if the more common low pressure lamp is chosen) will have to be tolerated. The radiation from a storage ring is a broad continuum (Fig. 3.1) which can be easily scanned by means of a monochromator.

3.5.4 Spectral resolution of emission

The extent to which the spectral bandwidth of the fluorescence can be narrowed is directly dependent on the intensity of the excitation source. For this reason in SPC instruments with laser excitation sources a high emission monochromaticity can be attained. If optical systems in flash lamp instruments are carefully designed, monochromators can be placed in both excitation and emission light paths.

3.5.5 Pulse width

Flash lamp pulse widths are of the order of 0.5 ns. Synchronously pumped mode-locked dye lasers furnish pulses with typical widths less than 10 ps. The advantage of this type of laser is obvious. Mode-locked ion laser pulses have widths of 100–200 ps whereas cavity-dumping from a CW laser furnishes 10 ns pulses. Pulses of 3 ns can be obtained from a Pockels cell in conjunction with a CW laser. If very short decay times are to be measured the mode-locked laser is clearly preferable, especially in the light of recent advances in reducing detector response times (see Section 4.5). Pulse widths in storage ring sources vary considerably from one installation to another, but are generally in the range 100 ps–1 ns.

3.5.6 Pulse stability

We have emphasized that accuracy in lifetime measurements depends to a large extent on pulse-to-pulse reproducibility. If pulse stability is perfect the pulse width imposes no limitation on the time resolution of the experiment. However, no source is perfectly stable and pulse jitter, which is usually proportional to pulse width, will be smaller in absolute terms in a narrow pulse. There is no doubt that all excitation sources have periods of instability. Most of the "rising edge effects" that hinder successful deconvolution over the entire decay curve (see Chapter 6) probably have their origin in excitation pulse jitter.

Acceptable stability is difficult to attain in air-filled lamps owing to electrode corrosion by the discharge. Similarly a great deal of attention is required initially in order to obtain stable operation of mode-locked lasers. CW lasers, on the other hand, from which pulses are selected by means of a cavity-dumper, provide extremely reproducible pulses without a great deal of effort on the part of the operator. Deuterium, hydrogen or nitrogen are less difficult flash lamp gases than air to work with but are subject to intermittent instability effects.

It appears, therefore, that at the present stage of development neither flash lamps nor mode-locked lasers can be guaranteed to provide, without some effort, excitation sources of acceptable stability. It should be emphasized, however, that operating conditions for all sources can be found such that stable excitation pulses (as judged by the absence of rising edge effects in deconvolution) are provided over relatively long (hours) times. Long-term drift is probably ineradicable, so that shorter data acquisition times are preferable.

3.5.7 Instrument response function

There is thought to be little change in the pulse shape from hydrogen- or deuterium-filled flash lamps over the wavelength range 200 to 450 nm. While this belief has not as yet received direct experimental support, it is justified by the success in deconvolution (at least over the decaying portion of the decay curve) achieved with instrument response functions collected at the emission wavelength. However, some contribution to rising edge distortions may originate in a slight wavelength dependence of the pulse shape. Until it is regular practice to use the entire decay curve in the deconvolution procedure, no definite conclusion can be reached. Our reluctance to accept the hypothesis of unchanging pulse shape stems in part from the high likelihood of some air contamination of hydrogen- or deuterium-filled flash lamps. Pulse shape is known to change from one spectral line to another in the nitrogen $C^3\pi u \rightarrow B^3\pi g$ transition. Similarly, for laser sources the instrument response function must be recorded at the excitation wavelength. The pulse shape in storage ring sources is believed to be constant over a wide wavelength range.

It will be clear that as far as the correct determination of the true instrument response function is concerned deuterium- or hydrogen-filled flash lamps appear to have the advantage over pulsed lasers – an advantage that must be considered in conjunction with the measurements of and corrections for wavelength effects discussed in detail in Sections 2.3, 3.3.2(c), 4.4 and 6.3.

3.5.8 Polarization

Exciting radiation from laser and storage ring sources is completely polarized and is ideal for experiments designed to take advantage of polarization or depolarization of fluorescence. Light output from flash lamps has some but by no means complete polarization and is considerably attenuated upon passage through a polarizer.

3.5.9 Scattered light

Stray exciting light resulting from reflections in the sample cell will be greater for the more intense laser sources. On the other hand the laser beam is almost perfectly focussed in comparison to flash lamp radiation and can therefore be easily prevented, by means of baffles from reaching the detector directly. In addition, since the laser radiation is concentrated into a narrow spectral bandwidth it can be filtered much more effectively than flash lamp or storage ring scattered radiation.

3.5.10 Radio frequency noise generation

The amplitude of the r.f. pulses to cavity-dumpers and mode-lockers is not large enough to cause any pick-up effects in the detection electronics. Pockels cell drivers generate very powerful r.f. voltages from which the detection system must be screened carefully. Some flash lamps generate detectable r.f. signals while others are quite free of this source of distortion. If the manufacturers' specifications are to be believed, in commercial SPC instruments r.f. interference is absent.

3.5.11 Expense

The home built flash lamp is the least expensive excitation source: it can be constructed from materials available in most laboratories. Perhaps the most expensive part is the monochromator. Clearly, commercial flash lamps will cost more. Most costly are lasers, in particular synchronously pumped dye lasers. In addition to the cost of the laser (and mode-locker) it may be necessary to incur expenses for a cooling water supply, deionizer for cooling water, fast photodiode and oscilloscope, replacement ion laser tubes, optics and dyes for wavelength conversion, cavity-dumper, Pockels cell, service by the manufacturer, etc. Cavity-dumpers are very much more expensive than Pockels cells, but are probably preferable in all other respects in the SPC experiment.

References

Andre, J. C., Lopez-Delgado, R., Lyke, R. L. and Ware, W. R. (1979). *Appl. Optics* **18**, 1355–1359.

Badea, M. G. and Brand, L. (1979). *Methods Enzymol.* **61** 378–425.

Bennett Jr., W. R., Kindlmann, P. J. and Mercer, G. N. (1965). *Appl. Optics, Suppl. 02* 34–54.

Berlman, I. B., Steingraber, O. J. and Benson, M. J. (1968). *Rev. Sci. Instrum.* **39**, 54–61.

Birch, D. J. S. (1977). *J. Phys. E* **10**, 1044–1049.

Birch, D. J. S. and Imhof, R. E. (1981). *Rev. Sci. Instrum.* **52**, 1206–1212.

Birks, J. B., King, T. A. and Munro, I. H. (1962). *Proc. Phys. Soc.* **80**, 355–361.

Brody, S. S. (1957). *Rev. Sci. Instrum.* **28**, 1021–1026.

Bushaw, T. H., Lytle, T. E. and Tobias, R. S. (1978). *Appl. Spectrosc.* **32**, 585–587.

Chan, C. K. and Sari, S. O. (1974). *Appl. Phys. Lett.* **25**, 403–406.

Coolidge, A. S. and James, H. M. (1939). *Phys. Rev.* **55**, 184–190.

D'Alessio, J. T., Ludwig, P. K. and Burton, M. (1964). *Rev. Sci. Instrum.* **35**, 1015–1017.

Franken, P. A. and Ward, J. F. (1963). *Rev. Mod. Phys.* **35**, 23–39.

Garbury, M., Vogl, T. P. and Hansen, J. R. (1957). *Rev. Sci. Instrum.* **28**, 826–827.

Grinvald, A. (1976). *Anal. Biochem.* **75**, 260–280.

Halpern, A. M. (1974). *J. Amer. Chem. Soc.* **96**, 7655–7661.

Harris, J. M., Gray, L. M., Pelletier, M. J. and Lytle, F. E. (1977). *Mol. Photochem.* **8**, 161–174.

Hatchard, C. J. and Parker, C. A. (1956). *Proc. Roy. Soc.* **A235**, 518–536.

Haugen, G. R. and Lytle, F. E. (1979). *Opt. Commun.* **29**, 204–208.

Henry, M. S. and Helman, W. P. (1972). *J. Chem. Phys.* **56**, 5734–5736.

Hundley, L., Coburn, T., Garwin, E. and Stryer, L. (1967). *Rev. Sci. Instrum.* **38**, 488–492.

Imasaka, T., Kawabata, Y. and Ishibashi, N. (1981). *Rev. Sci. Instrum.* **52**, 1473–1477.

Johnson, R. H. (1973). *IEEE J. Quant. Elect.* **QE9**, 255–283.

Jones, O. C. (1964). *J. Sci. Instrum.* **41**, 653–663.

Kinoshita, S., Ohta, H. and Kushida, T. (1981). *Rev. Sci. Instrum.* **52**, 195–199.

Knight, A. E. W. and Selinger, B. K. (1973). *Aust. J. Chem.* **26**, 1–27.

Leskovar, B., Lo., C. C., Hartig, P. R. and Sauer, K. (1976). *Rev. Sci. Instrum.* **47**, 1113–1121.

Lewis, C., Ware, W. R., Doemeny, L. J. and Nemzek, T. L. (1973). *Rev. Sci. Instrum.* **44**, 107–114.

Lopez-Delgado, R., Tramer, A. and Munro, I. H. (1974). *Chem. Phys.* **5**, 72–83.

Lyke, R. L. (1976). Personal communication.

Lyke, R. L. and Ware, W. R. (1977). *Rev. Sci. Instrum.* **48**, 320–326.

Malmberg, J. H. (1957). *Rev. Sci. Instrum.* **28**, 1027–1029.

Maydan, D. (1970). *J. Appl. Phys.* **41**, 1552–1559.

Morrison, R. (1967). "Grounding and Shielding Techniques in Instrumentation". Wiley, New York.

O'Connor, D. V. (1977). Thesis, p. 37. University of Western Ontario.

Phillips, H. B. and Swank, R. K. (1953). *Rev. Sci. Instrum.* **24**, 611–616.

Phillips, D. and O'Connor, D. V. (1981). Unpublished results.

Rayner, D. M., McKinnon, A. E., Szabo, A. G. and Hackett, P. A. (1976). *Canad. J. Chem.* **54**, 3246–3259.

Roberts, A. J., Ghiggino, K. P. and Phillips, D. (1980). *J. Phys. E.* **13**, 446–450.

Roberts, A. J. (1980). Thesis. University of Southampton.

Rumbles, G. (1982). *In* "Photosynthesis of Synthetic Polymers" (Phillips, D. and Roberts, A. J., eds) pp. 5–25. Science Reviews Ltd.
Sadkowski, P. (1981). Thesis. University of London.
Schuyler, R. and Isenberg, I. (1971). *Rev. Sci. Instrum.* **42**, 813–817.
Smith, P. W. (1970). *Proc. IEEE* **58**, 1342–1357.
Soutar, I., Phillips, D., Roberts, A. J. and Rumbles, G. (1982). *J. Polym. Sci.* **20**, 1759–1770.
Spears, K. G. and Hoffland, L. (1977). *J. Chem. Phys.* **66**, 1755–1756.
Spears, K. G. and Larsen, J. (1977). *Rev. Sci. Instrum.* **48**, 472–475.
Spears, K. G. and Cramer, L. E. (1978). *Chem. Phys.* **30**, 1–8.
Spears, K. G., Cramer, L. E. and Hoffland, L. D. (1978). *Rev. Sci. Instrum.* **49**, 255–262.
Stuart, D. C. and Kirk, A. D. (1977). *Rev. Sci. Instrum.* **48**, 186–189.
Thomaz, M. F., Barradas, I. and Ferreria, J. A. (1975). *J. Luminescence* **11**, 55–75.
Wahl, P., Auchet, J. C. and Donzel, B. (1974). *Rev. Sci. I* **45**, 28–xx.
Ware, W. R. and O'Connor, Y. (1977). Unpublished results.
Ware, W. R., Pratinidhi, M. and Bauer, R. K. (1983). *Rev. Sci. Instrum.* **54**, 1148–1156.
West, M. A. (1981). Personal communication.
Wild, U. P., Holzworth, A. R. and Good, H. P. (1977). *Rev. Sci. Instrum.* **48**, 1621–1627.
Yagi, S., Hikida, T. and Mori, Y. (1978). *Chem. Phys. Lett.* **56**, 113–116.
Yamazaki, I. (1982). Personal communication.
Yguerabide, J. (1965). *Rev. Sci. Instrum.* **36**, 1734–1742.
Yguerabide, J. (1972). *Methods Enzymol.* **26**, 498–578.

Appendix 3.A1

SPC instrument manufacturers

Applied Photophysics Limited, 20 Albemarle Street, London W1X 4BS, UK.

Low pressure gated flash lamp (capable of operation at 200 kHz). Glass-bodied lamp. PM tube: Philips XP2020Q. ORTEC electronics. Digital Corp PDPII microcomputer. Software capable of single, double and triple exponential analysis with automatic shift (analytic method) and scattered light correction, and option to analyse with fixed shift and lifetimes.

Edinburgh Instruments Limited, Riccarton, Currie, Edinburgh EH14 4AP, Scotland.

Low pressure gated lamp capable of operation at 150 kHz but optimized performance is at 25 kHz. Metal lamp of coaxial design. PM tube: Philips XP2020Q. ORTEC electronics. Lamp design illustrated in Fig. 3.11.

Photochemical Research Associates, Meg Drive, London, Ontario, Canada.

Low pressure gated flash lamp capable of operation at 60 kHz. Similar to the design in Fig. 3.10. Side on Hamamatsu PM. ORTEC electronics (but

ELSCINT snap-off timing single photon discriminator). PDP 11 microcomputer with software capable of single, double and triple exponential analysis.

Horiba Limited, Miyanohigashi Kisshoin, Minami-ku, Kyoto 601, Japan.

High pressure free-running flash lamp capable of operation at 20 kHz. Horiba electronics including multichannel TAC (see Section 5.2.5(c)). Computer includes two CPUs so that simultaneous data analysis and acquisition is possible.

Synchronously pumped dye laser systems

Spectra Physics (USA)

Laser instruments Division, 1250 West Middlefield Road, Mountain View, CA 94042.
Tel. (415) 961 2550 TWX (910) 379 6941 Telex 348488

Albuquerque, NM
(505) 881 7577

Atlanta, GA
(800) 631 5693

Boston, MA
(800) 631 5693

Chicago, IL
(800) 323 0583
(312) 956 0882

Cleveland, OH
(800) 323 0583

Houston, TX
(800) 231 3567
[in Texas (713) 688 9886]

Los Angeles, CA
(714) 770 8545

Piscataway, NJ
(800) 631 5693
[in NJ (201) 981 0390]

Washington, DC
(301) 345 7333

Worldwide

Spectra Physics France, 3 rue Leon Blum, Zone Industrielle de Glaises, 91120 Palaiseau France.
Tel.: 33 6920 2500 Telex: (842) 691183

Spectra Physics GmbH, Siemensstrasse 20, D-6100 Darmstadt, West Germany.
Tel.: 06151 7080 Telex: (841) 419 471

Spectra Physics Limited, 17 Brick Knoll Park, St Albans, Herts, AL1 5UF, UK.
Tel.: (0727) 30131 Telex: (851) 23578

Spectra Physics B.V., Kanaaldijk Noord 61, 5642 Eindhoven, The Netherlands.
Tel.: 040 81 45 55 Telex: 51668

Spectra Physics AG, Schwiezergasse 39, 4056 Basel, Switzerland.
Tel.: 061 380075 Telex: 064335

Canada, Latin America, Pacific Areas

Spectra Physics International, 2905 Stender Way, Santa Clara, CA 95051, USA.
Tel.: (408) 249 5200 TWX: (910) 338 0220 Telex: 357 460

Coherent Radiation (USA)

Coherent Laser Division, Corporate Headquarters, 3210 Porter Drive, Palo Alto, CA 94304.
Tel.: (415) 493 211 Telex: 34 8304

Coherent: Eastern USA, 1000 West 9th Avenue, Suite A, King of Prussia, PA 19406.
Tel.: (215) 337 3035 Telex: 84 6179

Coherent: Midwest USA, 870 East Higgins Road, Suite 142, Schaumburg, IL 60195.
Tel.: (312) 843 1650 Telex: 28 3592

Worldwide

Coherent limited, Science Park, Milton Road, Cambridge CB4 4BH, UK.
Tel.: 0223 68501 Telex: 817466

Coherent GmbH, Hermanstrasse 54-56, D 6078 Neu-Isenburg, West Germany.
Tel.: 06102 27073 77 Telex: 4 185656

Coherent SARL, 55 Rue Boussingault, 75013 Paris, France.
Tel.: 1 5898939 Telex: 842 204 900

Coherent BV, Meenthof 15, 1241 C.P. Kortenhoef, The Netherlands.
Tel.: 35 62504 Telex: 43514

Coherent Srl, Residenza Mestieri, Milano 2, 1 20090 Segrate, Italy.
Tel.: (02) 2138905 Telex: 843320

4 Photomultipliers

4.1 Introduction

The width of the instrument response function in the single photon counting experiment is due to a combination of three principal factors, namely, the duration of the excitation pulse, the spread in transit time of photoelectrons in the fluorescence photomultiplier tube and timing jitter in the electronic components. The fluorescence photomultiplier (PM) is therefore a crucial component of the apparatus and in this chapter we deal briefly with its operation.

As was stated in Chapter 2, the PM serves merely to time individual photons emitted by the sample, and does not produce an analogue representation of the decay of emission. Consequently many of the dark noise problems associated with current pulse operation of PMs are avoided. Since the PM is exposed only to a very low level of radiation occurring in well separated pulses, photomultiplier fatigue (Lopez and Rebolledo, 1981) is not important in SPC measurements (Coates and Andrews, 1981). In addition the rise time of the PM anode pulse is in itself unimportant since the discriminator merely times the arrival of this pulse. What is important in the measurement is not, strictly speaking, the transit time spread of photoelectrons, but the statistical dispersion associated with this spread. That is, time resolution is dependent on the uncertainty in the delay time between the emission of a photoelectron from the cathode and the arrival of the corresponding electron pulse at the anode (Birks and Munro, 1967). As the transit time and corresponding spread become shorter the dispersion decreases and time resolution increases. It must be remembered, however, that the time resolution in the SPC technique is not equal to the total instrumental time uncertainty. The technique achieves its superior time resolution because it is a statistical technique deriving its results from a very large number of independent experiments (individual excitation pulses).

It is not intended in the following sections to give a technical description of the mechanisms governing photoelectron ejection and secondary emission. Rather we concentrate on pointing out the aspects of the PM operation that affect the time-correlated SPC experiment. Brief details are also tabulated concerning some commonly used PM tubes. It should be noted, however, that not all photomultipliers are suitable for timing operation and the prospective purchaser should verify with the manufacturer that the PM he wishes to purchase will suit his needs. Incidentally, the early drawbacks in this regard associated with group III–IV elements as photocathode materials (Longworth, 1977) seem to have been overcome. A more complete account of single photon photomultiplier detection has been given by, for example, Poultney (1972). Much useful information can also be found in the Application Manuals published by the manufacturers whose addresses are listed in Appendix 4.A2. For details related to the possible incorporation of a PM tube on the trigger channel of the instrument the reader is referred to Section 3.

Fast PM tubes which are suitable for single photon detection can be classified into four main groups:

(1) Side-on tubes with Venetian blind dynode structure. These tubes usually have a small number of multiplication stages (e.g. 9) and consequently a low gain. They are by far the least expensive of the four types.
(2) End-on tubes with linear focussed dynode structure, a large number of stages (e.g., 12 or 14) and a high gain. They are quite expensive, particularly when specially selected for low dark noise. This type of PM is standard in SPC instruments.
(3) Crossed-field photomultipliers, the electron focussing in which is achieved by means of crossed electric and magnetic fields. These tubes have very fast rise times and are reported to be free of the after-pulsing commonly observed in the electrostatically focussed tubes listed in (1) and (2). They have a low gain however and are quite expensive.
(4) Channel photomultipliers in which a continuous secondary emission coating within a cylindrical microchannel carries out the multiplication (Timothy and Bybee, 1977). Again these devices have a low gain and are quite expensive. They have, however, extremely fast rise times and seem likely to supplant conventional PM tubes as detectors in picosecond single photon counting instruments.

To date the most common PM tubes in SPC experiments have probably been the Philips (Amperex) 56DUVP/03 and the RCA 8850. Both of these tubes have given very satisfactory performance but are now being superseded by faster devices. They are not particularly suited for detection of long wavelength (>650 nm) emission, for which purpose red-sensitive equivalents are available. For detection of vacuum u.v. radiation solar-blind PM tubes with appropriate window materials, available for example from the EMR company*, have proved satisfactory. Readers who may wish to detect radiation of even higher energy are referred to the discussion of windowless

*EMR Photoelectric (see Appendix 4.A2).

PM tubes in the book by Samson (1967). In recent years inexpensive side-on tubes have gained in popularity; in one commercial SPC instrument* this type of tube is standard. With the increasing proliferation of mode-locked lasers as excitation sources, however, there will be a corresponding demand for PM tubes with time resolution of the order of picoseconds. Microchannel plates, while suffering from the serious disadvantage of low gain, appear now to offer a time resolution increased by a factor of 5 or 10 over the other varieties of PM. In the following sections we shall outline some of the important properties of the single photon detector with reference to specific PM tubes, the performance of which has been described in the SPC literature. Some of the photomultipliers and their characteristics are listed in Appendix 4.A1.

4.2 Characteristics

4.2.1 Spectral response

The spectral sensitivity of the PM depends on the cathode material and in a somewhat trivial sense, on the window material. Quartz or sapphire windows are necessary for radiation from 300 nm to 180 nm (the cut-off of fused silica) and are thought to be less fluorescent than pyrex windows. There are a large number of cathode materials, the responses of which can be seen in the manufacturers' catalogues. The standard bialkali (KCsSb) response is from 250 to about 650 nm. Red sensitive PMs, suitable for single photon counting, are available from most PM tube manufacturers and extend this wavelength range to 850–900 nm.

Of this type the RCA C31034 has attained some popularity (Spears and Hoffland, 1977) and the inexpensive side-on type, Hamamatsu R928 has also been reported as satisfactory (Kinoshita *et al.*, 1981). It will usually be necessary to cool the red-sensitive photocathode in order to reduce excessive thermionic emission (see Section 4.2.3) although the red-sensitive EMI 9863 QB has an acceptably low level of dark noise at room temperature (Lampert, 1981). Work in the vacuum u.v. as far as 110 nm can be accomplished with PM tubes having MgF_2 or LiF windows. For instance Lyke and Ware (1977) described the performance of an EMR 561F-09-13 tube with an MgF_2 window in this wavelength region. Dark current in such solar-blind tubes is extremely low, an advantage offset to some extent by low gain arising from low photocathode quantum efficiency.

*Photochemical Research Associates (see Appendix 3.A1).

4.2.2 Gain

Detected light levels in single photon counting are very weak; as a consequence it has been traditional to detect single photons with high gain PM tubes. Since high gain is generally achieved by increasing the number of multiplication stages (dynodes) the price paid has usually been a decrease in time resolution. For instance it will be seen in Appendix 4.A1 that the Philips 56DUVP/03 (Lewis *et al.*, 1973), which has 14 stages, achieves a gain of 2 $\times 10^8$ (the highest quoted gain, incidentally, of all SPC PM tubes known to us) with a photoelectron transit time of 43 ns. The manufacturer, wishing to develop an equivalent of this tube with a shorter transit time reduced the number of stages to 12 in the XP2020 (Lampert *et al.*, 1983) thus reducing the transit time to 28 ns but at the same time decreasing the gain by almost a factor of 10.

Low gain can be compensated for by amplifying the PM anode pulses before discrimination, by discriminating at a low amplitude level or by operating the photocathode at a higher potential relative to the anode. As a consequence there is a steadily increasing tendency to employ PM tubes with a small number of dynode stages such as the red-sensitive RCAC31034, an end-on type with 11 stages, or the side-on 9 stage types, e.g., RCA1P28 and Hamamatsu R928. The latter has a surprisingly high gain but has a response that seems to depend strongly on the area of the photocathode illuminated (see Section 4.2.4).

Gains specified in the manufacturers' data sheets can usually be improved upon by operation of the photocathode at a higher relative potential than the common operating potential. Tube performance, especially as regards dark current, may degrade more rapidly if the manufacturer's maximum rated voltage is exceeded, but it is likely that this rating is conservative in most cases. We have operated an RCA 1P28 at 1500 V (the manufacturer's maximum rating is 1250 V) for a number of months without any apparent degradation in performance. (Actual arcing between dynodes does not occur in this tube until the interstage potential reaches 460 V (Beck, 1976).) Damage to inexpensive tubes of course is not serious; it would be perhaps more prudent not to exceed the maximum rating for the expensive types.

Owing to fluctuations in the multiplication process the total charge developed is subject to a statistical uncertainty (Jones *et al.*, 1971; Wright, 1981). As a result the anode pulses generated usually show large variations in pulse height, which require complicated discriminator circuitry for correct timing (see Section 5.2.2). This feature is less pronounced in side-on tubes when operated at a high voltage (Kinoshita *et al.*, 1981).

Discrimination of PM anode pulses with levels as low as 5 mV can be accomplished with many, but not all, constant fraction timing discriminators.

A low discriminator level is advantageous from the point of view of acceptance of relatively more single photon pulses (see Section 5.3), but may result in a somewhat inferior signal-to-noise ratio and increased interference from spurious r.f. signals. If this source of distortion is absent and the PM tube is not too noisy, low gain may be satisfactorily tolerated without having recourse to an amplifier or preamplifier. These, however, are in common use and may in fact lead to improved performance. With microchannel plate PMs in particular, e.g., the Hamamatsu R1294U-01, which have extremely low gains but extremely rapid transit times, amplification of the anode pulses will probably be necessary. Marconi Inc., Hewlett Packard Inc. and Ortec Inc. (see Appendix 5.A1) market amplifiers and preamplifiers that are reported to give satisfactory performance.

4.2.3 Dark current

A high dark current is not usually a critical problem in single photon counting experiments, both because the background is easily subtracted from the observed data and also because dark counts occurring outside the TAC range are ignored. Nevertheless it is desirable to maintain the dark counts as low as possible, especially if weakly fluorescing samples are under investigation or if the emission is longlived. Dark count levels, even among PM tubes of the same type, seem to vary enormously from one tube to the next. Consequently it has been customary to have the manufacturer select from his stock, at an extra cost, a tube with low dark count level. While the extra outlay is probably justified the tube so selected should be checked before final purchase, since the selection might be based on a count rate obtained at a low operating voltage.

Dark current can be permanently increased if high levels of light are allowed to impinge on the PM tube, when polarized. Even when unpolarized it should not be exposed to room light for long (> 1 min) periods. Initial dark count rates can be substantially reduced through polarization of the tube with no incident light; in fact it is customary, in instruments that are always shielded from room light and when safety is not at risk, to leave the high voltage permanently on. It is common practice to operate the PM with the photocathode at a negative potential with respect to earth and the anode at earth potential. Operation with the anode at high positive potential and the cathode earthed reduces the number of noise pulses, presumably by suppressing electroluminescence from the glass envelope which is maintained at cathode potential (Leskovar *et al.*, 1976). An unacceptable dark current level cannot be definitely specified. A rough guide to performance however is our observation that, at a discriminator level of about 70 mV, common tubes have a dark rate of certainly less than 500 Hz and more usually in the range 50–250 Hz. This is not to say that if the dark count exceeds this level the tube should be rejected. It should also be remembered that water vapour round the PM pins and condensed on the tube envelope increases the noise. In humid conditions, therefore, it is advisable to keep the PM housing continually flushed with dry nitrogen. Operation with the cathode at earth potential is reported to be particularly effective for noise reduction in a humid atmosphere (Halpern, 1974).

Undoubtedly the most effective means by which dark counts can be reduced is by cooling the photocathode. As we stated previously, red-sensitive photocathode materials generally have an intolerably high level of thermionic emission at room temperature, and many red-sensitive PMs must be cooled. Usually bialkali photocathodes (e.g. KCsSb) show no decrease in dark count when cooled below −10°C and should not in any case be cooled below −30°C. On the other hand, red-sensitive photocathode materials such as AgOCs (S1 response) and trialkali (S20), may be cooled to −180°C and in general show increased dark counts above −30°C.

Conveniently compact thermoelectrically cooled PM housings are available commercially* for both side-on and end-on tubes. Alternatively a cooled liquid can be circulated through coils wound round the housing. If the cathode is operated at a negative potential with respect to earth the tube envelope is at the same potential, and coils cannot be allowed to touch it. When the tube and housing are cooled below 0°C care must be taken to ensure that water vapour condensed on windows and dynode pins does not limit intensity or increase the dark count rate. Dry nitrogen flushed continuously through the housing (if the PM chamber is not sealed) and passed over the housing window should eliminate any condensation.

4.2.4 Transit time

An electron ejected from the cathode causes the transmission of an electron bunch along the dynode chain of the photomultiplier. The transit time is the elapsed time between the moment of ejection of the cathode photoelectron and the arrival of the corresponding electron bunch at the anode. Since electrons ejected in the multiplication process will have a range of velocities and may also travel along different paths there will be a spread in transit times of the order, in a conventional PM, of 1 or 2 ns. In the 12-stage RCA 8850, Leskovar *et al.* (1976) report a transit time spread of 0.41 ns. Transit time spreads are not listed in Appendix 4.A1 because of lack of data. The

*From Products for Research (see Appendix 4.A2).

available data indicate that the spread is roughly proportional to the transit time and in very many cases equal to the anode pulse rise time (Lytle, 1974). Consequently a reduced transit time yields a reduced transit time spread and, by implication, a reduced statistical dispersion (see Section 4.1) with an ensuing increase in the time resolution of the experiment.

The spread in transit times depends on, amongst other things, the energy of the primary photoelectron, the point on the surface from which it is emitted, and the extent to which the resulting time spread can be minimized by electrostatic focussing. An estimate of transit time spread can be determined with the assumption that the dynodes are parallel plates distance d apart (Beck, 1976). If the p.d. between them is V volts an electron will travel a distance d in a time t given by

$$t = d\left(\frac{2m}{eV}\right)^{1/2}, \tag{4.1}$$

where m and e are the mass and charge of the electron, respectively. The decrease in transit time for an electron starting with an energy V_0 is roughly

$$\Delta t \approx \left(\frac{V_0}{V}\right)^{1/2} t \tag{4.2}$$

$$\approx \left(\frac{d}{V}\right)\left[\left(V_0\frac{2m}{e}\right)\right]^{1/2}. \tag{4.3}$$

Similarly the transit time spread per stage resulting from a path difference Δd is roughly

$$\Delta t = \Delta d\left(\frac{2m}{eV}\right)^{1/2}. \tag{4.4}$$

In fact transit time spreads determined from these equations are too high. Nevertheless the inverse dependence on interdynode potential is clear. Wavelength dependence in transit time spreads will be considered in Section 4.4 and dynode chain voltage dividers in Section 4.3.

Present day end-on PM tubes and their dynode chains are designed so as to minimize the time difference of travel from different points on the photocathode to the first dynode. A focus or grid electrode is incorporated for this purpose. It is nevertheless observed that the shape of the measured decay curve and the position of its maximum vary with the area of the cathode illuminated, especially if the voltage divider network is improperly adjusted (Lewis *et al.*, 1973; Leskovar *et al.*, 1976). Unless the scattered photons for the instrument response function measurement and the fluorescence photons strike the same area of the photocathode difficulties in deconvolution will ensue. In order to reduce the effects of this artefact the fluorescence and scatter should, if possible, emanate from the same volume (see Section 2.2.4). It also helps to mask the PM windows with a black screen with a small hole in the centre. The screen, which will undoubtedly also reduce the count rate, should be positioned as close as possible to the window without touching it.

Side-on PM tubes, which were not originally designed for timing purposes, are more prone to transit time variations resulting from path length difference between the cathode and first dynode than end-on tubes. Since there are no focussing electrodes these path differences, the possibilities for which can be clearly seen in the diagram of Fig. 4.1, are not equalized. Figure 4.1 applies to PM tubes like the RCA 1P28 with a viewing window that can be as much as 3.5 cm long by 1 cm broad. The effect of illumination of different points on this surface on the transit time spread is shown in Fig. 4.2, which is adapted from the work of Kinoshita *et al.* (1981). It can be seen that the principal effect of varying transit time spread is a shift in the position of the instrument response function, although a change in shape also occurs. It is important therefore to illuminate the same area of the photocathode throughout the SPC experiment. Since transit time spread also depends to some extent on the operating voltage as indicated by Equations 4.3 and 4.4, it is customary to operate SPC PM tubes at the maximum voltage recommended by the manufacturer.

4.2.5 Single photon resolution

In some PM tubes designed for timing applications, anode pulses corresponding to single photon events have a voltage distribution that is easily distinguishable from pulses corresponding to two and three photon events. This single photon pulse height resolution (Foord *et al.*, 1969) is very well defined in two commonly used RCA PM tubes, namely the 8850 (Morton *et al.*, 1968) and the C31034 (Gulari and Chu, 1977). The pulse height

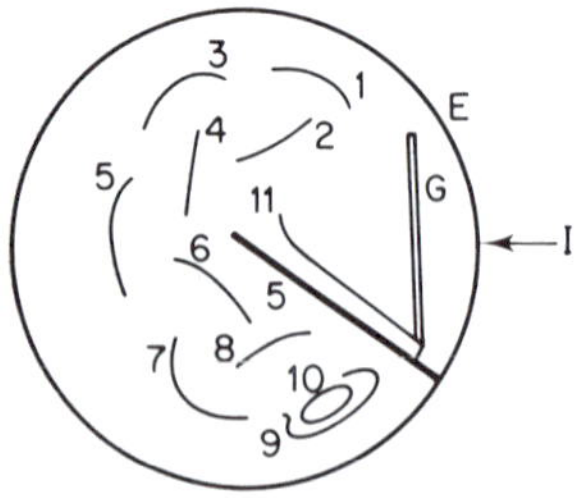

Figure 4.1 Dynode configuration of common side-on type PM tube (top view). 1–9, dynode numbers; 10, anode; 11, photocathode; G, grill; I, incident light; S, screen; E, glass or quartz envelope.

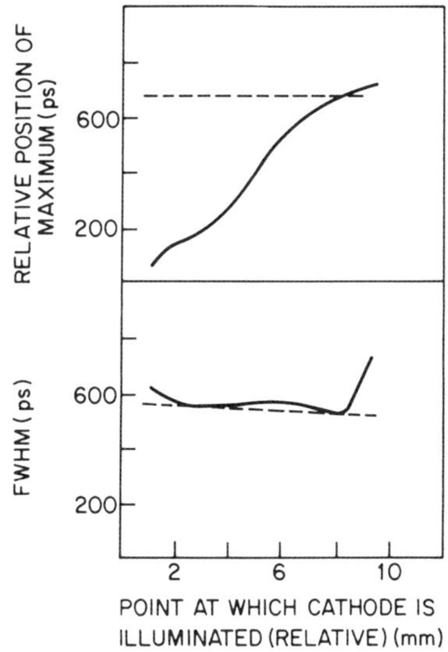

Figure 4.2 Variations of position and FWHM of instrument response function with point of illumination of side-on PM photocathode along short side (——) and long side (- - - - -) (after Kinoshita *et al.*, 1982).

spectrum of the RCA 8850 PM is illustrated in Fig. 4.3. Single photoelectron distributions for some EMI tubes have also been measured (Wright, 1981). The pulse height spectrum for the PM tube is obtained as follows (Morton, 1968; Kouyama, 1978). A low level of light is allowed to impinge on the photocathode. The signal from the anode or one of the later dynodes is routed to an integrating preamplifier, which essentially sums the amplitudes of all pulses input during a fixed time interval (say 50 μs) and outputs a pulse the amplitude of which is proportional to this sum. (The Ortec 113

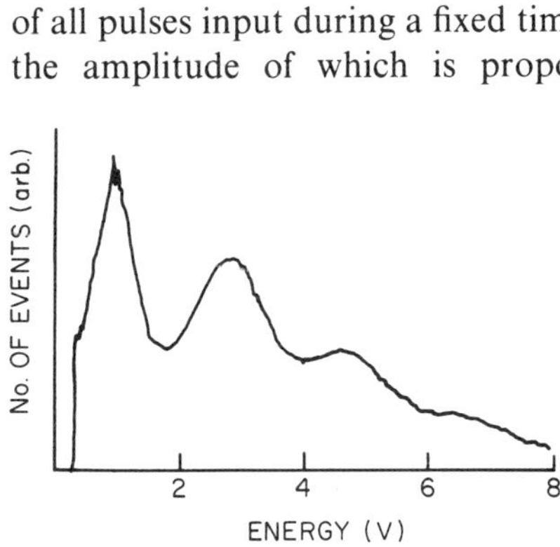

Figure 4.3 Pulse height spectrum for RCA 8850 PM tube operated at 2400 V (after Schuyler and Isenberg, 1971).

preamplifier is the usual choice.) It is generally necessary to amplify the PM pulses by a factor of *c.* 1000; therefore the pulses from the preamplifier may have to be routed to a second amplifier. However, the multichannel analyser, in which the spectrum is recorded, may possess sufficient amplification. The signals are routed to the AMP input of the MCA (see Section 5.2.7) and counts are accumulated in pulse height analysis mode. Channels now represent increments in energy so that the accumulated counts will show peaks corresponding to one-, two-, three- etc., photon events. If the level of light striking the photocathode is increased the peaks corresponding to more than one photon should grow in intensity (Kouyama, 1978). If necessary, the analyser channels may be calibrated with signals of known energy (Foord *et al.*, 1969).

On the basis of the pulse height spectrum a pile-up inspector may be designed, but only if the spectrum has a fairly narrow peak corresponding to single photoelectron events. For this purpose the signal to be inspected is taken from the last dynode, since the anode pulse is routed to the TAC. Therefore for the spectrum on which the energy window in the single channel analyser is based (see Section 5.2.5(b)), signals from this dynode should be processed. Pile-up inspection can reduce data collection times and may be advisable if thermally or photochemically unstable materials are under investigation. It depends for its success, however, on a small overlap between the energy distributions for one- and two-photoelectron events. Kouyama (1978) resolved the spectrum for the RCA 8850 into distinct distributions corresponding to one photon, two-photon and three-photon events, and showed that a suitable energy window can be chosen for the output pulses of this tube. In Section 5.2.5(b) the disadvantages attending pile-up inspection are discussed.

4.2.6 Magnetic focussing and screening

Transit time fluctuations can be reduced by focussing the electron bunch with a magnetic field. Magnetic focussing in conventional PM tubes, however, may create fringe areas in the photocathode and cause after-pulsing (Coates, 1975). Consequently it is rarely employed as a means of improving the timing capabilities in SPC experiments of conventional PM tubes. On the other hand, the recently-developed crossed-field photomultiplier, which employs crossed electric and magnetic fields for focussing purposes, is now commercially available and is suitable for single photon counting (see Section 4.5).

Accidental magnetic fields may interfere with the electron focussing and increase dark counts. It is therefore common to screen the PM tube from external magnetic fields by surrounding the tube with a cylinder of so-called mu-metal. Cylinders of this type, available from most PM tube manufacturers, should be about 1 cm longer than the length of the PM tube as illustrated in Fig. 4.4. The cylinder of mu-metal is preferably maintained at cathode potential and if the latter is at negative high voltage the cathode pin is attached through a high resistor (e.g., 6 MΩ) to the mu-metal. This connection should be made without subjecting the metal to undue strain, which might degrade its magnetic properties. Since the screen is now at high negative potential it must be insulated from the housing. A simple unscaled design for a housing, including magnetic screen, for an end-on PM tube is illustrated in Fig. 4.4.

4.2.7 Anomalous pulses

A cursory glance through the SPC literature will reveal that instrument response functions measured in most laboratories contain a secondary peak or shoulder that is thought not to be present in the pump pulse profile. It is now generally believed that this peak is a result of some photomultiplier process and it is usually displaced from the main peak by a fixed amount that depends on the particular PM tube. It seems, however, that this secondary peak is not present in instrument response functions measured with side-on PM tubes. A peak preceding the main peak is also sometimes observed.

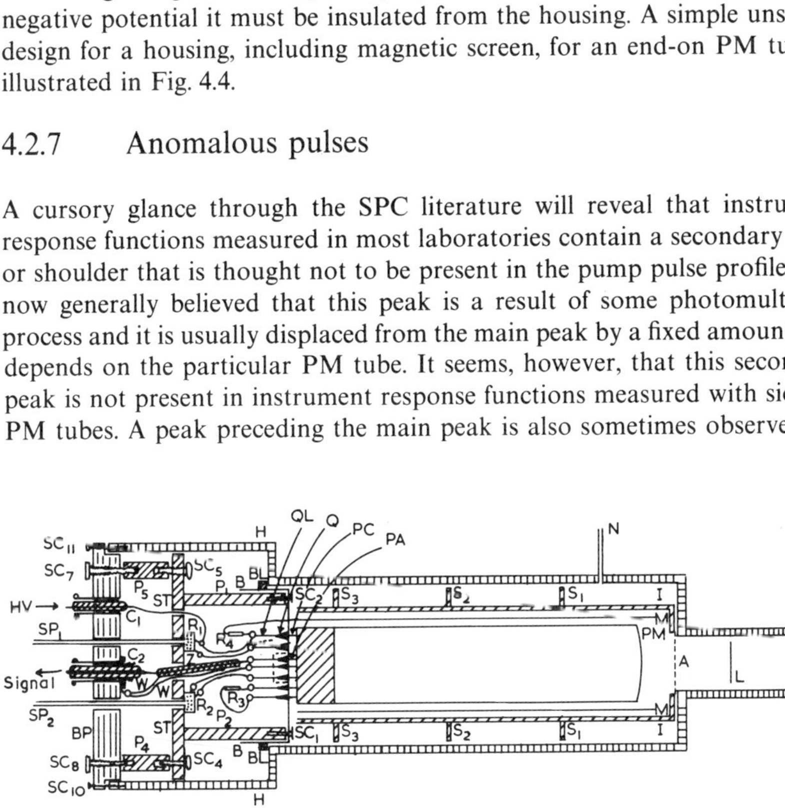

Figure 4.4 Design of housing for end-on PM tube (not drawn to scale) H, aluminium housing; N, port for N_2 flushing; L, lens or iris or quartz plate; A, aperture (if desired); I, plastic insulating cylinder; S_1–S_3, plastic lips on insulating cylinder; M, cylinder of mu-metal; B, commercial base with pin sockets (Q); QL, socket lugs; BL, lip on base; PC, cathode pin; PA, anode pin; R_4, resistor between mu-metal and cathode (*c.* 6 MΩ); R_3, fixed resistance; R_1,R_2, variable resistors (rheostats) mounted on stage (ST) with spindles SP_1, SP_2, ... protruding through back plate (BP); P_1,P_2(P_3 not shown), plastic posts connecting stage to base and attached with plastic (nylon) screws, SC_1,SC_2(SC_3); Z, coaxial cable; W, braid in coaxial cable; P_4,P_5(P_6 not shown), plastic posts attaching stage to back plate and connected with screws SC_4, SC_5(SC_6), SC_7, SC_8(SC_9); BP, back plate stepped so that r.f. seal (solder gasket) may be inserted; C_1, high voltage bulkhead plug; C_2, BNC type bulkhead plug.

The mechanisms that give rise to these spurious peaks are not known with certainty. After-pulsing is thought to occur as a result of an internal electron reflection, possibly from the first dynode to the photocathode (Stevens and Longworth, 1972). Whether a reflection over so short a length could cause (for the RCA 8850) a peak shifted by as much as 22 ns (Schuyler and Isenberg, 1971) is open to question. Whatever the cause, this peak can be minimized by careful adjustment of the interdynode voltages (see Section 4.3) and seems not to affect the success of deconvolution provided that it can be reduced to less than 1% of the height of the main peak. It should be borne in mind that flash lamps and pulsed lasers may produce real optical peaks which should not be confused with a peak resulting from a PM artefact (see Section 3.3). Unlike the peak resulting from after-pulsing, the early-time spurious peak seems to interfere seriously with successful deconvolution. It has been surmised that this peak results from photons that are not absorbed by the cathode but eject a photoelectron from the first dynode (Hartig *et al.*, 1976). It can be eliminated entirely through correct choice of interstage voltages. While a slight shoulder on the rising edge of the instrument response function will probably be tolerable in deconvolution, the intensity in this peak should be reduced as much as possible. In theory anomalous peaks should not affect mathematical analysis if they are produced in the same way in the measurements of instrument response function and decay curve. That they do implies perhaps that the mechanisms by which they are produced are influenced by some factor that changes between the two measurements, possibly the area of the cathode illuminated or the angle of illumination. Voltage divider adjustments might then be minimizing the effect of this factor on transit time spread while at the same time inhibiting the process by which the anomalous peaks are formed.

4.3 The Voltage Divider

Multiplication of the electrons in the PM tube occurs because the potential of each dynode is aranged so that it attracts the electrons ejected from the dynode preceding it. The resistance network that accomplishes this distribution of potentials is called the voltage divider, and adjustment of the resistors so as to achieve optimum PM performance is sometimes termed "tuning the PM tube".

If the PM power supply does not incorporate an internal voltage regulator the mains supply to it should be regulated. The anode pulse shape depends strongly on the applied voltage. Some PM tubes can be purchased ready-wired from the manufacturers. Recommended wiring diagrams can be obtained for all others. The wiring diagram is designed to divide the voltage so as to ensure linear performance by preventing space charge saturation and by providing an average current far in excess (10–100 times) of that formed by the anode. If these criteria are fulfilled the voltage divider will in most cases be suitable for the SPC experiment. However, it is usually found that the performance of the PM, as regards the FWHM and secondary peaks of the instrument response function as well as the signal-to-dark count ratio, can be improved by slight departures from the manufacturer's recommendations. As a consequence, in SPC instruments the PM is generally wired with a number of variable resistors, which can be "tuned" to give the best overall operation.

A suitable voltage divider is most conveniently chosen by following one of the manufacturer's designs and allowing at least three (and perhaps more) resistors to be variable. These should be, in standard end-on tubes, between the cathode (K) and grid (G), grid and first dynode (D_1), and last dynode and earth. Possible voltage divider designs for the popular 56 DUVP and XP2020 PM tubes are illustrated in Fig. 4.5 and 4.6, respectively. In the former there is provision for seven variable resistors, but it is usually found that all reasonable values of R_3–R_6 do not affect the measured instrument response function. Since variable resistors are relatively bulky it is preferable to employ only as many of them as can be accommodated in a compact r.f.-tight housing. Long leads from the PM pins to a separate box holding the voltage divider may function as antennae and should be avoided. Fixed resistors are of the precision carbon film variety since wirewounds have a high inductance and are not suitable. Capacitors, of the ceramic disc (low inductance) variety, are connected in parallel with the resistors over the last few stages in order to prevent space charge saturation at the pulse peaks. Moulded track miniature potentiometers with insulated spindles are satisfactory as variable resistors. If they are wired as rheostats, as illustrated in Fig. 4.5 and 4.6, the resistance (but not the potential) between separate stages can be altered more or less independently. It should be remembered that the potentiometer casings are not at earth potential and should therefore be mounted on an insulating platform, as illustrated in Fig. 4.4 (A diagram of a voltage divider suitable for a 9 stage side-on PM tube can be found in the recent publication of Ware *et al.* (1983)).

Direct wiring of the photomultiplier pins to the voltage divider mounted on a circuit board is reported to improve the noise counts (Lytle, 1974) but it is more usual to make use of the base with mounting sockets supplied by the manufacturer. Cable lengths should be as short as possible and solder joints clean and true. It is usual to use a 50 Ω load resistor from the anode to earth in order to attain a fast rise time in the associated electronics (Lytle, 1974) but we have found that any value for this resistor up to 10 kΩ leads to similar performance. In fact the 50 Ω input impedance of the following amplifier or discriminator can serve as the load resistor for the anode, but if no bleed

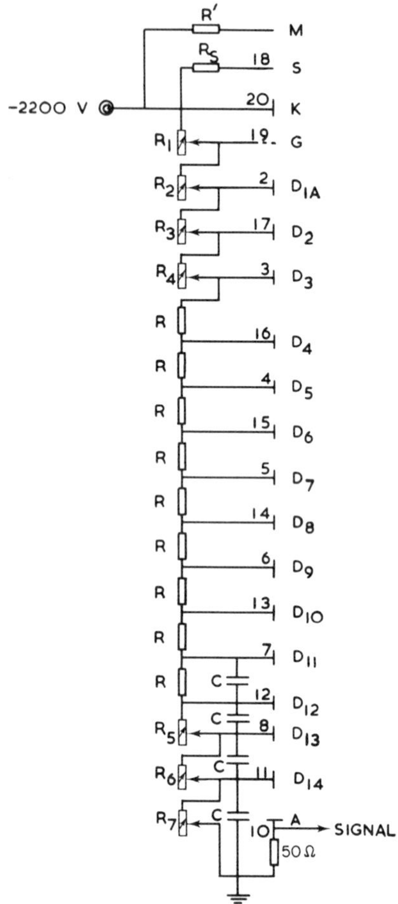

Figure 4.5 Suggested design of voltage divider for 14-stage Philips 56 DUVP photomultiplier. M, mu-metal screen; S, aquedag shield (internal connection); K, photocathode; G, grid (focussing electrode); D_{1A}, accelerating electrode and dynode 1; D_2–D_{14}, dynodes; A, anode; R_S, internal resistance (*c.* 16 MΩ); R_1–R_7, variable resistors (rheostats); R′,R, fixed resistors; C, capacitors. Suggested values: R′, 6 MΩ; R_1(0–20 kΩ), 6 kΩ; R_2(0–200 kΩ), 100 kΩ; R_3(0–60 kΩ), 40 kΩ; R_4(0–150 kΩ), 80 kΩ; R, 40 kΩ; R_5(0–100 kΩ), 40 kΩ; R_6(0–100 kΩ), 40 kΩ; R_7(0–100 kΩ), 40 kΩ; C, 10 nF.

resistor is employed great care must be taken to avoid damage to the electronics caused by the build up and sudden discharge of a high charge on the anode. Fifty ohm coaxial cable should be attached directly to the anode pin and earth connections made to a common earth. In order to seal the PM from r.f. interference all apertures should be as small as possible and the bulkhead connectors on the back plate should be of the type that can be

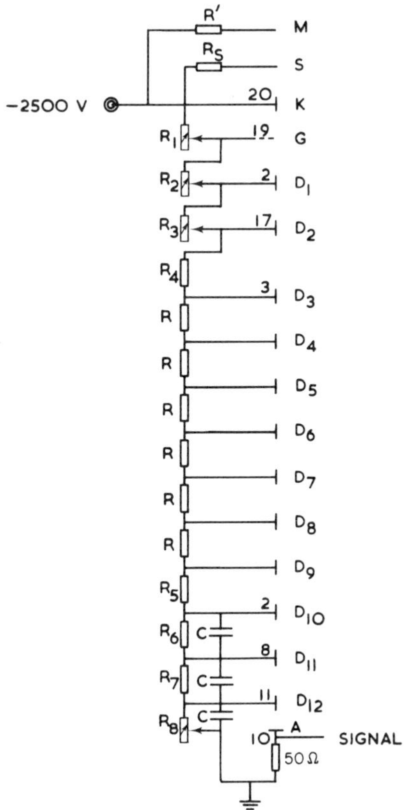

Figure 4.6 Suggested voltage divider design for 12-stage Philips XP2020 photomultiplier. M, mu-metal screen; S, aquedag shield (internal connection); K, photocathode; G, grid; D_1–D_{12}, dynodes; A, anode; C, capacitors; R_S, internal resistance; R′, R, R_4–R_7, fixed resistors; R_1–R_3, R_8, variable resistors (rheostats). Suggested values. R′, 6 MΩ; R_1(0–2 MΩ), 1.19 MΩ; R_2(0–2 MΩ), 940 kΩ; R_3(0–2 MΩ), 222 kΩ; R_4, 470 kΩ; R, 270 kΩ; R_5,R_6, 370 kΩ; R_7, 470 kΩ; R_8(0–2 MΩ), 600 kΩ; C, 10 nF.

sealed with a solder gasket (see Section 3.3.6). As an added precaution the potentiometer spindles can be taken out of the housing through waveguides attached to the back plate and sealed with solder gaskets.

Dark current may be reduced, as was stated in Section 4.2.3, if the pins are connected so that the cathode is at earth potential (Halpern, 1974; Leskovar *et al.*, 1976). This type of wiring is not common since it is necessary to couple the anode capacitively to the rest of the apparatus through an appropriately rated blocking capacitor. A positive high voltage is then applied to the last dynode. In addition to the hazard associated with degradation of the blocking capacitor there is also a risk, when the PM power supply voltage is adjusted, of inducing fast transient voltages of sufficient amplitude to damage electronic components. Nevertheless it is in theory reasonable to maintain the tube envelope at the same potential as the surrounding housing.

With reference to Fig. 4.6 a suggested procedure for choosing resistor values and tuning the divider can be outlined. The data sheet for the XP2020 PM tube gives the maximum current formed by the anode as 0.2 mA. This is the maximum current capable of being generated by the cascade of electrons from dynode to dynode. In single photon counting detection the light levels are so low that this figure can be reduced by some orders of magnitude, say, to 10^{-6} A. As was stated previously the current along the chain should be about 100 times greater (10^{-4} A minimum), setting a maximum value for the total chain resistance, at an applied potential of 2.5×10^3 V, at 25 MΩ, well above the 6.25 MΩ of the chain in Fig. 4.6. If the current is too high the power dissipation ($V \times I$) may generate too much heat. Consequently the resistors are kept fairly large. Since the current in the chain is ($2.5 \times 10^3/6.25 \times 10^6$) A $= 0.4$ mA, the power dissipated in the stage with the maximum voltage drop (K–G) is $I^2R = 0.19$ W. Therefore resistors rated at 0.25 W are satisfactory and are, moreover, conveniently small.

On the data sheet for the PM are also specified upper limits to the potentials to be applied between the various dynodes. While these recommendations are probably somewhat conservative it is found that too large a voltage drop between K and D_1 tends to induce a higher dark current. A large voltage drop, however, usually decreases the transit time spread. Hence a potential difference close to that specified as an upper limit by the manufacturer is probably a good compromise. For the XP2020 PM tube the recommendation is 800 V and the voltage divider suggested in Fig. 4.6 would provide 806 V at an applied voltage of 2500 V.

When the base has been wired and the resistances checked, pointers can be attached to the projecting spindles, and dials written on to the outside of the back plate so that the resistance produced by an adjustment of the spindle can be read off. In a rigorous adjustment procedure each resistance change should be accompanied by a change either in the applied high voltage or in the discriminator level in the CFTD. However, a purely empirical procedure in which only the PM resistors are altered will probably suffice.

The PM high voltage, which should be close to the highest value recommended by the manufacturers, is chosen together with a fairly low CFTD level, say 30–40 mV. An instrument response function is then collected, after the variable resistors have been set so as to achieve maximum counts. A note is then made of count rate, dark rate, and the shape of the instrument response function, particularly its FWHM and the intensities in any secondary peaks. Each rheostat is now adjusted in turn, but within a range in which the signal-to-dark count ratio is close to its highest value, and a note made of the shape of the instrument response function for each adjustment. It will probably be noticed that the most satisfactory shape is obtained with resistor settings that also yield close to the highest count rate, although some decrease from the maximum is to be expected. A secondary peak preceding the main peak in the instrument response function should be eliminated entirely, but it will probably be necessary to strike a compromise between minimum FWHM and minimum intensity in the secondary following the main peak. If the latter can be eliminated without a significant increase in the FWHM, so much the better. When the settings that yield the best compromise between high count rate, low dark current, narrow FWHM and minimum intensity in secondary peaks have been determined, a calculation is made to ensure that the resistances chosen do not give an excessive potential drop between any two stages. Finally the decay curve of a standard sample is collected and deconvolved. If the result is not satisfactory, and other sources of error can be ruled out, the resistors are re-tuned so that the compromise is biased in a different direction and deconvolution repeated. Once satisfactory settings have been attained re-tuning should not be necessary for a matter of months or even years.

The foregoing procedure refers primarily to end-on type PM tubes. Our experience with side-on types indicates that constant interdynode voltage all along the chain with capacitors in parallel with resistors across the last few stages leads to the best results.

4.4 Wavelength Dependence of Transit Time

The energy of the electron ejected from the photocathode depends on the work function of the cathode material and the energy of the incident photon. It has in the past been observed that, in spite of the developments in electrostatic focussing designed to minimize the effect of velocity variations in primary photoelectrons on the transit time through the dynode chain, the shape of the observed instrument response function depended on the wavelength at which it was measured (Lewis *et al.*, 1973; Wahl *et al.*, 1974; Andre *et al.*, 1979; Peterson *et al.*, 1979). In fact fast PM tubes are reported to show up to a four-fold increase in dispersion when the wavelength of the incident photons is reduced from 600 nm to 300 nm (Pietri and Nussli, 1968), with the main difference occurring in the cathode to first dynode region (Moszynski and Vacher, 1977). At the same time the actual transit time is expected to decrease in accordance with Equation 4.3. There have also been reports that some PM tubes with red-sensitive photocathodes such as the Philips XP1023 (Grinvald, 1976), the RCAC31034 (Spears *et al.*, 1978) and the Hamamatsu R928 (Ware *et al.*, 1983) show few wavelength effects.

A number of attempts have been made to ascertain to what extent observed wavelength variations in instrument response functions arise from PM effects. The difficulty arises because when the instrument response function is measured at different wavelengths observed variations may also result from pump pulse wavelength variations. In order to resolve this difficulty Ware and co-workers (Lewis *et al.*, 1974) measured instrument response functions at different wavelengths, both with and without a quantum counter placed in front of the photomultiplier. Light in the wavelength range of interest is totally absorbed by the quantum counter, the emission characteristics of which are assumed not to vary with excitation wavelength. Therefore the PM is exposed to photons of the same energy irrespective of the wavelength of the lamp's radiation. The broadening due to the time response (fluorescence lifetime) of the quantum counter can be incorporated in the photomultiplier response function. That is, the observed function, $P'(\lambda_e,t)$, is given by:

$$P'(\lambda_E,t) = E(\lambda_E,t) \otimes G(\lambda_Q,t) \otimes H(\lambda_Q,t) \quad (4.5)$$

$$= E(\lambda_E,t) \otimes H'(\lambda_Q,t) \quad (4.6)$$

where $G(\lambda,t)$ is the decay function of the quantum counter, λ_Q is the wavelength of quantum counter fluorescence, and the other symbols are as explained in Section 2.3. Of course the lifetime of quantum counter fluorescence should be as short as possible so that changes in $P'(\lambda_e,t)$ are not obscured by the extra convolution. It will now be understood that if $P(\lambda_E,t)$, given previously by Equation 2.20 as:

$$P(\lambda_E,t) = E(\lambda_E,t) \otimes H(\lambda_E,t)$$

shows a variation with λ_E that is not present in the function $P'(\lambda_E,t)$, the variation is a result of PM effects. The authors were able to conclude that both lamp and PM contributed to the overall effect but that the PM was mainly responsible. [Insertion of the quantum counter, sometimes called a wavelength shifter, in *all* SPC measurements could also serve to correct for the wavelength effects (Upton and Cline Love, 1979; Peterson *et al.*, 1979). It is not recommended because it reduces the sensitivity of the instrument and depends too critically upon the invariance of quantum counter fluorescence decay function on excitation wavelength.]

A further investigation of this problem, with wavelength independent pulses from a storage ring source, was carried out, again by Ware and co-workers (Andre *et al.*, 1979). In this experiment wavelength variations in the instrument response function must arise from PM effects. Since the shape of the storage ring pulse can be calculated the response functions of the PM tube can be derived. This function, for the EMR 651F-09-13 PMT, is shown at two wavelengths in Fig. 4.7. While the difference in FWHM between the two functions is not significant (FWHM is expected to, and is usually observed to, increase at shorter wavelength) the change in shape at longer times is quite marked.

On nanosecond timescales the pulses that are derived from synchronously pumped dye lasers (see Section 3.4.4) can be regarded as δ-functions since their FWHM is less than 10 ps. Consequently changes in shape within the time scale of 10 ps, as for instance upon second harmonic generation, will have only a minimal effect on instrument response functions measured on a nanosecond timescale. The change in the instrument response function between measurements at the doubled and undoubled light wavelengths should therefore be, to all intents and purposes, the change in the PM response function between these two wavelengths. These changes are illustrated by the functions shown in Figs 4.8, 4.9 and 4.10. In Fig. 4.8 are shown instrument response functions collected at 600 nm and 300 nm for an XP2020Q photomultiplier with two different voltage dividers. Clearly, the function shifts with wavelength. In addition in (b) a gross shape change is evident, whereas in (a) the change is not so obvious. In Fig. 4.9 the curves in Fig. 4.8(b) have been drawn with peak positions coincident in order to compare more closely the shape of the functions. Apart from some broadening in the wings at 600 nm the curves are not very different. It is worthy of note that the 600 nm function in Fig. 4.8(a) is considerably narrower than the

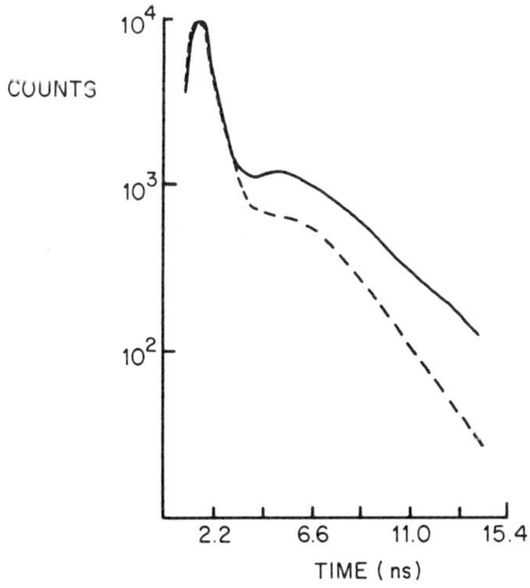

Figure 4.7 Response function of EMR 561F-09-13 photomultiplier at 220 nm (——) and 170 nm (-----). Reported FWHM at 220 nm was 1.31 ns and at 170 nm 1.21 ns (after Andre *et al.*, 1979).

600 nm function in (b), indicating that, when tuning the voltage divider, the FWHM of the instrument response function should not be the only guide. As a matter of fact both of the u.v. functions in Fig. 4.8 could be deconvolved quite satisfactorily from decay curves so that from this point of view it would be misleading to state that the voltage divider for (a) was maladjusted. Nevertheless, we feel that the PM tuning that produces the least shape change with wavelength is to be preferred.

Instrumental methods of correcting for PM wavelength effects are discussed in Section 2.3 and mathematical corrections in Section 6.3. Among the latter, the most common rests on the assumption that a wavelength-dependent PM response results merely in a time shift in the measured curve. In effect it is assumed that the photoelectron transit time varies but that the transit time spread remains constant. This assumption is perhaps a simplification but finds some support in the curves illustrated in Fig. 4.8(a) and 4.9. The shift to be seen in Fig. 4.8(a) is illustrated more clearly by the log plots in Fig. 4.10. Here there is again a slight change of shape but the major difference between the functions, again measured at 300 nm and 600 nm, is the shift along the time axis. The displacement is about 400 ps, in fair agreement with the figure of 1 ps per nm reported for this PM tube by Robbins *et al.* (1980). In Table 6.1 we show that mathematically a shift is an adequate correction for the wavelength effect. That it is may be explained by the mode of signal processing employed in the SPC measurement. Although the PM anode

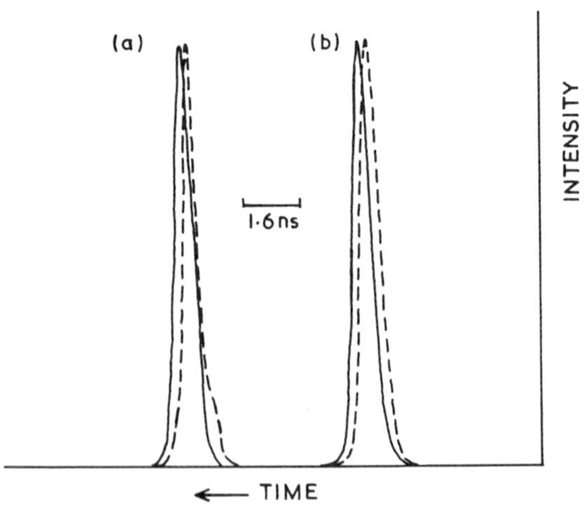

Figure 4.8 Instrument response functions of the fundamental, 600 nm (-----) and frequency-doubled, 300 nm (——) output of a synchronously pumped dye laser. The laser pulses, extracted with a cavity-dumper, have an FWHM of *c.* 6 ps. (a) and (b) represent the results of measurements with slightly different PM voltage divider.

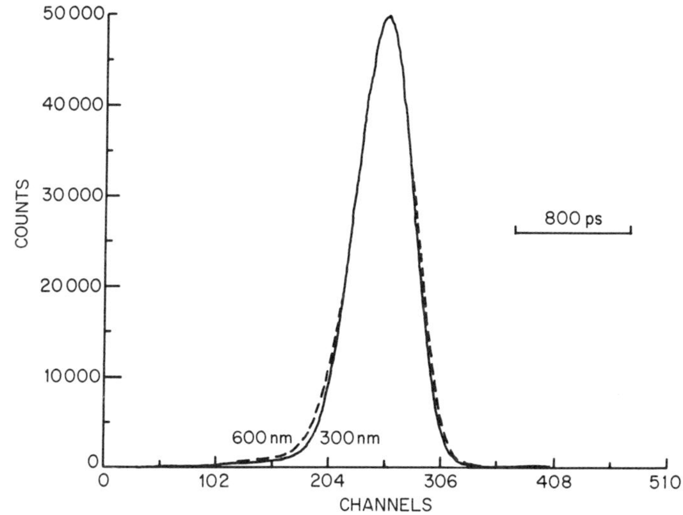

Figure 4.9 Comparison of the shape change of an instrument response function measured at 600 nm and 300 nm. Conditions are as in Fig. 4.8(b).

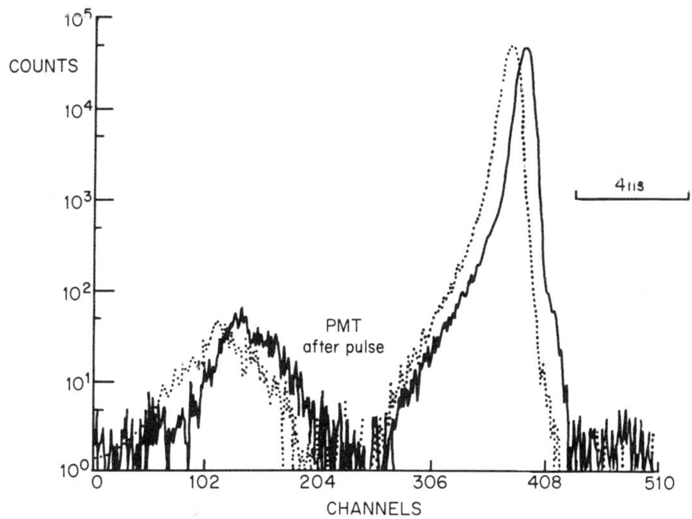

Figure 4.10 Comparison on a vertical log scale and expanded horizontal scale of functions measured under identical conditions as those in Fig. 4.8(b) (-----) 600 nm (——) 300 nm. Inverted TAC configuration.

pulse shape may vary with wavelength, the discriminator, which times from some point on the leading edge of the pulse, sees the variation merely as a time shift. In the light of the increased dispersion in fast PM tubes at shorter wavelengths this explanation seems somewhat simplistic. Nevertheless, the shift correction appears to be satisfactory in the analysis of nanosecond decays; whether it will remain so for picosecond decays remains to be seen. There is some evidence however that modern developments in PM technology are reducing these potentially serious wavelength affects. Ancillary difficulties associated with a mathematical shift routine are discussed in Chapter 6.

4.5 New Developments

With the increasing proliferation of picosecond lasers as excitation sources in single photon counting instruments the demand for PM tubes with transit time spreads, and therefore δ-function responses, in the tens rather than hundreds of picoseconds has risen. The most promising development has been the microchannel plate PM tube. It was also hoped to improve timing characteristics by replacing the purely electrostatic electron focussing of conventional PM tubes by focussing achieved with crossed electric and magnetic fields.

The performance of a static crossed-field PM, the Varian VPM-154, has been described by Koester (1979). Since the excitation source in his instrument was a picosecond laser, the duration of the pump pulse made a negligible contribution to the width of the instrument response function. The crossed-field PM has a low gain and must be coupled to an amplifier. Although the transit time dispersion was reported to be less than 30 ps the instrument response function had a FWHM of 228 ps, compared with 760 ps for an RCA 31034 and 360 ps for the RCA 8850. It is likely that the amplifier made a large contribution to this FWHM so that the improvement in PM performance is quite significant. Nevertheless instrument response functions of similar FWHM have been obtained with much less expensive PM tubes (Robbins *et al.*, 1980).

More promising than the crossed-field PM is the microchannel plate PM (Barker and Weston Jr., 1976; Yamazaki *et al.*, 1982). A microchannel plate is a secondary electron multiplier consisting of an array of millions of glass capillaries (called channels) with internal diameters 10 μm to 20 μm. These capillaries are fused into the form of a thin (<1 mm) disc and the inside walls are coated with a secondary electron emissive material. Both ends of the channels are covered with a thin metal film which acts as the electrode. A schematic representation of electron amplification in a single channel is given in Fig. 4.11(a) and of a microchannel plate PM tube in Fig. 4.11(b). It can be seen that the channels are sliced at an angle in order to prevent passage of the primary electron through the channel without striking the wall.

Among the drawbacks associated with microchannel plate PM tubes are ion feedback causing reduced photocathode lifetime and after-pulsing, and reduced counting rate resulting from the "dead time" required to neutralize the residual positive charge remaining in the channel after passage of the electron current. It is also reported that data distortions introduced by this type of PM are progagated differently in instrument response function and decay curve, i.e., that successful data deconvolution is difficult to achieve (Fleming, 1983). On the other hand these PM tubes are extremely fast with a very small transit time spread, have good single photon resolution and can achieve relatively high gains through incorporation of two or three microchannel plates. In addition they are reported to be subject to very small or negligible wavelength effects (Yamazaki, 1982). Dark current in microchannel PM tubes with red-sensitive photocathodes can be quite high, however; there appears to be no cooled housing available commercially at the time of writing.

In Fig. 4.12 is illustrated the instrument response function obtained for a picosecond pulse from a synchronously pumped dye laser with a Hamamatsu

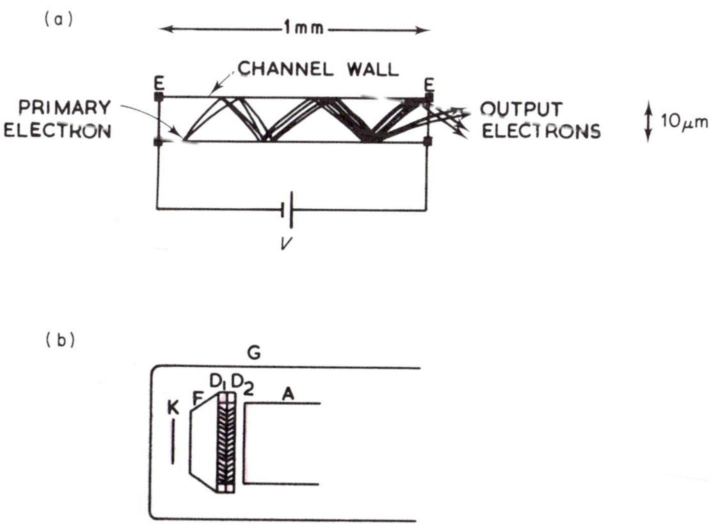

Figure 4.11 Schematic representation of (a) electron amplification in a single microchannel. E, electrodes. Parabolic trajectories determined by initial velocity. (b) PM tube with two microchannel plate discs D_1 and D_2. G, glass envelope; F, focus; K, semi-transparent photocathode; A, anode.

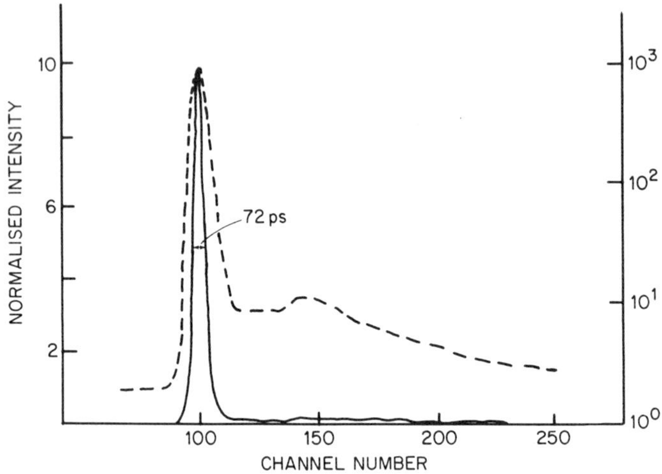

Figure 4.12 Instrument response function for laser pulse (FWHM 6 ps) obtained with Hamamatsu R1294U PM operated at 3000 V. (——) linear plot (- - - - -) log plot. 12.8 ps per channel. (Yamazaki, 1982).

R1294U two-stage microchannel plate PM tube operated at 3000 V. The FWHM of this function is 72 ps, considerably shorter than has been reported for conventional or crossed-field PM tubes. It should be noted, however, that electronic processing of the PM signals was carried out with the latest generations of electronic components, specifically a Hewlett Packard 8447D 1 GHz preamplifier, Ortec 583 CFTD incorporating an upper level discriminator (allowing selection of single photon pulses), and Ortec 467 TAC. Because signal transmission from PM tube to amplifier and discriminator can contribute substantially to the width of measured instrument response functions (Robbins *et al.*, 1980) the extremely narrow function in Fig. 4.12 is a result partly of decreased electronic jitter. Nevertheless, the performance of the PM is quite impressive. Recently Hamamatsu have produced a new "proximity type" two-stage microchannel PM tube, which is reported to yield an even narrower δ-pulse response function (Yamazaki, 1982). It will be noted that this PM produces a long tail, probably owing to ion feedback effects. It is to be hoped that in the development of these PM tubes, at present at a relatively early stage, this problem can be eliminated. At the time of writing the two-stage Hamamatsu microchannel PM tube costs about twice as much as conventional 12- or 14-stage end-on type PMs. It is possible that the cost will be reduced once production increases but even at the present price this type of PM must be seriously considered by the SPC user who intends to measure ps lifetimes. Further details can be found in Appendix 4.A1.

References

Andre, J. C., Lopez-Delgado, R., Lyke, R. L. and Ware, W. R. (1979). *Appl. Optics* **18**, 1355–1359.

Barker, J. R. and Weston Jr., R. E. (1976). *J. Chem. Phys.* **65**, 1427–1442.

Beck, G. (1976). *Rev. Sci. Instrum.* **47**, 537–541.

Birks, J. B. and Munro, I. H. (1967). *Prog. React. Kinet.* **4**, 239–303.

Coates, P. B. (1975). *J. Phys. E.* **8**, 614–617.

Coates, P. B. and Andrews, J. W. (1981). *J. Phys. E.* **14**, 1164–1166.

Fenster, A., LeBlanc, J. C., Taylor, W. B. and Johns, H. E. (1973). *Rev. Sci. Instrum.* **44**, 689–694.

Fleming, G. R. (1983). Private communication.

Foord, R., Jones, R., Oliver, C. J. and Pike, E. R. (1969). *Appl. Optics* **8**, 1975–1989.

Grinvald, A. (1976). *Anal. Biochem.* **75**, 260–280.

Gulari, E. and Chu, B. (1977). *Rev. Sci. Instrum.* **48**, 1560–1567.

Halpern, A. M. (1974). *J. Amer. Chem. Soc.* **96**, 7655–7661.

Hartig, P. R., Sauer, K., Lo, C. C. and Leskovar, B. (1976). *Rev. Sci Instrum.* **47**, 1122–1129.

Jones, R., Oliver, C. J. and Pike, E. R. (1971). *Appl. Optics* **10**, 1673–1680.

Kinoshita, S., Ohta, H. and Kushida, T. (1981). *Rev. Sci. Instrum.* **52**, 572–575.

Koester, V. J. (1979). *Anal. Chem.* **51**, 458–459.

Kouyama, T. (1978). *Jpn. J. Appl. Phys.* **17**, 1409–1418.

Lampert, R. A. (1981). Personal communication.

Lampert, R. A., Chewter, L. A., Phillips, D., O'Connor, D. V., Roberts, A. J. and Meech, S. R. (1983). *Anal. Chem.* **55**, 68–72.

Leskovar, B., Lo, C. C., Hartig, P. R. and Sauer, K. (1976). *Rev. Sci. Instrum.* **47**, 1113–1121.

Lewis, C., Ware, W. R., Doemeny, L. J. and Nemzek, T. L. (1973). *Rev. Sci. Instrum.* **44**, 107–114.

Longworth, J. W. (1977). *Photochem. Photobiol.* **26**, 665–668.

Lopez, R. J. and Rebolledo, M. A. (1981). *Rev. Sci. Instrum.* **52**, 1852–1854.

Lyke, R. L. and Ware, W. R. (1977) *Rev. Sci. Instrum.* **48**, 320–326.

Lytle, F. E. (1974). *Anal. Chem.* **46**, 545A-557A.

Morton, G. A. (1968). *Appl. Optics* **7**, 1–10.

Morton, G. A., Smith, H. M. and Krall, H. R. (1968). *Appl. Phys. Lett.* **13**, 356–357

Moszynski, M. and Vacher, J. (1977). *Nucl. Instrum. Meth.* **141**, 319–323.

Petersen, S. H., Demas, J. N., Kennelly, T., Gafney, H. and Novak, D. P. (1979). *J. Phys. Chem.* **83**, 2991–2996.

Pietri, G. and Nussli, J. (1968). *Philips Tech. Rev.* **29**, 267–287.

Poultney, S. K. (1972). *Adv. Elect. Elect. Phys.* **31**, 39–117.

Robbins, R. J., Fleming, G. R., Beddard, G. S., Robinson, G. W., Thistlethwaite, P. J. and Wolfe, G. J. (1980). *J. Amer. Chem. Soc.* **102**, 6271–6279.

Samson, J. A. R. (1967). "Techniques of Vacuum Ultra-violet Spectroscopy", pp. 188–205. John Wiley, New York.

Schuyler, R. and Isenberg, I. (1971). *Rev. Sci. Instrum.* **42**, 813–817.

Spears, K. G. and Hoffland, L. D. (1977). *J. Chem. Phys.* **66**, 1755–1758.

Spears, K. G., Cramer, L. E. and Hoffland, L. D. (1978). *Rev. Sci. Instrum.* **49**, 255–262.

Stevens, S. S. and Longworth, J. W. (1972). *IEEE Trans. Nucl. Sci.* **NS-19**, 356–359.

Swords, M. D. (1977). Thesis. University of Southampton.

Timothy, J. G. and Bybee, R. L. (1977). *Rev. Sci. Instrum.* **48**, 292–299.

Upton, L. M. and Cline Love, L. J. (1979). *Anal. Chem.* **51**, 1941–1945.
Wahl, Ph., Auchet, J. C. and Donzel, B. (1974). *Rev. Sci. Instrum.* **45**, 28–32.
Ware, W. R., Pratinidhi, M. and Bauer, R. K. (1983). *Rev. Sci. Instrum.* **54**, 1148–1156.
Wright, A. G. (1981). *J. Phys. E* **14**, 851–855.
Yamazaki, I. (1982). Personal communication.
Yamazaki, I., Murao, T. and Yoshihara, K. (1982). *Chem. Phys. Lett.* **87**, 384–388.
Zimmerman, H. E., Werthmann, D. P. and Kamm, K. S. (1974). *J. Amer. Chem. Soc.* **96**, 439–449.

Appendix 4.A1

In Table 4.A1 we give details of PM tubes that appear to have been most commonly used in SPC measurements. Also listed are some less popular tubes that appear to have some unique advantages. We need hardly stress that the list is by no means exhaustive. Nor is it meant to be assumed that we believe PM tubes in this list are more suitable for SPC instruments than others that we do not refer to.

The figures for cost are approximate and are given in pounds (£) sterling. Users of all the tubes listed have already been referred to in this chapter unless indicated by a superscript.

Figures for gain, transit time and anode pulse rise time have been taken, for the most part, from the manufacturers' literature. As regards transit time and anode pulse rise time the figures are not of great use since it is generally true that the manufacturers specifications can be improved considerably by voltage divider tuning. In addition the gain can be increased by operating the PM at a higher voltage than that applied in the manufacturers' tests.

Appendix 4.A2

The following companies manufacture photomultiplier tubes that are suitable for single photon counting and timing. In addition to their normal catalogues some manufacturers publish applications manuals containing a wealth of theoretical and practical information. Some of these publications are given with the address of the corresponding manufacturer.

(1) Philips, Eindhoven, The Netherlands.
[Amperex Fast Response Photomultipliers.]

(2) RCA, Harrison, New Jersey, USA.
[RCA Photomultiplier Manual.]

(3) Thorn/EMI, Hayes, Middlesex, UK.
[EMI Photomultiplier Manual and Application Notes.]

(4) Hamamatsu TV Company Limited,
1126 Ichino-cho, Hamamatsu, Japan.

(5) Varian Instrument Division,
611 Hansen Way, Palo Alto, California 94303, USA.

(6) ITT Electronic Services,

(7) EMR Photoelectric, P.O. Box 44, Princeton, New Jersey 08540, USA.

For PM Housings

(8) Products for Research Incorporated
78 Holton Street, Danvers, Massachusetts 01923, USA.

Table 4.A1

Manufacturer and Model No.	Spectral response (cathode material)	Gain (at applied voltage)	Anode pulse rise time/ns	Transit time/ns	Remarks	Cost £ sterling
EMI 9594QB[a]	200–750 nm S-20 ($CsNa_2KSb$)	1×10^7 (2900 V)	2	50	14 stages. Equivalent of 56DUVP	[h]
EMI 9863 QB	S-20	2.5×10^7 (2600 V)	2.5	45	Red sensitive. 1 cm diameter cathode[b]	997
Philips 56DUVP/03	200–600 nm (KCsSb)	2×10^8 (2200 V)	2	43	14 stages specially selected for low dark counts	[i]
Philips XP2020Q	200–600 nm (KCsSb)	3×10^7 (2200 V)	1.5	28	12 stages	450
RCA 8850	200–600 nm (KCsSb)	7×10^6 (2000 V)	2.1	31	12 stages. Good single photon resolution	640
RCA 8575[c]	200–600 nm (KCsSb)	1×10^7 (2000 V)	2.5	30	As for 8850	297
RCA C31034	200–950 nm (GaAs)	6×10^5 (1500 V)	2.5[d]	30	11 stages. Good SP resolution very flat response	827

RCA 1P28	250–600 nm (S-5)	2.5×10^6 (1000 V)	1.6	20	Side-on inexpensive	41
Hamamatsu R928	200–930 nm (NaKCsSb)	2×10^7[e] (1000 V)	2.2	22	As for 1P28 but red sensitive	112
EMR 561F-09-13	120–320 nm (CsTe)	1×10^7[f] (3000 V)	*c*. 2	*c*. 20	Solar blind. 13 stages. MgF_2 window, very low dark count	[j]
Varian VPM 154A-6D	200–1100 nm (InGaAsP)	1×10^5 (3300 V)	0.12	~1	Crossed field. Very expensive	>5000
Hamamatsu R1294 U-01	300–850 nm (NaKCsSb)	1×10^7 (3000 V)	[g]	~1	Microchannel plate	2500

[a]Reference: Swords, 1977.
[b]Small photocathode area reduces sensitivity but increases time resolution.
[c]Reference: Zimmerman *et al.*, 1974.
[d]Given as 0.8 ns by Lytle, 1974.
[e]This value from Kinoshita *et al.*, 1981.
[f]This value from Lyke and Ware, 1977.
[g]See Section 4.5.
[h]No longer available. Nominal replacement is EMI 9813QB at £391.
[i]No longer available. Nominal replacement is XP2020.
[j]Must be ordered.

5 Electronics

5.1 Introduction

At the heart of the single photon counting instrument lies the time-to-amplitude converter, usually referred to as the TAC. This device achieves the time correlation between excitation and emission events that is the basis of the SPC experiment. It is common to measure this time difference in a commercial TAC NIM standard unit. In fact the electronic modules in most SPC instruments consist of little more than the basic TAC with, usually, one or two discriminators for noise reduction, and electronic delay lines for correct pulse synchronization. For data storing it is customary to have a pulse height multichannel analyzer (MCA) incorporating an analogue-to-digital converter and memory stores. Data analysis in a computer is now probably universal.

Other components may also be used if more sophisticated pulse processing is employed. Frequently the pulses from low gain PM tubes must be amplified, in which case an amplifier or preamplifier will be necessary. Dual discriminators, linear gates, single channel analysers, timing filter amplifiers, clock modules and many other components have been used in extensions or modifications of the basic experiment. Needless to say, the more components that are used the more difficulties are likely to be encountered in setting up the experiment. In this chapter we shall do little more than describe in some detail the basic equipment and give references to more complicated instruments.

5.2 Individual Electronic Components

5.2.1 NIM Bin

Most electronic components used in SPC equipment are standard Nuclear Instrument Modules (NIMs) and take their voltage, through a connection on the rear panel, from a NIM power supply which is normally incorporated in a mounting crate with a certain number of sockets (usually 12). The power supply and crate are called a NIM Bin. Sockets at each module position always provide four voltages, ± 12 V d.c. and ± 24 V d.c. In addition

± 6 V d.c. lines are usually wired but may have to be connected to a source external to the Bin. Since some modules do require either + or -6 V d.c. power a NIM Bin/Power Supply with ± 6 V d.c. direct wired is preferred. If a new NIM unit (particularly a discriminator) does not function correctly the NIM Bin should first be checked for the ± 6 V d.c. lines.

NIM Bins are not subject to many faults and do not require much attention. Occasionally, a Bin from one manufacturer may become overloaded if modules from different manufacturers are mounted in it. Test points are usually provided on the front panel of the Bin so that each power supply voltage level can be checked when all the modules are in place. Perhaps the only other serious fault associated with NIM Bins is overheating arising from the close proximity of modules in the crate. If the crate is mounted in a rack on which vacuum tube equipment is also mounted overheating may be a serious problem. Even without this extra heat source the modules themselves can generate enough heat to cause damage to resistors or capacitors. More commonly, intermittent malfunctioning of some component occurs; it can sometimes be corrected by placing the module in the Bin in a position with an unoccupied module position on each side.

Many manufacturers of electronic equipment supply NIM Bin/Power Supplies. The most popular manufacture is probably ORTEC Inc. but the components are also available from Canberra Industries, Inc., and from Nuclear Enterprises Ltd. Addresses for these companies can be found in Appendix 5.A1.

5.2.2 Discriminators

The output from the fluorescence photomultiplier consists of a broad distribution of pulse heights, some pulses being generated by dark noise, some by single photon events and others by multiple photon events. It is essential to route these pulses through a discriminator in order to improve the signal-to-noise ratio and to furnish the TAC with constant amplitude pulses that are independent of the PM pulse shapes. However, since photomultiplier pulses of different amplitudes will not cross a discriminator level at the same time, leading edge discrimination is not recommended for the fluorescence PM pulses. On the other hand, leading edge discriminators usually have less internal jitter than discriminators involving more complicated circuitry (Nemzek, 1975); they are therefore quite suitable for pulses with very little amplitude jitter, such as the trigger signals for the excitation pulse. Discrimination for the single photon pulses is usually of the constant fraction timing or snap-off timing variety.

Trigger pulses for the pump pulse are usually derived from a photomultiplier, photodiode, antenna or from a logic circuit in the pulsing electronics.

Logic pulses are generated with very little noise and are sometimes routed, without discrimination direct to the TAC. Many TACs however, accept only negative-going pulses so that the trigger pulses may have to be inverted. Feed-through inverting transformers of the BNC type are useful for this purpose. Elimination of the trigger channel discriminator, while removing one source of time jitter in the instrument, is not recommended for signals from photomultipliers, photodiodes or antennae which may have a significant noise component. Leading edge discriminators for this channel are available from many manufacturers and are sometimes called FAST or 100 MHz discriminators. They accept negative pulses of *c.* 20 mV amplitude and usually have the input connector terminated internally in 50 Ω. Output pulses from this discriminator should of course be compatible with the input pulse requirements of the TAC.

As Fig. 5.1 illustrates, the broad distribution of pulse heights in the PM signals can give rise to errors when timing is performed with a simple, leading edge discriminator. Pulses A and B may have their source in photons emitted the same time after excitation but are seen to cross the discriminator level at different times. This jitter with leading edge timing has been verified by Kinoshita *et al.* (1981). However, level timing goes some way towards correcting the timing error resulting from leading edge discrimination. In this type of discriminator, sometimes called a differential discriminator, timing is from a low voltage level where the time difference between pulses of different amplitudes is fairly small (see Fig. 5.1) but noise is excluded by timing only those pulses that cross an upper discriminator level $-(V_2+\Delta V)$. A rather

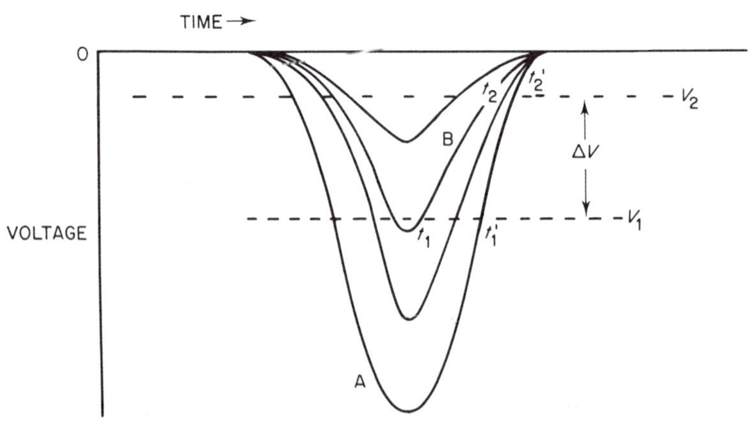

Figure 5.1 Schematic representation of timing errors associated with variable pulse amplitudes and leading edge discrimination. A, B, different amplitude pulses having the same time origin; $t_1-t_1^1=\Delta t_1$, timing error at high discriminator level, $-V_1$; $t_2-t_2^1=\Delta t_2$, smaller timing error at lower discriminator level, $-V_2$.

elaborate extension of this principle is described by Wild *et al.* (1977), who split the PM pulse into two channels, routing one channel to a leading edge discriminator and the other to a differential discriminator. Only pulses with amplitudes falling within a narrow range (set on the differential discriminator) are accepted and are timed at a low discriminator level by the leading edge discriminator. Although the jitter associated with this procedure is quite low the pulse rejection in the differential discriminator is probably so severe that the method can be used only for experiments in which the fluorescence count rate is extremely hgih.

In order to eliminate the timing errors occurring even in differential discriminators, discriminating with constant fraction timing is usually applied to the single photon pulses. As the name implies, with this type of discriminator the pulses are timed from a point on the leading edge that is a fixed fraction of the pulse height. For pulses of similar shape but varying amplitudes this timing point is constant. Figure 5.2 illustrates the way in which the constant fraction timing is achieved (Gedcke and McDonald, 1967, 1968). Suppose the input pulse has amplitude V_a and is to be timed at an amplitude fV_a on the leading edge. In the discriminator the output of the input buffer is fed through two paths, in one of which the pulse is delayed by an amount δ and inverted [Fig. 5.2(a)], while in the other the undelayed pulse is attenuated to a maximum amplitude fV_a. [Fig. 5.2(b)]. Both signals are then added to form the zero crossing signal [Fig. 5.2(c)]. Only pulses with amplitudes greater than a given threshold, set on the front panel, are timed.

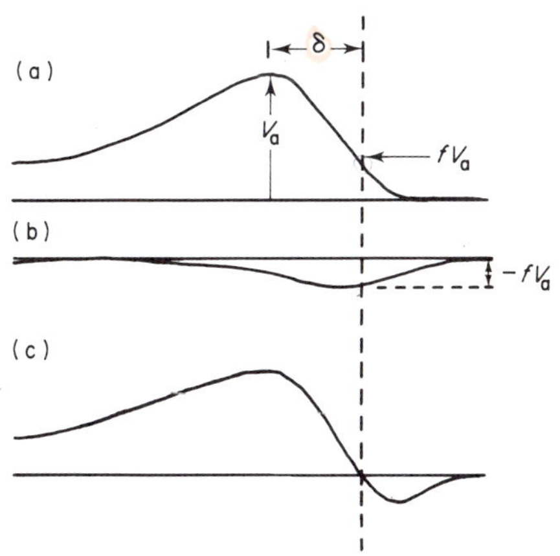

Figure 5.2 Schematic representation of constant fraction timing discrimination. Timing is from a position on the leading edge that is always a fixed fraction f of the input pulse height. Input pulse not shown. (a) Input pulse (amplitude $-V_a$) inverted and delayed by time δ. (b) Undelayed input pulse attenuated to maximum amplitude $-fV_a$. (c) Zero-crossing pulse (sum of (a) and (b)).

In Section 5.3 the setting of this threshold will be discussed. The most recent constant fraction timing discriminators (Bedwell and Paulus, 1976) seem to have much less internal jitter than earlier models and are therefore recommended. For example, Leskovar *et al.* (1976) developed a modified CFTD having a measured timing walk of only 35 ps. Commercially available components are now capable of equally good time resolution. Recently the ORTEC company has introduced a Model 934 Quad discriminator which includes four separate constant fraction discriminators in one NIM module. The Model 583 also incorporates an upper level discriminator and would appear to have excellent timing precision (Yamazaki *et al.*, 1982). It can be used for both the trigger and the fluorescence channels of the SPC instrument. Its main disadvantage is that it requires LEMO rather than BNC connectors, but as BNC to LEMO adapters are readily available this discriminator has become quite popular because of its good jitter characteristics and compactness. Simpler CFTD discriminators are also available from ORTEC Ltd and also from CANBERRA Ltd.

Instead of the CFTD discriminator a snap-off timing discriminator can be employed (Arbel *et al.*, 1974). This type, which is incorporated in one commercial SPC apparatus (Photochemical Research Associates), is available from the Elscint company.

5.2.3 Amplifiers

Amplifiers in SPC electronic arrangements are most commonly incorporated in three places, as illustrated in Fig. 5.3. With reference to the notation in this figure, A1, a preamplifier, may be essential if PM or trigger pulses are not of sufficient amplitude to trigger the discriminator. A2 is a timing-filter amplifier (TFA). It was once believed that the TFA, while improving signal-to-noise ratios, served as a filter for radio frequency interference (Lewis *et al.*, 1973). A3 is a bias amplifier (BA), and may be useful in the selection of the correct timescale. Otherwise it is not essential.

If possible, gain in the single photon pulses should be achieved in the PM, since the preamplifier (A1) may act as an antenna for r.f. interference. By introducing an additional time jitter it may also decrease the overall time

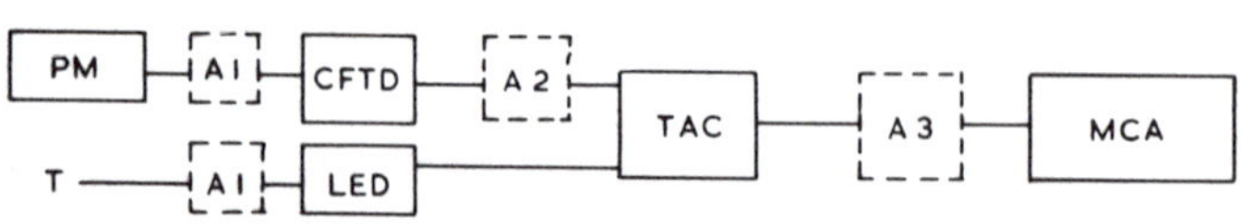

Figure 5.3 Position of amplifiers in SPC electronics. A1, A2, A3, amplifiers; T, excitation trigger. For other symbols, see Fig. 2.1.

resolution of the instrument, as Koester and Dowben (1978) have reported. It is also well established that the gross signal-to-noise ratio is improved if the amplification can be achieved in the photomultiplier (Kingston, 1978). However, when the current multiplication in the PM is small it is usually coupled to a preamplifier, which should have a fast rise time and an output pulse compatible with the input requirements of the discriminator. It should also be positioned as close to the PM tube as possible in order to minimize r.f. pick-up. A NIM module may not be suitable in this regard. A small preamplifier, which can be attached without support direct to a PM housing (if necessary through a plug-to-plug BNC adapter), is marketed by ORTEC Inc. (Lyke and Ware, 1977). This type of preamplifier usually derives its power from a NIM Bin Power Supply via some intermediate module (e.g., a Spectroscopy Amplifier), but the modification for direct wiring to the mains is easy to carry out. An advantage of this component is that it is easy to screen electrically. Excellent timing performance has been achieved with the Hewlett Packard Model HP 8447D 1 GHz preamplifier, which has outclassed the Ortec preamplifier when coupled to a microchannel plate PM tube (Yamazaki *et al.*, 1982). Another useful amplifier, available from Marconi Inc. (Robbins *et al.*, 1980), is powered without modification direct from the mains supply and is reported to have virtually insignificant time jitter and to improve instrument performance through correct pulse shaping. We have used this amplifier on the output pulses of an XP2020 Q photomultiplier and observed an improvement in performance of the instrument. However it should be recognized that spurious r.f. signals will also be amplified with such a system, and particular care must therefore be taken with screening when using a preamplifier. Our advice is that unless it is definitely necessary for pulse amplification a preamplifier may be more of a disadvantage than a benefit. The same remarks do not apply to the trigger side of the electronics where the discriminators threshold can be set high enough to exclude any r.f. noise picked up in the amplifier. Amplification of the trigger pulses is usually not necessary. Nor is there a need to amplify the output pulses from the single photon discriminator with a timing filter amplifier (TFA). It is an expensive and uncertain means of attenuating r.f. signals and can be difficult to adjust. Furthermore, there appears to be an advantage as regards r.f. suppression if gain is accomplished in the PM tube rather than in the electronics (Lyke and O'Connor, 1977), contrary to the belief of Lewis *et al.* (1973). When assembling a new SPC instrument, therefore, we would suggest that the added expense of purchasing a TFA is of doubtful value. As far as we know only one of three commercially available instruments incorporates this component (Applied Photophysics Ltd). In passing we should point out that the TFA can also be used on the PM side of the discriminator, where it serves as a preamplifier (Haugen *et al.*, 1979).

Although the correct placement of the decay curve in the analyser channels can be achieved by means of variable delay lines, a biased amplifier (BA) greatly facilities this operation and also provides in effect an increase in the number of time ranges available in the TAC. The mode of operation of this device will be described in conjunction with the details of the TAC.

Note: Some commercial manufacturers market photomultiplier tube bases that incorporate a preamplifier or a constant fraction timing discriminator. The latter is a very attractive feature, eliminating the cable connection between PM and discriminator, which may be an antenna for r.f. interference. However, we are aware of no reports of their use in a laboratory SPC instrument and are reluctant to make further comments beyond stating that this close coupling of preamplifier or discriminator to the PM anode will probably be a standard feature of future SPC instruments. At present these bases do not accommodate adjustments of the interdynode voltage divider and require some more development.

5.2.4 Delay lines

A variable delay line is an essential feature of every SPC instrument; it serves as a means of triggering the TAC–MCA combination is such a way as to achieve optimum placement of the decay curve in the MCA channels. In addition, if the TAC is operated in "reverse" mode it is preferable to STOP the voltage sweep with a trigger pulse corresponding to the excitation cycle in which the signal that STARTED the sweep occurred (see Section 2.2.6). This synchronization of STOP and START pulses is best achieved by means of a variable delay line. Furthermore, coaxial cable delays have been reported to act as high frequency cut-off filters and to reduce the oscillations arising from r.f. pick-up in the photomultiplier (Lewis *et al.*, 1973). Our experience, on the contrary, inclines us to the opinion that long lengths of cable furnish rather effective antennae for r.f. signals.

Calibrated variable delays are available in NIM single-width modules from a number of manufacturers. They fit conveniently into standard NIM Bins, but because they usually consist merely of lengths of coaxial cable they require no power. In these modules the maximum delay attainable is about 70 ns and can be reached in 1 ns or 0.5 ns steps. Longer delays require a bigger housing but, if no power is required, the large housing should not be inconvenient. Accuracy in commercial delay boxes is quoted at ± 100 ps for each delay section, so that if such delays are used to calibrate the TAC–MCA combination an independent measurement of the delay should first be made.

Home-built delay lines should be made from 50 Ω coaxial cable, e.g. RG-58C/U, and very careful soldering or crimping carried out. BNC bulkhead connectors can serve to carry the signals into the delay lines which must be housed in an r.f. tight metal box. BNC connectors are open to criticism as regards poor r.f. filtering, earthing continuity and transmission of high frequency signals (Lytle, 1974), but since most of the NIM modules accept only BNC connectors there is little to be gained by using a different type on the delay line housing. Lengths of cable yielding delays of say 10, 25, 50, 100 and 200 ns, measured on the approximation that signals travel at 30 cm per ns, can be housed in a box that could be accommodated in a conventional NIM Bin rack.

All delay lines, whether home-built or commercial, should be checked for pulse reflections since multiple pulsing, which can seriously distort the collected data, sometimes results from impedance mismatches and faulty solder joints at connectors. We emphasize that r.f. tight housings are essential for delay cables since, although serving as attenuators for r.f. picked up in the PM, the coaxial cable is itself an antenna. As an added precaution the delay line should be placed between the PM or trigger pulse source and the discriminator, so that further r.f. suppression may be achieved.

When the required delay is too long for a convenient length of coaxial cable certain types of pulse generator can perform the same function (Haugen *et al.*, 1979).

5.2.5 The TAC

5.2.5(a) *Function*

In this instrument the time correlation between excitation and emission events is carried out as depicted in Fig. 5.4. Upon receipt of a START signal,

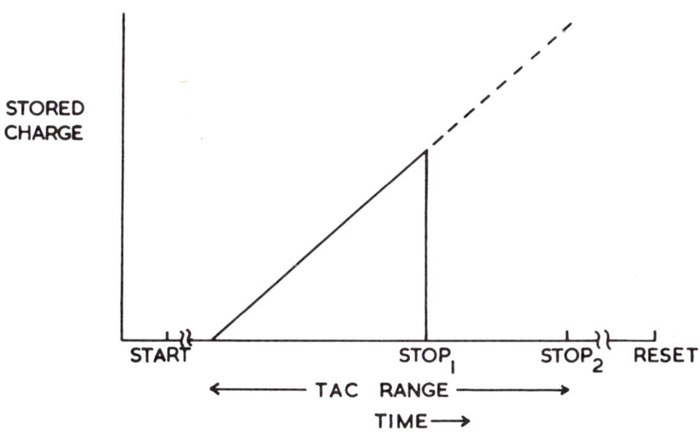

Figure 5.4 Schematic representation of TAC operation.

and after a certain fixed delay, a timing capacitor is charged linearly from a constant current source. The charging is discontinued ($STOP_1$) upon acceptance of a STOP pulse and an output pulse is generated with an amplitude derived from the final charge in the capacitor. Therefore the output pulse height is proportional to the time difference between START and STOP pulses. If no STOP pulse is received after a time called the TAC range, charging is automatically discontinued ($STOP_2$). Again a fixed time elapses before the capacitor is RESET, at which time the instrument is ready for another START pulse.

TAC ranges, usually from 50 ns to 80 μs, are switch selectable on the front panel. Hence if a range of 100 ns is selected the maximum output pulse amplitude, generally 10 V, corresponds to a 100 ns time difference between START and STOP signals.

In addition to the START and STOP inputs and the output (called the CONVERSION output) a standard TAC unit will generally possess a second CONVERSION, a TRUE START output, GATE and STROBE inputs and an INTERNAL STROBE facility. The TRUE START output can be used to monitor the number of signals that actually start the voltage ramp. A more useful monitor is that of the number of START signals converted to output signals (measured at one of the CONVERSION outputs) since this number can be taken as the number of "detected" photons and is adjusted to achieve single photon conditions (see Section 2.2.1). The GATE input in conjunction with the EXTERNAL STROBE is sometimes made use of to block TAC conversions that cannot be processed by the analogue-to-digital converter. In addition, pulses can be furnished through the EXTERNAL STROBE input to inhibit a converted output under certain conditions, e.g., when a multiple photon event has been detected. There may also be a BUSY output, from which pulses can be routed to other electronic equipment, for example, the repetition rate driver of the excitation source.

As discussed in Section 2.2.6, the TAC may be operated either in the conventional or the "inverted" configuration. If, as in most simple instruments, none of the additional options are chosen, the GATE is set at ANTICOINCIDENCE with no GATE input signal, the BUSY output is not used, and the STROBE option is set at INTERNAL. On some modules it is advised to terminate unused output connectors in 50 Ω. The relative advantages of conventional and inverted configurations have been thoroughly investigated by Haugen *et al.* (1979). They show that if the experiment is carried out under single photon conditions, the conventional configuration should not be used above START rates of *c.* 150 kHz. At higher repetition rates the "inverted" mode is necessary, and should possibly be used also at low repetition rates since it will always lead to some reduction in data acquisition times. However, in instruments that incorporate a pulse pile-up inspector the inverted mode of operation is not suitable (Haugen *et al.*, 1979). We have already stated that when operating the TAC in the "inverted" mode correct synchronization of START and STOP signals is advisable (Section 5.2.4). This synchronization is achieved by varying the delay between the START and STOP signals until the position of the curve in the MCA channels is independent of the excitation pulse repetition rate.

Perhaps the most serious source of distortion in the TAC is non-linearity associated with the sweep of the voltage ramp. Since the non-linearity is most severe at times corresponding to small charges on the capacitor the distorted decay curve is likely to show spurious oscillation in low-numbered MCA channels. Every TAC should be checked for differential non-linearity, if possible before purchase. One simple check entails the processing of uncorrelated signals; the START signals are taken from the excitation source or a pulse generator and random STOPs can be supplied by the PM, exposed if necessary to a small amount of room light. Counts should build up in the MCA in a horizontal line. With a 50 ns TAC range any non-linearity will be clearly evident when 2000–3000 counts have been accumulated in each channel. Alternatively, an experiment similar to time calibration with a variable delay line (*vide infra*) can be performed and the amplitude of the TAC conversion output pulse measured on an oscilloscope. A plot of delay against pulse height should be a straight line.

In our experience many commercial TAC modules suffer from severe non-linearity in certain time ranges. One instrument tested was actually non-linear over its entire range and was unusable (Roberts, 1982). It is therefore essential to characterize this fault and to determine which channels should be avoided when collecting (or analysing) the decays. For this purpose the variable delay lines are indispensable. Modern TACs, while apparently less subject to non-linearity associated with small timing capacitor charges than older instruments, may have a small non-random oscillation over the complete range. Usually this oscillation is of too low an amplitude to interfere with data analysis and can therefore be tolerated.

5.2.5(b) Pile-up inspection

Since pulse pile-up distortions are intimately connected with the mode of operation of the TAC we shall briefly discuss these effects in this section. Instrumental pile-up correction is based on one of two techniques – pulse discrimination or energy discrimination. In the former, the number of fluorescence-generated signals arriving at the TAC during a TAC cycle is monitored; the arrival of a second signal inhibits the transmission of the pulse associated with the first signal to the MCA. Energy discrimination assumes that a pulse height spectrum measured at a late dynode of the PM tube has a well-defined peak that corresponds to one photon events. If the sum of the amplitudes of all the pulses from this dynode during one excitation cycle exceeds the appropriate single photon level the TAC output for that cycle is inhibited.

An ingenious, if somewhat awkward, experiment that corrects for pile-up with pulse counting has been devised by Davis and King (1970). In it the PM pulse is split into two channels. One channel starts the TAC; the other is delayed and [because there elapses a fixed time before the TAC, having accepted a START pulse begins its voltage sweep (see Fig. 5.4)] can be timed so as almost but not quite to stop the TAC. An additional PM pulse occurring during a cycle will therefore stop the TAC. Also routed to the STOP input of the TAC are the excitation trigger pulses delayed by the TAC range. Data are now collected first for 10 seconds, say, with the MCA in ADD mode and both the PM and trigger pulses connected to the STOP input; then the trigger pulses are disconnected and data "collected" with the MCA in SUBTRACT mode. The MCA is cycled in this way between ADD and SUBTRACT until the desired number of counts have been accumulated. Random starting of the TAC is prevented by maintaining it in an inhibit mode until the beginning of each cycle. Techniques of pile-up correction, such as this, are difficult to implement correctly because of dead time in the inhibit circuit and precise timing requirements (Coates, 1972; Haugen *et al.*, 1979). Commercial components are available but are reported to be inefficient (Knight and Selinger, 1973). However, modifications that increase the data collection rate have been described (Knight and Selinger, 1973; Williams and Sandle, 1970).

Pile-up inspection employing energy discrimination depends for its success on a negligible or at least very small overlap between the peak corresponding to one and the peaks corresponding to multiple photon events in the pulse height spectrum of the photomultiplier tube. Some RCA photomultiplier tubes are particularly useful in this application (see Fig. 4.3). A typical instrument employing energy discrimination would include a PM with a well-resolved pulse height spectrum, a preamplifier with μs decay time, a spectroscopy amplifier and a timing single channel analyser. During the decay time of the preamplifier, multiple pulses from the dynode of the PM tube are converted to a single pulse with an amplitude proportional to the sum of the amplitudes of the original PM pulses. The spectroscopy amplifier shapes this pulse so that it is suitable for input to the single channel analyser, which is preset to output a gating pulse only when the input pulse is within the range of the single photon energy. Because of the rather slow decay time of the preamplifier (50 μs), this type of pile-up rejection is unsuitable for fast repetition rate sources.

Pile-up correction has been thoroughly investigated by Harris and Selinger

(1979). They concluded that avoidance of pile-up distortions is best achieved by reduction of the count rate (see Section 2.2.1). If a definite reduction in data collection times can be safely obtained by pile-up inspection, they recommend energy discrimination for excitation pulses with lifetimes less than 80 ns and pulse discrimination for other excitation sources. Therefore for the measurement of nanosecond lifetimes the former seems to be the method of choice.

A limitation on the maximum efficient count rate attainable with pile-up inspection is set by statistical considerations. Following the notation of Section 2.2.1 let $p_1(i)$ be the probability of emission of one photoelectron in the time interval $t_{i-1/2}$ to $t_{i+1/2}$ and let w_i be the average number of photoelectrons ejected from the photocathode in this time interval. Assume that all cycles corresponding to the emission of more than 1 photoelectron are excluded from the counting. The probability, $P_1(i)$ of detecting a pulse in the ith time interval is given by

$$P_1(i)=p_1(i)\prod_{j\neq i}^{\infty} p_0(j) \tag{5.1}$$

From Equations 2.4 and 2.5

$$P_1(i)=w_i e^{-w_i}\left[\exp\left(-\sum_{j\neq i}^{\infty} x_j\right)\right] \tag{5.2}$$

$$=w_i e^{-w}, \tag{5.3}$$

where

$$w=\sum_i w_i,$$

the average number of photoelectrons ejected during an excitation cycle. After N_E excitation cycles the number of counts in the ith channel is

$$N_i=N_E\beta z_i e^{-\beta z}, \tag{5.4}$$

where the notation is as in Section 2.2.1. It can be seen that this number is proportional to the intensity at time t_i, z_i, without any counting rate restrictions. If n is the number of analyser channels, the total number of useful counts, N_D, is given by

$$N_D=\sum_{i=1}^{n} N_i=N_E\beta\left(\sum_i z_i\right)e^{-\beta z}$$

$$=N_E\beta z e^{-\beta z} \tag{5.5}$$

For $\beta z=1$, N_D is a maximum, so that $F_D=N_D/N_E=e^{-1}=0.37$.

[Mathematical correction of pile-up distorted curves is described in Section 6.4.]

Commercially available SPC instruments do not offer pile-up inspection as a standard feature. Manufacturers would doubtless include the option if the customer requested it. However, with a well-aligned flash lamp apparatus and a ratio of converted counts to excitation events of 0.01 a decay curve should be measurable in at most 30 min. Considerable reduction in this time is standard of course, in instruments using a high repetition rate laser. Since data transfer and analysis must also be taken into account in an evaluation of time-saving, the actual reduction in data processing achieved by a pile-up inspector is likely to be quite marginal. In experiments where collection times are abnormally long the reason is usually a low count rate. A pile-up inspector might perhaps be suitable for experiments with high count rates in which the excitation source stability is not high or for experiments with biologically or chemically labile samples. Otherwise careful consideration of the actual time-saving expected is recommended.

5.2.5(c) Multichannel TAC

This component has been developed by the Research Institute of Applied Electricity, Hokkaido University, for Horiba Ltd (Hara, 1983). A schematic diagram is illustrated in Fig. 5.5. It consists essentially of several TAC units (channels) whose voltage sweeps are initiated by the same start

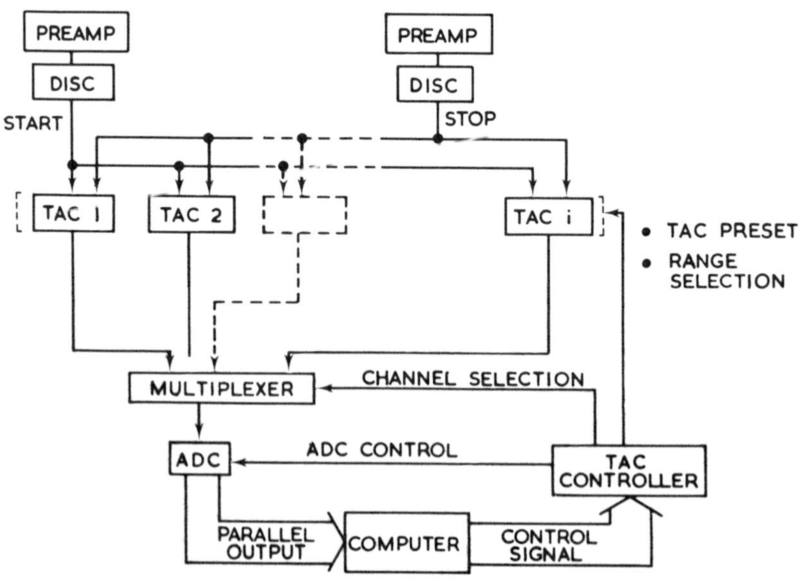

Figure 5.5 Schematic diagram of Horiba multichannel TAC operation (from Hara, 1983).

pulse. This allows a high stop count rate because the electronics can detect and time several photons for each exciting flash. With this component a count rate of 50% is possible for a 100 ns lifetime (Hara, 1983). The multichannel TAC has not yet been tested critically but it appears to offer a decided improvement over conventional TACs.

5.2.6 Biased amplifier

As was stated in Section 5.2.3, it is convenient to route the output pulses from the TAC to the MCA through a biased amplifier. This component amplifies the TAC output pulses yielding increased resolution and, through its ability to bias out time delays, helps to achieve suitable placement of the decay curve in the MCA channels. The effect of bias level and gain on the voltage of the TAC conversion output pulse is illustrated in Fig. 5.6. It can be seen that

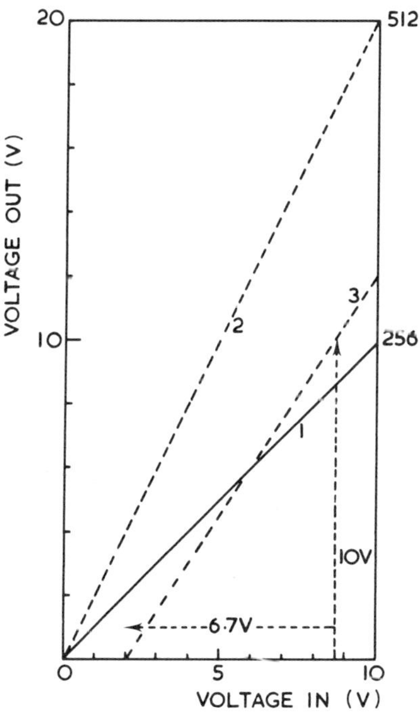

Figure 5.6 Effect of bias level and gain in biased amplifier used in conjunction with TAC:

(1) bias level 0, gain 1 $V_{out} = V_{in}$

(2) bias level 0, gain 2 $V_{out} = 2\ V_{in}$

(3) bias level 2 V, gain 1.5 $V_{out} = 1.5\ V_{in}$

voltages corresponding to short times can be biased out and a specific region of TAC voltage say 2 V–8.6 V (line 3) corresponding to say 20 ns–87 ns on a 100 ns TAC range, can be expanded to fill the N MCA channels (V_{out} = 0–10 V corresponds to 0–N channels). Gains of 15 above the bias level are standard in these components so that the usual minimum TAC range of 50 ns can be effectively reduced to 3.3 ns, allowing greater resolution of picosecond decays. Biased amplifier components may be purchased separately or be incorporated in the TAC. In either case time jitter in this amplifier appears to be negligible, TAC–BA combinations having less than 20 ps jitter (Ware, 1971; Koester, 1979). The BA is not an essential component of SPC electronics since even very short lifetimes are easily resolvable on a 50 ns timescale and delay generators or coaxial cables can be used to position the decay in the MCA channels. Nevertheless it can be a convenient component and the extra expense of a biased time-to-amplitude converter (TAC and BA combined in one NIM unit) over a standard TAC will probably be justified.

5.2.7 Multichannel analyser (MCA)

The variety of MCAs on the market is so large that we cannot hope to describe them all in this short section. We shall therefore give a brief description of a standard instrument and mention the features that we consider desirable for the SPC experiment.

A basic MCA has an analogue-to-digital converter, a memory consisting of channels for storing data and data input and output facilities. A standard instrument generally incorporates lower and upper level discriminators and two modes of data collection – pulse height analysis mode or multichannel scaling mode. It is usual and, in our opinion, essential to be able to display the contents of the memory store on a CRT screen, which may be a standard feature of the MCA, or an oscilloscope dedicated to the SPC experiment. Live 'scope display during data acquisition is preferable. MCAs may also possess the following features: PRESET TIME or PRESET COUNT which enables the operator to collect data to a certain peak height or for a certain time (useful but not essential); variable display scale (highly desirable); log display (useful but not essential); digital zero offset (sometimes useful); choice of data ADD or SUBTRACT (normally not used but can be convenient for background correction or pulse pile-up correction); internal clock (very useful); % dead time indicator (useful as a diagnostic); cursor for monitoring the counts in a particular channel (useful); region of interest integrator (useful for background estimation); multiple memory groups (very useful); AUTOMATIC ERASE (after readout) mode (useful in a computer controlled instrument but could be a disadvantage otherwise).

For pulse height analysis experiments 200 to 600 channels are sufficient in a decay curve. A smaller number may give less than satisfactory resolution in double exponential decays, especially if some channels are reserved for collection of background counts. Time resolution is increased with increasing number of channels but so also are data collection times. Five hundred to 600 channels are perhaps most common. Optical resolution in time-resolved spectroscopy is also dependent on the number of channels. Perhaps a suitable number for all types of experiment is about 1000 channels divided into 4 memory groups; in a single data collection one could use 250, 500 or 1000 channels. Calibration of the analyser channels is discussed in Section 5.4 and the effect of channel width on time resolution in Chapter 6.

Oscilloscope display of the accumulating data is necessary in order to judge visually whether the experimental conditions are correct. A log display is useful for judging exponentiality, but since convolution tends to mask the simple kinetics of the sample lack of this feature is not serious.

Data in the analyser must be transferred to a computer. Most analysers READOUT in several modes. For instance, standard MCAs will drive a paper tape perforator or cassette deck or will output data serially to an appropriate interface. Data transfer from analysers is sometimes tricky, especially when the instrument is hard wired to a computer. Most of the problems, however, can be solved with a reasonable knowledge of the appropriate electronics.

Many analysers have what the uninitiated might regard as a bewildering array of switches and knobs on the front panel. It is of course important to adjust the settings correctly, but this adjustment is quite simple to accomplish by consulting the user's manual, and the MCA is as easy to use as any of the other electronic components. Dead time in the MCA, which occurs while it is processing the converted output from the TAC, is not a limitation on data collection rates since it can accept input pulses at a rate faster than the TAC can supply them. Dead time in MCAs is discussed by Harris and Selinger (1979) and Bennett *et al.* (1965) among others. MCAs may be NIM units or independently powered components.

5.2.8 Rate meter

It is always essential to monitor the converted TAC signals so that a correct ratio, F_D, of detected photons to excitation events may be maintained (Section 2.2.1). This measurement (before decay curve collection) can be achieved by operating the MCA in multichannel scaling mode and by advancing a channel every second. For this purpose a cursor is useful. More convenient however is a rate meter. A counter, as such, merely accumulates counts; hence a counter that accumulates for one second, displays the total, resets to zero and repeats the cycle is necessary. Digital readout is more convenient although a dial readout may be less expensive. While a rate meter that can accept both positive and negative polarity input signals is most convenient, acceptance of only one polarity, that of the signals from the second CONVERSION output of the TAC, is quite satisfactory. This output signal is usually positive unipolar or positive bipolar. If it must be inverted to comply with the input requirements of the rate meter excessive attenuation may occur in a simple BNC-type inverting transformer. However, negative-going pulses may be taken from the discriminator in the fluorescence channel. If the excitation source repetition rate is very high (100 kHz–5 MHz) this count will be close to the TAC rate. If there is only one CONVERSION output from the TAC the signal to the MCA can be monitored before data collection. A rate meter for the trigger signals is convenient if the excitation pulse repetition rate is not displayed on the driving electronics for the excitation source (e.g., in a free-running flash lamp). This count can be used, with all sources, to monitor source stability. Input pulse polarity requirements are not critical for this rate meter since either positive-going or negative-going signals can be derived from the discriminator in the trigger channel.

Some rate meters incorporate a discriminator with adjustable threshold on the input. If the threshold is set too high or too low an incorrect count will be registered. The correct threshold can sometimes be set by monitoring the known trigger pulse count rate. Alternatively the signals to be monitored may be counted in the MCA, as outlined in the preceding paragraph, and the rate meter discriminator threshold adjusted until the count rate agrees with that measured in the MCA. In fact it is advisable to calibrate any counter before use since it is found that these components frequently perform less than satisfactorily.

5.2.9 Cables and connections

Electronic signals are usually transferred from one component to another along coaxial cable. Cable impedance is almost invariably 50 Ω. For standard BNC connectors RG 58C/U cable is suitable. Other connectors may accept coaxial cable of a different thickness. For instance, between LEMO connectors RG 174/U narrow width cable is satisfactory. Doubly-screened cable (e.g. RG 55/U, RG 71B/U) may act as an attenuator of r.f. signals. Signal transmittance has been discussed by Lytle (1974). Since pulse rise time increases with the square of the length of cable, lengths of greater than 1 m are not recommended. Soldering of cable cores to connector pins and crimping of screens to the connector bodies should be examined frequently, as dry solder joints are apt to break.

Inputs on NIM components are normally BNC bulkhead sockets (also called jacks), which accept BNC free plugs. Straight BNC adaptors (socket-to-socket or plug-to-plug) are very useful as also are, though less frequently, BNC elbow plugs. At least one BNC T-adaptor (with one plug and two sockets) should be available, especially if the TAC–MCA combination is to be calibrated by the simple method outlined in Section 5.4. Unused inputs on some components should be terminated in 50 Ω for which purpose BNC termination plugs are available; BNC 50 Ω THROUGH terminators will also suffice for this purpose. The latter are useful for signal transmission to a device whose input is not internally terminated in 50 Ω (inputs on some TACs require this type of termination). Trigger pulses are frequently of positive polarity; if so they must be inverted before input to most amplifiers, discriminators and TACs. A high frequency BNC type inverting transformer effects this polarity change in a most convenient way. BNC 50 Ω attenuators are useful when signals are monitored on an oscilloscope but may also be used to attenuate signals from, say, a fixed gain amplifier.

The 4-module CFT discriminator (Ortec Model 934) referred to in Section 5.2.2 possess LEMO type sockets for all input, output and internal connections. In this unit each of the four discriminators actually used requires, as an internal delay, about 10 cm of coaxial cable of external diameter 2.5 mm connected to two front panel sockets by 2 plugs of SUHNER type 11 QLA 01-2-2c. Adapters, BNC free socket to LEMO free plug, SUHNER type 33 QLA-BNC 01-1, can be obtained with which standard RG 58C/U cable can be connected to the input and output sockets of each discriminator. Details of these connectors, cables and adaptors can be had from the addresses given in Appendix 5 A1.

It is important to avoid earth loops when signals are transmitted from devices that are earthed independently. Good interdevice earthing should also help in the suppression of r.f. interference. Lytle (1974) advised connecting the chasis of the separate devices with a hollow wire braid threaded with a solid copper conductor. Conductor and braid are soldered together every 30 cm or so. With this technique a common d.c. and r.f. earth should be attained. Further discussion of r.f. attenuation can be found in the book by Morrison (1967).

5.2.10 Other components

Initial optimization of flash lamp performance is greatly aided by use of an oscilloscope with which the trigger pulse (and thyratron gating pulse, if used) is monitored while high voltage, arc gap etc., are adjusted. Pulse instability is easily observable by this means. The fairly common and relatively inexpensive Tektronix type 454 oscilloscope can be used for this purpose. Electronic components can also be tested on such an oscilloscope if a pulse generator is available. Pulses, of a type specified in the instruction manual for the component, are input to the component and its output is monitored on the 'scope. Frequently it is sufficient to monitor the response of the component to the pulses actually processed in the experiment thus dispensing with the need for a pulse generator. For many components this check is unnecessary until the instrument performance deteriorates, due to some unknown cause.

A convenient method of calibrating the TAC–MCA combination is by means of a commercial time calibrator. This is a NIM standard module with an internal crystal oscillator specifically designed for use with a TAC. It outputs START and STOP pulses, separated by an integral multiple of a precise time period, which are routed to the appropriate inputs of the TAC. One common model (ORTEC Model 462) has a facility (called a Dispersion Amplifier) for dispersing the peak over a number of channels so that its centroid can be calculated exactly. Some years ago a time calibrator of this type in our laboratory yielded different results when the Dispersion Amplifier was operated. For this reason, and also as a normal precaution, we recommend that time calibrators should themselves be calibrated against some known time period.

For time resolved spectral studies it is necessary, if the MCA does not possess an internal clock, to time the channel advance with an external clock. NIM standard modules are available that output a pulse after some chosen time interval. There is generally an input connection at the rear panel of the MCA to which the pulse can be routed so that the channel is advanced by one whenever a pulse is received.

As was stated in Section 5.6 commercial pile-up inspectors are now available. Before purchase, however, it is best to ascertain the efficiency of these devices. As Knight and Selinger (1973) pointed out, discussing earlier models, unless the device is gated to inspect pulses only during the TAC open period, it can be quite inefficient. For the other method of pile-up inspection, energy discrimination, a timing single channel analyser (SCA) can be employed. This device is also available as a NIM standard module. Pulses from the PM are routed to an integrating amplifier and, usually, a spectroscopy amplifier before input to the SCA. The integrating amplifier is also necessary in the determination of the pulse height spectrum of the PM. Suitable commercial components are (all ORTEC Models): 113 preamplifier, 451 spectroscopy amplifier and 420-A timing SCA.

Schuyler and Isenberg (1971) have described an SPC instrument which records two decay curves simultaneously through the use of COINCIDENCE modules and a LINEAR GATE. A block diagram of their instrument is shown in Fig. 5.7. It will be seen that energy discrimination pile-up inspection is employed in the two fluorescence channels while only a single

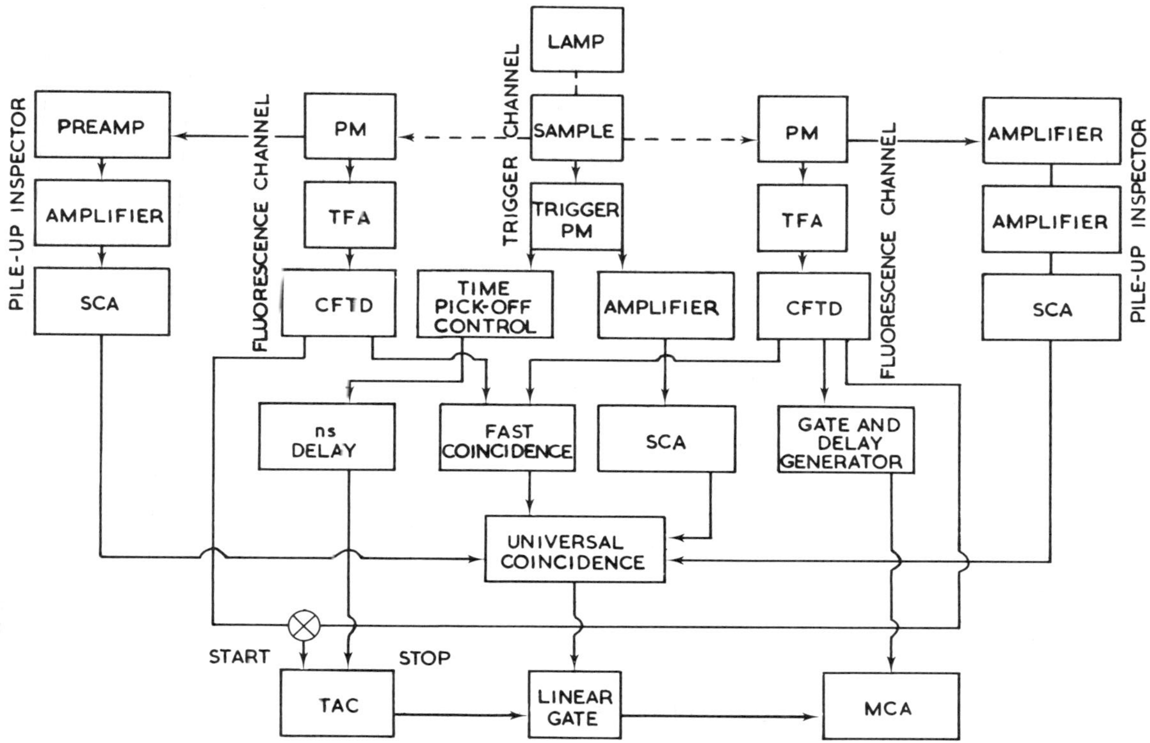

Figure 5.7 Block diagram of SPC instrument employing pulse pile-up inspection and alternation of data collection between two fluorescence channels. Symbols as in Fig. 2.1 and text (adapted from Schuyler and Isenburg, 1977).

TAC–MCA is necessary. The UNIVERSAL COINCIDENCE gates the timing information to the MCA while the FAST COINCIDENCE blocks data collection if the two PMs output anode pulses within 100 ns of each other. While the advantage of collecting two decay curves simultaneously is obvious in some experiments (e.g., when fluorescence polarized parallel and perpendicular to the direction of polarization of the exciting light is to be time-resolved), for most users the expense of assembling an instrument so sophisticated is possibly prohibitive. Close attention to the actual reduction in data collection and processing times should be paid before the components are purchased.

5.3 Setting Discriminator Levels

The discrimination level for the trigger pulses is not critical since these are usually of reasonably constant amplitude. The final setting is chosen by raising the level increasingly above threshold while the discriminator output is monitored with a rate meter. When noise has been discriminated against a region of constant output counts is reached. The operating level is then set about one third of the way along this plateau region.

Much more crucial in the success of the SPC experiment is the setting of the discrimination level on the fluorescence (usually constant fraction timing) discriminator. As the level is raised the FWHM of the instrument response function gets smaller, a distortion that must be clearly distinguished from increased time resolution. The effect has been explained on the basis of overlap of separate single photon anode pulses, which cause a multiphoton pulse with a much larger amplitude than a single photon pulse. Since this overlap is more severe at times close to the maximum of the photon distribution function, a high discriminator setting, which discriminates in favour of multiphoton events, biases the collected decay towards shorter times. The distortion is different for pump pulse and decay curve so that the convolution integral is no longer applicable.

Fluorescence PM pulses, as stated previously, have a wide amplitude distribution; there is no clear dividing line between pulses arising from dark noise and pulses arising from fluorescence photons. Increasing the discriminator level, however, seems to increase the signal-to-noise ratio and also to reduce somewhat the interference from spurious r.f. signals (Nemzek, 1975). Consequently the chosen discrimination level will be a compromise between acceptance of relatively more single photon pulses and rejection of dark noise and r.f. generated pulses. In order to adjust the level one should first try to find a signal plateau. If no clear plateau is observable the tedious but necessary adjust and deconvolve technique must be resorted to. A standard, preferably of short lifetime, is chosen and pump pulse and decay curves measured for various discrimination levels until the correct lifetime and acceptable fitting is obtained. An alternative to this procedure is collection of the instrument response function only and determination of its FWHM at different discriminator levels; an obvious narrowing of the function indicates excessive discrimination. While this procedure is certainly more convenient the final setting should always be tested by deconvolution using a compound with a known lifetime.

5.4 Time Calibration and Choice of Timescale

As discussed by Yguerabide (1972), the timescale in nanoseconds per channel can be estimated with a knowledge of the time range of the TAC, the gain of the BA, if used, and the input and output ranges of the MCA. Suppose the amplitude of the TAC output signal is V volts for a START–STOP time difference equal to the TAC range, TR ns, yielding a calibration of V/TR volts per nanosecond for the TAC pulse. The input range of the MCA, I volts, is determined by its upper and lower level discriminator settings and is usually about 10 volts for pulse height analysis. The output range, O, is sometimes referred to as the gain or conversion gain and is the number of channels corresponding to the input range. The calibration of the MCA range is therefore I/O volts per channel, yielding a time calibration C in nanoseconds per channel given by

$$C = \frac{TR}{V} \times \frac{1}{G} \times \frac{I}{O} \text{ ns per channel,} \tag{5.6}$$

where G is the gain of the BA (-1 if no BA is used). Equation 5.6 gives only a rough estimate of the timescale. Owing to slightly variable internal settings in the TAC and MCA the actual timescale will usually be somewhat different.

One method of experimental time calibration has been outlined in Section 5.2.10. A much simpler method, requiring only accurately calibrated delay lines, is illustrated in Fig. 5.8. Pulses of constant amplitude, such as those from a pulse generator or from the trigger for the excitation source, are split into two channels. One channel is connected to the START input on the TAC and the other is routed through two delay boxes. With D_F switched out, D_V is adjusted until a spike appears on the analyser screen. D_F is switched in and the channel difference between the two spikes, N_{DF}, is noted. Then the spikes are erased, D_F switched out, D_V changed and a spike recorded. Again D_F is switched in and the resulting channel difference between the spikes noted. This process is repeated for various values of D_V and the resulting values of N_{DF} averaged, giving a time calibration of D_F/N_{DF} nanoseconds per channel.

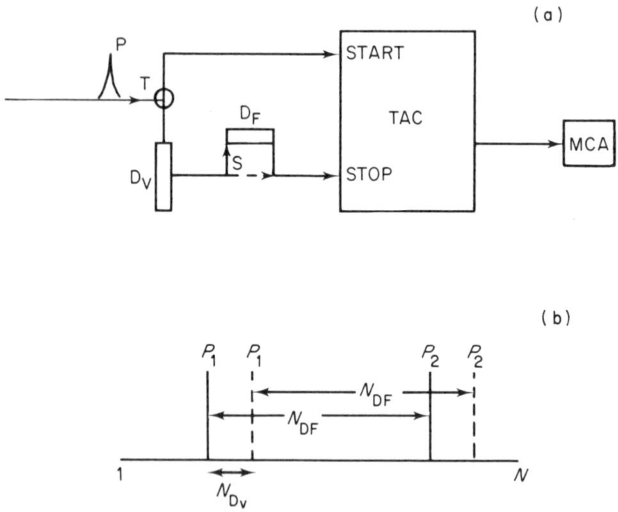

Figure 5.8 Illustration of time calibration of TAC–MCA combination. (a) Experimental arrangement P, constant amplitude pulse; T, BNC T-piece (pulse splitter); D_V, variable delay; S, switch; D_F, fixed delay.

(b) Peaks in MCA channels P_1, P_2, counts accumulated in narrow peaks; N, total number of channels; N_{DF}, number of channels corresponding to fixed delay, D_F; N_{DV}, number of channels corresponding to variable delay, D_V.

The advantage of this method is that if the spikes fall in a single channel the uncertainty in the position is reduced by the averaging. An alternative is to hold D_V constant and shift P_2 progressively across the channel range by increasing D_F. By drawing a straight line through a plot of channel number against delay added the time calibration is obtained. Whichever method of calibration is used, channels corresponding to a non-linear region of the TAC ramp should be avoided.

Calibration of longer TAC timescales cannot be achieved with small variable delays; perhaps the most satisfactory method of such calibration is a commercial time calibrator. Other methods employ a pulse generator with a precision variable delay (Ware, 1971) and, less satisfactorily, a fluorescence standard with a long lifetime. Hexafluoroacetone at 25°C and a pressure of 100 Torr has been suggested for this purpose (Knight and Selinger, 1973). Its lifetime is quoted as 84 ± 0.5 ns. This value should be confirmed before acceptance. We would recommend a method based on an accurately known and easily generated delay in preference to a decay time measurement.

Single exponential decay curves should be collected on a range that allows the curve to decay over 2 to 3 decades of intensity in the available channels. Multiexponentials are more difficult and may require two experiments on different time scales for adequate resolution. Although equations for choosing the TAC range have been suggested (Hall and Selinger, 1981) we feel that they introduce an unnecessary level of complication into what is really a matter of common sense.

References

Arbel, A., Klein, I. and Yarom, A. (1974). *IEEE Trans. Nucl. Sci.* **NS-21**, 3–8.

Bedwell, M. O. and Paulus, T. J. (1976). *IEEE Trans. Nucl. Sci.* **NS-23**, 234–243.

Bennett, Jr., W. R., Kindlmann, P. J. and Mercer, G. N. (1965). *Appl. Optics*, Supplement 02: "Chemical Lasers", 34–57.

Coates, P. B. (1972). *J. Phys. E.* **5**, 148–150.

Davis, C. C. and King, J. A. (1970). *J. Phys. A.* **3**, 101–109.

Gedcke, D. A. and McDonald, W. J. (1967). *Nucl. Instrum. Meth.* **55**, 377–385.

Gedcke, D. A. and McDonald, W. J. (1968). *Nucl. Instrum. Meth.* **58**, 253–260.

Hall, P. and Selinger, B. (1981). *J. Phys. Chem.* **89**, 2941–2946.

Hara, K. (1983). Personal communication.

Harris, C. M. and Selinger, B. K. (1979). *Aust. J. Chem.* **32**, 2111–2129.

Haugen, G. R., Wallin, B. W. and Lytle, J. E. (1979). *Rev. Sci. Instrum.* **50**, 64–72.

Kingston, R. H. (1978). "Detection of Optical and Infra-Red Radiation", p. 47. Springer-Verlag, Berlin.

Kinoshita, S., Ohta, H. and Kushida, J. (1981). *Rev. Sci. Instrum.* **52**, 572–575.

Knight, A. E. W. and Selinger, B. K. (1973). *Aust. J. Chem.* **26**, 1–27.

Koester, V. J. (1979). *Anal. Chem.* **51**, 458–459.

Koester, V. J. and Dowben, R. J. (1978). *Rev. Sci. Instr.* **49**, 1186–1191.

Leskovar, B., Lo, C. C., Harig, P. R. and Sauer, K. (1976). *Rev. Sci. Instrum.* **47**, 1113–1121.

Lewis, C., Ware, W. R., Doemeny, L. J. and Nemzek, T. L. (1973). *Rev. Sci. Instrum.* **44**, 107–114.

Lyke, R. L. and O'Connor, D. V. (1977). Unpublished results.

Lyke, R. L. and Ware, W. R. (1977). *Rev. Sci. Instrum.* **48**, 320–326.

Lytle, F. E. (1974). *Anal. Chem.* **46**, 817A–830A.

Morrison, R. (1967). "Grounding and Shielding Techniques in Instrumentation", *passim*. Wiley, New York.

Nemzek, T. L. (1975). Thesis. University of Minnesota.

Robbins, R. J., Fleming, G. R., Beddard, G. S., Robinson, G. W., Thistlethwaite, P. J. and Woolfe, G. J. (1980). *J. Amer. Chem. Soc.* **102**, 6271–6279.

Roberts, A. J. (1982). Personal communication.

Schuyler, R. and Isenberg, J. (1971). *Rev. Sci. Instrum.* **42**, 813–817.

Ware, W. R. (1971). In "Creation and Detection of the Excited State", Vol. 1A (Lamola, A. A., ed.) p. 246. Marcel-Dekker, New York.

Wild, U. P., Holzwarth, A. R. and Good, H. P. (1977). *Rev. Sci. Instrum.* **48**, 1621–1627.

Williams, O. M. and Sandle, W. J. (1970). *J. Sci. Instrum.* **3**, 741–743.

Yamazaki, J., Murao, J. and Yoshihara, K. (1982). *Appl. Optics* **21**, 2297–2298.

Yguerabide, J. (1972). *Meth. Enzymol.* **26**, 498–578.

Appendix 5.A1

In this appendix we list names and addresses of manufacturers of SPC electronic modules. We would emphasize that the list is by no means exhaustive; only companies commonly referred to in the literature are listed. In order to conserve space only the address of the international headquarters of the company is given. If that is not known the address of the representative known to us is substituted.

A. General SPC equipment

(1) (EG and G) ORTEC Incorporated, 100 Midland Road, Oak Ridge, Tennessee 37830, USA.

(2) Canberra Industries Incorporated, 45 Gracey Avenue, Meriden, Connecticut 06450, USA.

(3) Nuclear Enterprises Limited, Sight Hill, Edinburgh EH11 4EY, Scotland.

(4) Elscint Incorporated, P.O. Box 297, Palisades Park, New Jersey 07650, USA.

B. Preamplifiers (as in A, also)

(5) Marconi Instruments Division, Marconi Electronics Incorporated, 100 Stonehurst Court, Northvale, New Jersey 07647, USA.

C. Calibrated delay units (as in A, also)

(6) Hamamatsu TV Company Limited (see Appendix 4.A2).

D. Time-to-amplitude converters (as in A, also)

(7) Societa Industriale per L'Elettronica Avanzata (SILENA), Via Negroli, 10/A, 20133 Milano, Italy.

(8) Le Croy Research Systems Corporation, 126 North Route 303, West Nyack, New York 10994, USA.

E. Multichannel Analysers (as in A, also)

(9) Hewlett Packard Company, 150 Page Mill Road, Palo Alto, California 94304, USA.

F. Connectors

BNC cables and connectors from any electronic components supply company. Lemo connectors and adaptors from (*inter alia*):

(10) Lemo S.A. Electrotechnique, 1110 Marges, Switzerland.

(11) Huber and Suhner AG, 9100 Herisau, Switzerland.

(12) Kings Electronic Company Incorporated, Tuckahoe, New York 10707, USA.

In-line inverting transformers, attenuators and terminators:

(13) Tektronix Incorporated, P.O. Box 500, Beaverton, Oregon 97077, USA.

G. Oscilloscopes

Tektronix Incorporated (see 13)
Hewlett–Packard Company (see 9)

(14) Fabri–Tek Incorporated, 5901 South County Road 18, Minneapolis, Minnesota 55436, USA.

H. Time calibrator

(EG and G) ORTEC Incorporated (see 1)

(15) Tennelec Incorporated, P.O. Box 601, Oak Ridge Turnpike, Oak Ridge, Tennessee 37830, USA.

6 Data Analysis

6.1 Introduction

It is beyond the scope of this short monograph to examine critically all the mathematical techniques that have been applied to the analysis of single photon counting lifetime data. Our intention is briefly to outline the more commonly used methods, reserving detailed description for the technique of our choice (reiterative least-squares fitting and convolution). In addition, we shall give some attention to the extension of this technique to treat data for which *a priori* assumptions about the functional form of the decay law cannot be made. Detailed critical comparisons of the common techniques in their application to both real and synthetic data have been carried out by McKinnon *et al.* (1977) and O'Connor *et al.* (1979).

6.2 The Data

In the SPC pulse height analysis experiment, outlined in Chapter 2, for each converted TAC START signal a count is added to the appropriate location in the data storage device. We shall in this chapter assume that data are stored in the channels of a multichannel pulse height analyser (MCA). When the counts in the decay curve have reached a suitable precision, the number of counts in each channel $I(t_i)$, follows a Poisson distribution with a standard deviation, σ_i, given by (Equation 2.11):

$$\sigma_i = \sqrt{I(t_i)}.$$

In addition to the error resulting from this counting uncertainty, decay curve data are contaminated with spurious counts usually having their origin in photomultiplier dark noise and leakage room light. The level of background counts caused by these uncorrelated signals is sometimes treated as a variable parameter in least-squares fitting routines (Robbins *et al.*, 1980) but it is more usual and, we believe, better practice to estimate an average dark count per channel in the way described in Chapter 2 and to subtract this number from the collected data channel by channel in accordance with Equation 6.1 (see Equation 2.20):

$$I_0(t_i) = I(t_i) - B. \tag{6.1}$$

The same procedure is applied to the data in the instrument response function, $P(t_i)$ to yield the corrected data, $P_0(t_i)$.

Given, then, the two functions, $P_0(t)$ and $I_0(t)$, corrected for constant background, it is the purpose of the mathematical analysis to determine the decay law of the sample $G(t)$. For extremely long lifetimes deconvolution is not necessary and hence the function, $P_0(t)$, is not required. Usually, however, $P_0(t)$ is convolved to a greater or lesser degree with $G(t)$ so that the convolution Equation 2.19 (rewritten here with subscripts denoting background correction) or some variant thereof must be solved in the data analysis:

$$I_0(t) = \int_0^t P_0(t')G(t-t')\,\mathrm{d}t'. \tag{6.2}$$

It will be realized that Equation 6.2 is an approximation since the decay curve is a histogram rather than a continuous distribution. To account for the fact that each channel registers not the decay at an instantaneous time t_i, but an integrated decay over channel width, Δt, the number of counts in channel i should be represented by:

$$C(t_i) = \int_{t_i}^{t_i+1} I_0(t)\mathrm{d}t. \tag{6.3}$$

But if we approximate this integral according to the law of the mean we can write

$$C(t_i) \approx \Delta t I_0\left(t_i + \frac{\Delta t}{2}\right). \tag{6.4}$$

Similarly, the observed counts in the pump pulse profile can be written:

$$R(t_i) \approx \Delta t P_0\left(t_i + \frac{\Delta t}{2}\right). \tag{6.5}$$

Substituting Equation 6.2 in Equation 6.4 gives

$$C(t_i) = \Delta t \int_0^{t_i+(\Delta t/2)} P_0(t')G\left(t_i + \frac{\Delta t}{2} - t'\right)\mathrm{d}t', \tag{6.6}$$

and approximating using the law of the mean again,

$$C(t_i) = \Delta t^2\left[P_0\left(\frac{\Delta t}{2}\right)G(t_i) + P_0\left(t_1 + \frac{\Delta t}{2}\right)G(t_i - t_1) + \ldots\right.$$
$$\left. + P_0\left(t_{i-1} + \frac{\Delta t}{2}\right)G(t_i - t_{i-1}) + \frac{1}{2}P_0\left(t_i + \frac{\Delta t}{2}\right)\right] \tag{6.7}$$

where it is assumed

$$\int_{t_i}^{t_i+(\Delta t/2)} P_0(t')G(t-t')\,\mathrm{d}t' = \frac{1}{2}\int_{t_i}^{t_i+\Delta t} P_0(t')G\left(t_i+\frac{\Delta t}{2}-t'\right)\mathrm{d}t'$$

and $G(0)=1$ for a normalized function. With the aid of Equation 6.5 we get [with $R(t_0)=0$]

$$\begin{aligned} C(t_i) &\approx \Delta t\left[R(t_i)G(t_i-t_1)\ldots + R(t_{i-1})G\left[t_i-t_{i-1}+\frac{1}{2}R(t_i)\right]\right] \\ &\propto \int_0^{t_i} R(t')G(t_i-t')\,\mathrm{d}t'. \end{aligned} \tag{6.8}$$

Therefore we are justified in using the observed histograms of instrument response function and decay curve in the deconvolution procedure; care must be taken however, if an unnormalized preexponential factor is to be used for kinetic information. The problem of data quantization is discussed in greater detail and with more rigour by Bouchy (1982).

6.3 Distortions

Data that do not conform to Equation 6.2 can sometimes be analysed successfully if the nature of the distortion is known and easily characterized. Nevertheless, we should like to repeat that, wherever possible, distortions should be eliminated at source by adjusting the operating conditions of the instrument. In this section we consider some of the deviations from Equation 6.2 that can be treated mathematically, albeit with reduced confidence in the results.

It is usually possible rigorously to eliminate scattered exciting light by means of suitable filters and baffles in the optical path. Blackened Wood's horns as sample cells, particularly for gas phase studies, are quite effective in dissipating absorbed exciting radiation. In spite of the increasing number and sophistication of optical and spatial filters, however, contaminating stray light in the decay curve may have to be tolerated in some experiments, for instance when resonance or near resonance fluorescence is under investigation or when front surface viewing is employed for solid samples. It is unfortunately, but not surprisingly, true that scattered light contamination is usually most severe when the level of fluorescence is low and when, therefore, other interfering effects such as impurity fluorescence or pump pulse instability are likely to be present.

If the decay times under investigation are not too short, some deconvolution routines allow one to exclude short time portions of the decay from analysis, thus avoiding the scattered light. Frequently, however, this treatment is not possible. Another approach is to represent the distorted decay function, following the notation of Section 2.3, by the equation

$$I_0(\lambda_e,t) = E(\lambda_E,t) \otimes G(\lambda_e,t) \otimes H(\lambda_e,t) + CE(\lambda_E,t) \otimes H(\lambda_E,t). \quad (6.9)$$

If the difference between the photomultiplier response at the excitation and observation wavelengths, $H(\lambda_E,t)$ and $H(\lambda_e,t)$ respectively, can be ignored Equation 6.9 simplifies to

$$I_0(\lambda_e,t) = P_0(\lambda_E,t) \otimes G(\lambda_e,t) + CP_0(\lambda_E,t). \quad (6.10)$$

This equation can now be used, instead of Equation 6.2, for the analysis of contaminated data. The successful application depends on the requirement that the pump pulse, $E(\lambda_E,t)$ contributing to the measured instrument response function, $P(\lambda_E,t)$, should be the same as that contaminating the decay curve. In other words, exciting radiation reaching the fluorescence PM without passage through the sample cell or similar radiation multiply-reflected from the cell walls before entering the detector may render Equation 6.10 inapplicable. In addition, if $H(\lambda_e,t) \neq H(\lambda_E,t)$, there appears to be no suitable correction. In one case the excitation pulse shape is wavelength-independent and Equation 6.9 becomes

$$I_0(\lambda_e,t) = P_0(\lambda_e,t) \otimes G(\lambda_e,t) + CP_0(\lambda_E,t), \quad (6.11)$$

an equation that requires the collection of the instrument response function at two wavelengths for successful application. On the other hand, if the instrument response function cannot be measured at the observation wavelength the empirical shift correction (*vide infra*) should not be applied to both terms on the r.h.s. of Equation 6.9. Nevertheless, the variation with wavelength of the photomultiplier response may be slight, in which case Equation 6.10 or the variant

$$I_0(\lambda_e,t) = P_0(\lambda_e,t) \otimes G(\lambda_e,t) + CP_0(\lambda_e,t) \quad (6.12)$$

may in fact be successful. It should be realized that mild stray light contamination can be indistinguishable from some other common distortions. Consequently successful analysis with Equations 6.10 or 6.12 does not necessarily indicate the presence of stray light in the decay. Conversely, scattered light may indeed be present even if analysis with these equations fails, for the reasons already outlined. Every effort, therefore, should be made to eliminate stray light experimentally.

In Chapter 4 we stated that a shift routine in the data analysis was the usual means of correcting for a wavelength-dependent photomultiplier response, when the instrument response function cannot be measured at the observation wavelength. This correction is empirical, but is sometimes justified by a consideration of the signal processing in the SPC electronics. It is admitted that light of different energies striking the PM photocathode causes the consequent electron bunches travelling down the tube to have different electron distributions. Hence the anode pulse shapes vary with the energy of the incident photons. It is now argued that when these anode pulses are processed in a timing discriminator this instrument sees the different shapes merely as differently sloped rising edges. Therefore the changing photomultiplier response leads in fact only to a difference in the time of output of the discriminator timing pulse.

Our belief, explained in more detail in Chapter 4, is that this argument is too simplistic, as the curves presented in Figs 4.8 to 4.10 would seem to imply. Nevertheless it appears to be true that the PM interdynode voltages may be adjusted so as to yield an instrument response function that merely shifts, without changing shape, as the wavelength of the incident light is altered. For instance, in Table 6.1 are shown the results of analysing (with least-squares fitting) decay curves of the fluorescence of N,N-dimethyl-1-naphthylamine (DMNA) observed at different emission wavelengths but all excited at $\lambda_E = 300$ nm. It will be seen that shifting the zero-time of the instrument response function corrects, first, for some unknown distortion and then for the wavelength effect. Further comments on this table were given in Chapter 4. Equation 6.13 is a representation of the correction to be applied.

Table 6.1 Results of data analysis for decay of DMNA in CH_2Cl_2 at 26°C. $\lambda_E = 300$ nm (Meech, 1982).

λ_e/nm	Region of fit[a]	Shift, δ/ps	τ_0/ns	χ_ν^2 [b]	DW[c]
	D	–	2.396	1.1	2.05
375	W	–	2.405	1.4	–
	W	−8	2.397	1.1	2.08
	D	–	2.386	1.1	1.9
425	W	–	2.377	1.7	–
	W	+8	2.384	1.2	1.8
	D	–	2.424	1.1	1.9
475	W	–	2.388	9.6	–
	W	+41	2.423	1.1	1.9

[a] D, decay only; W, whole curve.
[b] See Section 6.7.2.
[c] See Section 6.7.7.

$$I_0(t) = P_0(t + \delta) \otimes G(t), \tag{6.13}$$

where δ represents the zero-time shift.

It will be noticed that this equation should also be effective in correcting for long-term drift in the excitation pulse time profile. Details of the application of Equation 6.13 will be found in the sections describing the various deconvolution techniques.

There is no reliable mathematical correction for r.f. interference, although small distortions of this sort may sometimes be corrected for with the zero-time shift. The effect of r.f. interference on the decay depends on the TAC range of the experiment but severe distortions give rise to a visible oscillation in the measured decay curve. Such an oscillation is illustrated in Fig. 6.1 in which the decay of gaseous *p*-fluorobenzotrifluoride is presented. For this decay, which also contains some contribution from scattered light, excitation was with a high pressure flash lamp. Instrumental techniques for dealing with r.f. noise are discussed in Chapter 3.

Impurity fluorescence is also difficult to treat mathematically. In the simplest case there is an additional component in the decay and if this lifetime

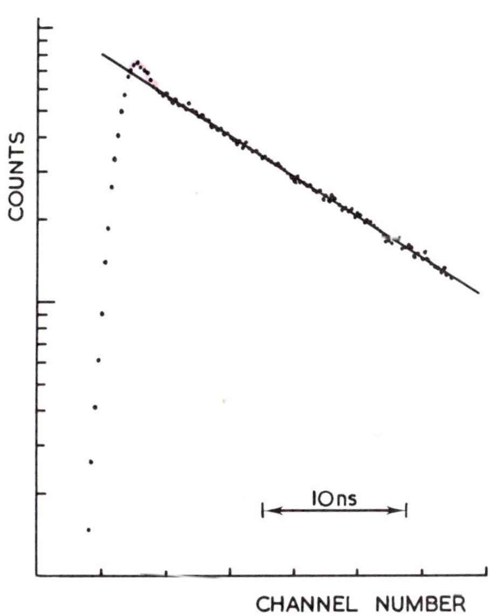

Figure 6.1 Decay of 2.0 Torr *p*-fluorobenzotrifluoride excited at 264 nm with a high pressure flash lamp. Method of analysis: straight line fit. Broken line, instrument response function; points, observed decay; full line, straight line fit. Some scattered light contamination is observable at the maximum of the decay curve. The slight oscillation of the experimental data with respect to the fitted line is caused by r.f. pick-up.

is quite different from the decay times of the components actually under investigation a correct allowance may be made for it. However, since analysis in terms of more than two exponential components is, generally speaking, very difficult, impurity fluorescence, even of the straightforward variety referred to, will render analysis of all but the simplest decays impossible to achieve. Furthermore, impurities may also quench the fluorescence of interest and in addition seem not to come singly but in battalions. In our experience contaminating fluorescence usually renders analysis of decay curves utterly impossible.

Data distorted by pulse pile-up are usually collected deliberately and corrected. This correction is now discussed.

6.4 Correction for Pulse Pile-up

Coates (1968) formulated mathematical expressions relating the number of counts in each channel of pile-up distorted data to the number in data collected under true single photon conditions. Suppose the decay curve has been collected at high count rates and let N_i be the observed number of counts in channel i. If P_i is the probability of occurrence of a count in channel i during one excitation cycle, then

$$P_i = \frac{N_i}{\left(E - \sum_{j=1}^{i-1} N_j\right)}, \tag{6.14}$$

where E is the total number of excitation cycles. It can be shown that, if S_i is the true probability of occurrence of an event in channel i, during one excitation cycle under single photon conditions

$$S_i = P_i + 0.5P_i^2. \tag{6.15}$$

Therefore, from the observed decay data and a knowledge of E, calculated from the known repetition rate of the excitation source and the total time of data collection (usually stored in channel 1 of the MCA), the undistorted decay, S_i, for each channel may be calculated.

The assumptions entailed in deriving Equations 6.14 and 6.15 led Coates to put an upper limit of 0.1 on the ratio of detected photons to excitation events, F_D, that could be tolerated without introduction of significant errors. However, coherence effects (Davis and King, 1970), and more seriously variations in the excitation source intensity (Coates, 1972a), lower the tolerable limit for F_D. In fact under conditions of variable source intensity (which does not in itself affect the normal decay curve measurement), the exact pile-up correction equations are difficult to formulate. When the source is stable collection at high F_D may lead to an increased signal-to-noise ratio (Gregory and Helman, 1972), but care must be taken to ensure that Equation 6.15 is valid under the prevailing experimental conditions (Coates, 1972b).

For a single exponential decay the exact effect of pile-up distortion has been derived (Holzapfel, 1974) and tested experimentally (Harris and Selinger, 1979). It now appears that the surest way to avoid errors due to pile-up is to adjust F_D to the single photon value (see Section 2.2.1). However, if high values of F_D are preferred, Harris and Selinger (1979) recommend correction by means of an electronic energy discrimination pile-up inspector (see Section 5.2.5(b)). We would add that if any form of pile-up correction is resorted to the effect of count rate should be investigated. In particular, if mathematical correction is applied the upper limit for F_D should be ascertained for each range of lifetimes.

6.5 Mathematical Analysis Techniques

6.5.1 Straight line fitting

Single exponential decay curves with long lifetimes may be analysed with a linear least-squares fitting technique. Double exponential decays can be fitted with a four parameter least-squares function. Alternatively, the tail of the decay can be fitted to a single exponential function and this component subtracted from the short time data. Our cautions concerning disregard of the instrument response function have already been given, to which we would merely add that, if it is definitely established that the width of the instrument response function (at say 1/20 maximum) is negligible compared to the lifetimes being measured, the decay data can indeed be assumed unconvolved and analysed accordingly. However, it is with data for which convolution cannot be ignored that we are principally concerned and we pass quickly on to deconvolution techniques.

6.5.2 Phase-plane plot

A second method, the success of which also depends on a fairly large difference between the width of the instrument response function and the lifetime of the decay, is the phase-plane method proposed by Demas and Adamson (1971). It has the advantage (now of reduced importance) of being easily programmable on a desk calculator. In order to outline the method we assume that $G(t)$ in Equation 6.2 is a single exponential function, that is,

$$G(t) = a_1 e^{-t/\tau_1}. \tag{6.16}$$

Equation 6.2 can be written

$$I_0(t)=a_1\mathrm{e}^{-t/\tau_1}\int_0^t P_0(t')\mathrm{e}^{t'/\tau_1}\,\mathrm{d}t'. \tag{6.17}$$

Both sides of the equation are integrated between the limits of 0 and t

$$\int_0^t I_0(y)\,\mathrm{d}y=a_1\int_0^t \mathrm{e}^{-y/\tau_1}\left[\int_0^y \mathrm{e}^{-t'/\tau_1}P_0(t')\,\mathrm{d}t'\right]\mathrm{d}y \tag{6.18}$$

$$=-a_1\tau_1\mathrm{e}^{-t/\tau_1}\int_0^t \mathrm{e}^{-t'/\tau_1}P_0(t')\,\mathrm{d}t'+a_1\tau_1\int_0^t P_0(y)\mathrm{d}y \tag{6.19}$$

$$=-\tau_1 I_0(t)+a_1\tau_1\int_0^t P_0(y)\,\mathrm{d}y. \tag{6.20}$$

Equation 6.20 is rearranged to yield

$$\frac{\int_0^t I_0(y)\,\mathrm{d}y}{\int_0^t P_0(y)\,\mathrm{d}y}=-\tau_1\frac{I_0(t)}{P_0(y)\,\mathrm{d}y}+a_1\tau_1. \tag{6.21}$$

Since the integrals are easily calculated from the observed $P(t)$ and $I(t)$ curves a straight line plot yields the value of τ_1 from the slope. It should be pointed out that the weighting to be applied in the least-squares fitting of Equation 6.21 is problematic. Unweighted fitting seems to give the best results (Greer *et al.*, 1981).

Recently this method has been extended to double exponential decay and to decays having a component of scattered light (Jezequel *et al.*, 1982). The equations from which the decay parameters are derived are extremely cumbersome and the solution involves least-squares fitting. It seems likely therefore that the method will continue to be less popular than the methods in common use. It is worth noting however that Equation 6.21 could serve as a useful diagnostic of non-exponentiality.

6.5.3 The method of moments

Although this method is widely used, especially by physical biochemists, no more than a brief outline of the essential details will be given here. The interested reader is referred to the two comprehensive papers by Isenberg (1973a, b) that set forth the theory and applications of the method and to the later papers by Isenberg and co-workers that deal with extensions and modifications (e.g., Small and Isenberg, 1977).

In normal applications the decay function, $G(t)$, of Equation 6.2 is an exponential or sum of exponentials, that is:

$$G(t) = \sum_{k=1}^{n} a_k e^{-t/\tau_k}. \tag{6.22}$$

The ath moments of the decay and instrument response function are defined, respectively, by

$$\mu_a = \int_0^\infty I_0(t) t^a \, dt \tag{6.23}$$

$$m_a = \int_0^\infty P_0(t) t^a \, dt. \tag{6.24}$$

When Equations 6.22 and 6.2 are substituted into Equation 6.23, integration by parts and some rearrangement yields the expression

$$\frac{\mu_a}{a!} = \sum_{s=1}^{a+1} \left\{ D_s \frac{m_{a+1-s}}{(a+1-s)!} \right\}, \tag{6.25}$$

where

$$D_s = \sum_{i=1}^{n} (a_i \tau_i^s). \tag{6.26}$$

With the calculated moments and Equation 6.25 the quantities D_s and hence the parameters a_i and τ_i can be determined.

It will be noticed that the integrations in Equations 6.23 and 6.24 extend to infinity whereas the observed data, particularly the decay curve, are truncated at a finite time. A "cut-off" correction has therefore been devised (Isenberg, 1973a) whereby the observed decay curve is extrapolated to infinity with the decay law calculated for no extrapolation. A new decay law is then determined, which leads to a refined extrapolation, and so on until some convergence criterion is satisfied. For double or multiple exponential decays the number of iterations may be quite large, to reduce which the data can be exponentially depressed. It is also maintained that results are improved if the raw data are smoothed (Small and Isenberg, 1977). Poisson noise is a definite obstacle to the success of the method of moments, and hence smoothing has at least a pragmatic justification. Scattered light contamination and a shift in the zero-time between instrument response function and decay curve, as defined in Equations 6.10 and 6.13 respectively, can be treated in the method of moments by means of a technique called moment index displacement (Isenberg, 1973a).

For single exponential decays the method is quite successful. The recovered decay parameters from double exponential decays, however, seem to depend upon the extent of exponential depression and smoothing and upon the amount of index displacement (O'Connor *et al.*, 1979). Moreover, the technique takes no explicit account of the known variance in the data [although this can be used to estimate the errors in the recovered decay parameters (Isenberg, 1973b)]. Knight and Selinger (1971) have criticized the technique on statistical grounds while McKinnon *et al.* (1977) and O'Connor *et al.* (1979) found it unsatisfactory in the analysis of real and synthetic data. Nevertheless it must be admitted that it is fairly widely used and is claimed to be successful even in the analysis of three component data (Isenberg and Dyson, 1969).

6.5.4 Laplace transforms

As was stated in Section 2.2.3, and can be easily proved by means of simple integration techniques, convolution is Laplace transformed into multiplication. Consequently, the Laplace transforms of the observed decay curve and instrumental response function, i.e.,

$$i(s) = \int_0^\infty I_0(t)e^{-st}\,dt \tag{6.27}$$

$$p(s) = \int_0^\infty P_0(t)e^{-st}\,dt \tag{6.28}$$

can be substituted in the rearranged transform of the convolution integral (Equation 2.18),

$$\frac{i(s)}{p(s)} = g(s), \tag{6.29}$$

yielding the Laplace transform of the decay function.

Earlier versions of this method of deconvolution were developed for the specific case of monomer–excimer kinetics (Helman, 1971; Almgren, 1973) but more recently Gafni *et al.* (1975) outlined a variant of general applicability to any decay consisting of an exponential or sum of exponentials. In this formulation both scattered light contamination and zero-time shifts can be treated explicitly. Although attractive in concept the method has a number of drawbacks, among the most severe of which are: integration of truncated data to infinity requiring a "cut-off" correction; no weighting of the data with the Poisson variance; and a rather restricted range of suitable s values which depends in a complex and as yet uncharacterized manner on the decay times under scrutiny. Ameloot and Hendrickx (1982) have recently developed a non-iterative cut-off correction that improves the performance of the technique, but the inherent disadvantages remain. While probably satisfactory for single exponential analysis, the method cannot be recommended for more complex decays at the present stage of its development.

6.5.5 Modulating functions

A modulating function, $\phi(t)$, is defined to be any function that, together with its first derivative goes to zero at time $t=0$ and some other time $t=T$. Functions of this kind have been applied by Valeur and Moirez (1973) and Valeur (1978) to decay curve analysis. Suppose the decay function $G(t)$ is a sum of two exponentials, that is:

$$G(t)=a_1 e^{-t/\tau_1}+a_2 e^{-t/\tau_2}. \tag{6.30}$$

It is easy to show that

$$G(t)+(\tau_1+\tau_2)\dot{G}(t)+\tau_1\tau_2\ddot{G}(t)=0, \tag{6.31}$$

where

$$\dot{G}(t)=\frac{\mathrm{d}G(t)}{\mathrm{d}t}, \text{ etc.}$$

Differentiation of Equation 6.2 leads to the relation

$$\begin{aligned} &I_0(t)+(\tau_1+\tau_2)\dot{I}_0(t)+\tau_1\tau_2\ddot{I}_0(t) \\ &\quad=\int_0^t [G(t-t')+(\tau_1+\tau_2)\dot{G}(t-t')+\tau_1\tau_2\ddot{G}(t-t')]\,\mathrm{d}t' \\ &\quad+[(\tau_1+\tau_2)G(0)+\tau_1\tau_2\dot{G}(0)]P_0(t)+\tau_1\tau_2 G(0)\dot{P}_0(t) \end{aligned} \tag{6.32}$$

which, upon substitution with Equation 6.31 reduces to

$$I_0(t)+(\tau_1+\tau_2)\dot{I}_0(t)+\tau_1\tau_2\ddot{I}_0(t)=A_1P_0(t)+A_2\dot{P}_0(t), \tag{6.33}$$

where

$$A_1=a_1\tau_1+a_2\tau_2 \tag{6.34}$$

$$A_2=(a_1+a_2)\tau_1\tau_2. \tag{6.35}$$

If Equation 6.33 is multiplied across by the modulating function $\phi(t)$, then integrated between $t=0$ and $t=T$, T being the time at which the curves are truncated, and then rearranged, it yields the equation

$$\begin{aligned} &(\tau_1+\tau_2)\int_0^T I_0(t)\dot{\phi}(t)\,\mathrm{d}t-\tau_1\tau_2\int_0^T I_0(t)\ddot{\phi}(t)\,\mathrm{d}t \\ &\quad+A_1\int_0^T P_0(t)\phi(t)\,\mathrm{d}t-A_2\int_0^T P_0(t)\dot{\phi}(t)\,\mathrm{d}t \\ &\quad=\int_0^T I_0(t)\phi(t)\,\mathrm{d}t. \end{aligned} \tag{6.36}$$

This equation can be solved for the four decay parameters if four modulating functions are substituted. Extension of the method to three or more exponentials is straightforward and it can also be used to correct for scattered light and zero-time shift distortions.

Valeur and Moirez (1973) reported that modulating functions of the form

$$\phi(t) = t^m (T - t)^n, \tag{6.37}$$

in which m and n are integers are most suitable for the application of the technique to fluorescence decay convolution. Again, no account is taken of Poisson noise but unlike the previous two techniques this method does not require a "cut-off" correction. The recovered decay parameters do not depend greatly on the values of m and n, which may be selected so as to modulate the decay in such a way that data with the highest number of counts are given greatest weight. Although not widely popular the method has been found effective in separating two (O'Connor *et al.*, 1979) and three (Striker, 1982) closely spaced exponential components. It is, moreover, quite rapid.

6.5.6 Fourier transforms

The Fourier transform, $f(\nu)$, of a function $F(t)$ is defined by the equation

$$f(\nu) = \int_{-\infty}^{\infty} F(t) e^{-2\pi i \nu t} \, dt. \tag{6.38}$$

As in Laplace transformation, convolution (Equation 6.2) is converted to multiplication in Fourier space, that is

$$i(\nu) = p(\nu) \times g(\nu). \tag{6.39}$$

Therefore, since the Fourier transforms of the instrument response function and decay curve can be calculated (notice that again a "cut-off" correction is necessary), the Fourier transform of the decay law can be determined. One of the advantages of Fourier over Laplace transformation is that in the former back-transformation to the time domain is easily achieved through the equation

$$G(t) = \int_{-\infty}^{\infty} g(\nu) e^{2\pi i \nu t} \, d\nu. \tag{6.40}$$

A number of methods based on the Fast Fourier Transform (Cooley and Tukey, 1965) have been developed in recent years for application to multi-exponential decays (Provencher, 1976a, b) and deconvolution (Wild *et al.*, 1977; Andre *et al.*, 1979). All are plagued, however, by the well-known accentuation of random (Poisson) noise upon Fourier transformation of the data. Perhaps the most successful is the method of Wild *et al.* (1977) in which the (single exponential) decay parameters are determined in Fourier space. This method is ideally suited to the excitation source employed by its proposers (a mode-locked Ar^+ ion laser with pulse repetition rate 115 MHz). A further development by Andre *et al.* (1979), in which spurious oscillations are avoided by extrapolating the function $g(v)$ by the Fourier transform of a single exponential, shows some promise, especially for decays in which the functional form of $G(t)$ is unknown. However it is probably true to say that until more progress in dealing with Poisson noise is achieved, deconvolution methods based on Fourier transformation will not attain wide acceptance.

6.5.7 Least-squares curve fitting

Perhaps the most widely used deconvolution technique for single photon counting data is curve fitting, especially the method involving linearization of the fitting function and least-squares fitting. The physical significance of the results of this procedure has been questioned by Isenberg (1973a) and its statistical validity by Hall and Selinger (1981). In spite of these criticisms the method yields good results and its success in analysing data can be judged by a number of well-defined criteria. For these reasons we fully recommend it and in this and the following sections describe its applications in some detail. In passing we should mention another curve fitting procedure, based on the maximum likelihood estimator, which has been proposed by Hall and Selinger (1981). This method is similar in many ways to the least-squares technique but has been developed only for one component decays and is therefore of limited applicability.

As we stated previously, comparisons of the advantages and disadvantages of the least squares method *vis-à-vis* the other common deconvolution techniques have already been carried out (McKinnon *et al.*, 1977; O'Connor *et al.*, 1979). In these publications the least-squares method was found to yield the most satisfactory results. It is a statistical technique, however, and is strictly speaking valid only for very large samples. Consequently in the analysis of the relatively small samples available in SPC data it is most important to have reliable criteria by which the success of the fit can be judged. In addition, we would stress that the decay parameters recovered from any deconvolution technique should always be checked for consistency with a physical model and with the results of other independent experiments.

In the least-squares formulation the optimum description of a set of data points is one that minimizes the weighted sum of the squares of the deviations of the experimental points $y(t_i)$ from the calculated fitting function $Y(t_i)$. That is, the quantity

$$\chi^2 = \sum_{i=1}^{n} W_i[y(t_i) - Y(t_i)]^2 \tag{6.41}$$

is minimized. In this equation W_i is the weighting factor for the ith data point and n is the total number of data points. In the application to SPC data W_i is the reciprocal of the expected value for $y(t_i)$, which is, until the fit is completed, unknown. Usually W_i is approximated by the reciprocal of the measured value $y(t_i)$. The alternative, approximation by the reciprocal of $[Y(t_i)+\text{background}]$, leads to less simple fitting equations but little difference in the final result (Irvin and Livingston, 1974). Therefore, reverting to our previous notation, we assume that the fitting function, $Y(t_i)$ is optimum when the expression

$$\chi^2 = \sum_{i=1}^{n} \left\{ \frac{[I_0(t_i) - Y(t_i)]^2}{I(t_i)} \right\} \tag{6.42}$$

is minimized*. It should be noticed that the uncorrected count $I(t_i)$ rather than $I_0(t_i)$ is the more accurate representation of the variance (Bevington, 1969, p. 111). Of course a rigorous expression for the weighting factor would include a contribution from the counting error associated with the instrument response function data (Powell and McDonald, 1972). Such an expression would, because of the convolution effect, be rather complex and it is usually assumed that the points $P_0(t_i)$ are error free. This assumption is completely justified empirically, since instrument response functions collected to vastly different total count numbers yield identical results, provided that a certain minimum count number is achieved (Phillips and O'Connor, 1982).

Following Bevington (1969) the fitting function $Y(t)$ is linearized by expansion in a curtailed Taylor series in the fitting parameters, i.e.,

$$Y(t_i) = Y^0(t_i) + \sum_{j=1}^{l} \left[\frac{\partial Y^0(t_i)}{\partial a_j} \delta a_j \right], \tag{6.43}$$

in which $Y^0(t_i)$ is expressed by the convolution integral given previously in Equation 6.2:

$$Y^0(t_i) = \int_0^{t_i} P_0(t')G(t_i - t')\,\mathrm{d}t',$$

and $G(t)$ contains the fitting parameters (a_j), e.g.

$$G(t) = a_1 \mathrm{e}^{-t/a_2} + a_3 \mathrm{e}^{-t/a_4}. \tag{6.44}$$

Substitution in Equation 6.42 yields

$$\chi^2 = \sum_{i=1}^{n} \left\{ \frac{\left[I_0(t_i) - Y^0(t_i) - \sum_{j=1}^{l} \left(\frac{\partial Y^0(t_i)}{\partial a_j} \delta a_j \right) \right]^2}{I(t_i)} \right\}. \tag{6.45}$$

*In this formulation $Y(t_i)$ is written in Equation 6.42. In practice when χ^2 is calculated the unexpanded function $Y^0(t)$ (see Equation 6.43) will always be used.

The condition for a minimum,

$$\frac{\partial \chi^2}{\partial(\delta a_k)} = 0, \tag{6.46}$$

leads to equations of the form

$$A_k = \sum_{j=1}^{l} (\delta a_j B_{jk}) \tag{6.47}$$

in which

$$A_k = \sum_{i=1}^{n} \left\{ \left(\frac{I_0(t_i) - Y^0(t_i)}{I(t_i)} \right) \frac{\partial Y^0(t_i)}{\partial a_k} \right\} \tag{6.48}$$

and

$$B_{jk} = \sum_{i=1}^{n} \frac{1}{I(t_i)} \left[\frac{\partial Y^0(t_i)}{\partial a_j} \times \frac{\partial Y^0(t_i)}{\partial a_k} \right] \tag{6.49}$$

Equation 6.47 is conveniently written in matrix form

$$\mathbf{A} = \boldsymbol{\delta}\mathbf{a} \cdot \mathbf{B} \tag{6.50}$$

and $\boldsymbol{\delta}\mathbf{a}$ is calculated by inverting **B**, i.e.,

$$\boldsymbol{\delta}\mathbf{a} = \mathbf{A} \cdot \boldsymbol{\beta} = \mathbf{A}\mathbf{B}^{-1} \tag{6.51}$$

or

$$\delta a_j = \sum_{k=1}^{l} A_k \beta_{jk} \tag{6.52}$$

In execution, a functional form is chosen for $G(t)$, e.g. Equation 6.44, and some or all of the characteristic quantities are chosen as the fitting parameters, a_k in $Y(t)$. When the fitting equation includes an expression for scattered light, Equation 6.10, the contribution from scatter, C, is also made a fitting parameter. The zero-time shift, Equation 6.13, is treated in the same way. Initial estimates, a_k^0, are chosen, $Y^0(t)$ is calculated with Equation 6.2 or 6.10, and χ^2 with Equation 6.42. Partial derivatives, $\partial Y^0(t_i)/\partial a_k$, are now determined, the matrix **B** is set up and inverted, and parameter increments calculated by means of Equation 6.52. A new function $Y^0(t)$ is calculated using the new parameters, $a_k = a_k^0 + \delta a_k$, in $G(t)$ and a new value for χ^2 determined. This process of parameter increment determination and reconvolution is repeated until the value calculated for χ^2 converges to a minimum.

An estimate of the standard deviation in the final value of each fitting parameter may be obtained on the assumption that the corresponding diagonal element of the inverse error matrix $\boldsymbol{\beta}$ expresses the variance i.e. $\sigma_{a_j}^2 = B_{jj}$.

The accuracy of this estimate is of some importance since quoted uncertainties in rate constants or thermodynamic properties derived from kinetic analysis of lifetime data are frequently based on the standard deviations as estimated. This point is fully discussed by Bevington (1969) from a theoretical standpoint. Our opinion is that for multi-exponential fits errors calculated in this way, while probably correct relative to each other are, as absolute errors, too low.

The reason for this is that in the method as outlined the inherent assumption is that deviations from the assumed form of $G(t)$ are statistical in origin, whereas it may be that systematic errors are the dominant errors in lifetime measurements (Behlen and Rice, 1981). Even if this contention seems to be overstating the case somewhat, there is no doubt that the safest way to estimate the errors on the recovered parameters is to repeat the experiment a large number of times and estimate the precision from the spread in the results. For example, Abramson *et al.* (1972) performed a particular lifetime measurement 52 times and found a standard deviation of 1.6% in the results, an error significantly higher than that predicted by the statistical analysis of a single experiment.

Marquardt (1963) has outlined a means by which the minimum in χ^2 may be rapidly attained through simultaneous variation of each parameter. In this procedure Equation 6.47 is modified to the form

$$A_k = \sum_{j=1}^{l} \delta a_j B'_{jk}, \tag{6.53}$$

where

$$\begin{aligned} B'_{jk} &= B_{jk}(1+\lambda) \qquad && \text{for } j = k \\ &= B_{jk} && \text{for } j \neq k \end{aligned} \tag{6.54}$$

and λ is a number that is initially set relatively large (*vide infra*) and is reduced as the minimum is approached.

Routines that calculate initial guesses for the fitting parameters, a_k^0, such as the method of moments or of modulating functions or a straight line fit to the tail of the decay, are sometimes incorporated in the computer programming. In our opinion these extra routines are unnecessary. On the one hand any reasonable estimates of a_k^0 will recover the same final parameters from undistorted data; if a secondary minimum exists it will be at so high a value of χ^2 as to be clearly recognizable. On the other hand, while many local minima may exist on the χ^2 surface for badly distorted data, causing the recovered values to depend on the initial estimates, there is probably no greater likelihood that the routines referred to will find better estimates for a_k^0 than guesses based on knowledge about the decay. For such data the minimum in

χ^2 may be very shallow; the solution is to eliminate as far as possible the source of distortion.

Partial derivatives of the fitting function with respect to the variable parameters $[\partial Y^0(t_i)/\partial a_k]$ may be calculated either analytically or numerically. Supposing $G(t)$ to be a sum of exponentials (Equation 6.44), the analytic expressions for the partial derivatives with respect to a_1 and a_2 are

$$\frac{\partial Y^0}{\partial a_1} = \int_0^{t_i} P_0(t')e^{-(t_i - t')/a_2}\,dt' \tag{6.55}$$

and

$$\frac{\partial Y^0(t_i)}{\partial a_2} = \frac{a_1}{a_2^2}\int_0^{t_i} (t_i - t')P_0(t')e^{-(t_i - t')/a_2}\,dt'. \tag{6.56}$$

These equations are easily calculated from the experimental curves. If it is not desired to calculate the partial derivatives in this way, non-analytical methods can be found in the appropriate text books (e.g., Bevington, 1969, p. 242).

The approximation employed in numerical integration is not critical, Simpson's rule, the trapezoidal rule and the law of the mean for integration yielding essentially identical results. Another simplifying technical point is that the matrix **A** in Equation 6.50 is symmetric ($A_{jk} = A_{kj}$).

Finally, we have found that the speed of the fit is influenced by the starting value for the Marquardt parameter λ in Equation 6.54. We set it at 10^{-2} for single exponential fitting and at 5×10^{-2} or 10^{-1} for double and triple exponential fitting. When a new value for a_k is determined λ is decreased by a factor of 10, but if the search is not finding the minimum in χ^2 (i.e., χ^2 is increasing upon recalculation), λ is increased by 10 and a new inverse matrix **B** determined.

A Fortran deconvolution program that fits decay data to a sum or difference of two exponentials is listed in Appendix 6.A2. In Appendix 6.A1 is listed a Fortran routine that convolves an experimental instrument response function with a decay function that may contain up to 30 exponentials. Gaussian noise (which for high counts is a close approximation to experimental Poisson noise) may be added to the calculated data (SUBROUTINE GLUSS, also listed). However, while synthetic data have much value, especially for testing newly written deconvolution routines and for determining the ability of an analysis technique to cope with certain types of decay data, they are no substitute for real data in the testing of deconvolution procedures.

Further discussion of the applications of least-squares fitting is given in Section 6.6. Here we would briefly mention a recent extension of the technique, which has been proposed by Knutson *et al.* (1983). It employs a so-called "global analysis", that is in the analysis of several decay curves a mode framework that encompasses them all is set up. In this way some of the model parameters may be head fixed, if the values are known from analysis of an individual decay, and fitting error may be examined. Some parameters can also be common to the different but related individual curves so that, overall, the number of fitting parameters, and, consequently, the fitting errors, can be reduced. The authors have developed an implementation of the technique employing matrix mapping, in which parameters pertaining to only one decay (local parameters) and those common to all decays (globals) can be conveniently separated. They have tested the method using both simulated and experimental heterogenous data with excellent results. An incidental advantage of the global mapping technique is that it can be applied to lifetime measurements employing plane and modulation techniques. The method in its application to single photon counting should be particularly suited to, for example, anisotropy or quenching data.

6.5.8 Exponential series

In view of the fact that quite a number of decay curves are measured for which the functional form of the decay law, $G(t)$, is not known, Ware and coworkers (Ware *et al.*, 1973; Hui and Ware, 1976) devised an extension of the least-squares fitting technique that is aimed at determining an accurate representation of $G(t)$ rather than at recovering individual decay parameters directly. It appears to possess fewer disadvantages than other "non-*a priori*" methods, such as Fourier transformation (see Section 6.5.6) or the less well-known method based on Lagrange multipliers (Andre *et al.*, 1982). For this method the decay function is represented by a sum of exponentials, that is:

$$G(t) = \sum_{j=1}^{l} a_j \exp(-t/\gamma_j), \tag{6.57}$$

where, provided that l is chosen large enough, a_j and γ_j have no physical or mechanistic significance. Once the representation of $G(t)$ has been deconvolved it can be examined from the point of view of various physical models or manipulated to construct, for example, time-resolved fluorescence spectra.

In the original development of the technique the γ_j were fixed and only the a_j allowed to vary, thus saving some computation time and avoiding the necessity of choosing initial estimates for γ_j. The number of exponential terms was 6, 10 or 15 (Ware *et al.*, 1973). In more recent applications (Easter *et al.*, 1976; Meech *et al.*, 1981) both the a_j and γ_j are allowed to vary and the number of exponential terms is reduced to 3 or 4. A significant advantage of a function with fewer exponentials is that spurious oscillations in the recovered function, sometimes observed when Equation 6.57 contained 6 or more terms, are eliminated. These fluctuations are introduced when regions of low signal-to-noise ratio (very short or fairly long times) in the data are fitted by a function that is inherently unstable; such a function is unconstrained at regions of low signal and tends to oscillate. However, as Ware *et al.* (1973) point out, these oscillations clearly result from the mathematical technique and are unlikely to be mistaken for real physical effects. Since they occur at the time extremes of the recovered decay function an intermediate segment may be chosen for further analysis.

When the representation of $G(t)$ has been chosen the exponential series method is implemented in the same way as the least-squares technique described in the preceding section. That is, the fitting function, (with variable parameters a_j alone or a_j and γ_j) is linearized and a Marquardt search is carried out for the minimum value of χ^2.

If it is decided to allow only the a_j in Equation 6.57 to serve as fitting parameters, a suitable set of γ_j must be chosen for the data under analysis. This choice is discussed by Ware *et al.* (1973), Meech *et al.* (1981) and Bouchy (1982). We would suggest that the γ_j should be evenly spaced about the lifetimes of the decay and that on short timescales at least one sub-nanosecond γ_j should be included in order to facilitate a good fit on the rising edge of the decay curve, where some real picosecond phenomena may be strongly manifest.

It will be realized that in many applications of the least-squares technique, discussed in Section 6.5.7, a curtailed exponential series, with both a_j and γ_j variable, is the form of the decay function. Since the number of terms is small the a_j and $\gamma_j(\tau_j)$ assume physical significance. The advantages of the present method, in which no *a priori* assumptions about the functional form of $G(t)$ are made, is that it is suitable in the study of photokinetic phenomena which are difficult or impossible to characterize by a definite decay function. If a representation of the decay function of such systems is required, undistorted by the optical pump and detector response, the exponential series method is at present the most satisfactory means of obtaining it. Perhaps the most useful application of the technique is in the construction of time-resolved spectra, a subject fully discussed in the following chapter.

6.6 Application of the Least-Squares Fitting Technique

In Section 6.5.7 we described least-squares fitting and reconvolution in some detail, with explicit reference to the convolution Equation 6.2 and the modified Equation 6.10. The difficulty of treating Equation 6.13, which takes account of a shift in the zero-time between instrument response function and decay curve, will be realized in a simple way if it is attempted, by a change of variable, to remove the quantity δ from the exponential function $P(t)$ into the decay function $G(t)$ where it can be treated analytically. It will then be found that the limits of integration in the convolution equation will include δ. An analytical solution for δ can be found which involves differentiation of the instrument response function. This is the procedure for determining the shift adopted in the program listed in Appendix 6.A2.

As we have already pointed out this shifting technique has attained some popularity as a means of correcting for a wavelength dependent photomultiplier response. For this purpose it appears to be a good empirical correction. Its most appropriate application, however, is the correction for a shift in the pump pulse profile leading to a shifted instrument response function. If the shift correction routine is applied it should be remembered that it could mask the presence of effects resulting from real picosecond processes such as rotational relaxation in moderately viscous solvents.

Spurious rising edge effects can often be corrected for with the shift routine. An alternative is to avoid the rising edge altogether in the analysis by fitting only over that portion of the data beginning at the maximum of the decay curve. Indeed, since any section of the decay may be selected for analysis, a region of scattered light contaminated data may also be avoided. The least-squares technique is particularly suited to truncation of the analysis region because the only modification required is an adjustment of the limits of summation in Equation 6.42 which may be rewritten as:

$$\chi^2 = \sum_{i=n_1}^{n_2} \frac{[I_0(t_i) - Y(t_i)]^2}{I(t_i)}, \tag{6.58}$$

where n_1 and n_2 are the first and last channels of the region chosen for analysis. Again, interesting short time effects may be missed when the analysis is curtailed in this way. The advantage of correcting for scattered light with Equation 6.10 is that short times are not avoided. It must be pointed out, however, that in the analysis based on Equation 6.10 some mathematical correlation between a scattered light intensity factor and a short lived decay is probably unavoidable (Grinvald, 1976).

The channel fitting range to be used in Equation 6.58 is a matter of much debate even among experienced users of the SPC technique. Definite rules are difficult to give with any degree of confidence; on the other hand, it would render comparison of results a much less arduous task if the fitting range were always specified. With the latest generation of PM tubes and provided that the observation wavelength is not grossly shifted from the excitation wavelength, there appears to be no reason why the entire decay should not be analysed, given a stable excitation pulse. Assuming the normal number of counts in the channel of maximum counts (see Chapter 2) analysis over the whole curve would mean analysis from t_{n_1} with $I_0(t_{n_1})=1000$ to t_{n_2} with $I_0(t_{n_2})=100$ or to t_n if the counts have not decayed to 100.

Occasionally it may be advisable to avoid analysis over channels corresponding to a non-linear region of the TAC range (see Chapter 5); if so those channels must be determined and the decay collected in such a way that the "non-linear" channels are not necessary for the deconvolution and subsequent analysis.

Sometimes ineradicable instabilities in the excitation pulse may cause rising edge distortions that are best avoided in deconvolution analysis. If so, the fitting region is chosen to begin at the channel of maximum counts. Little is to be gained in such cases by fitting over the whole curve and having recourse to the shift routine. A common case is one in which a short decay time can be recovered from analysis over the decay, but could be determined with greater accuracy from analysis of the whole curve. Our usual practice, therefore, is to fit first over the decay and then, with shift, over the whole curve. If the fit from the maximum is not successful (see the following section), or if significantly different decay parameters are recovered from the two fitting regions, we repeat the experiment after checking the stability in the excitation source. In Table 6.1 will be found an example of the type of agreement between the two fitting regions that should be observed.

Before concluding this section we briefly mention two little used techniques for "improving" the quality of the data, smoothing and selective data rejection. As regards smoothing of the raw data (Dromey and Morrison, 1970), Nemzek (1975) carried out experimental tests with a number of smoothing routines applied to SPC data and found that in all cases results of least-squares deconvolution were poorer after smoothing. Data smoothing cannot be recommended as a preliminary to least-squares fitting since the Poisson noise is actually an advantage in this technique. Moreover extreme care must be exercised so that loss of information does not result from smoothing. In one variant of the least-squares procedure, data can be selectively rejected (Irwin and Livingstone, 1974). Residuals (see Section 6.7.4) are calculated upon each iteration; if the ratio of any residual to the standard deviation of the residuals exceeds some arbitrary limit, the corresponding data point is removed. There appears to be no justification for rejecting SPC data points in this way. A rogue point probably results from an instrumental artefact, which should be tracked down and eliminated.

6.7 Evaluating the Success of the Fit

6.7.1 Preliminary

Some means should be available by which to judge the physical significance of the parameters recovered with any deconvolution technique. On a certain level this question is outside the scope of data analysis; the significance of the parameters can best be judged against knowledge about the system under investigation derived from independent experiments and by consistency with some reasonable physical model. Nevertheless knowledge of excited state kinetics in many areas is often so limited that proposed models may accommodate many different types of decay functions. For instance the concentration in the first excited singlet state (S_1) of simple aromatic molecules in room temperature solution usually decays strictly exponentially, even though a thermalized Boltzmann distribution populates a number of vibronic states in S_1. But is a single exponential decay always expected for such molecules at elevated temperatures in the vapour phase, even in the presence of a buffer gas? When questions like this occur in lifetime studies it is important to have complete confidence in the results of decay curve analysis so that subtle mechanistic points can be investigated in detail.

Data analysis of SPC decay curves assumes the simplest explanation of the observations (consistent with a proposed hypothesis). For example an aromatic molecule excited to S_1, in room temperature solution, would be expected to decay with a single exponential decay function. If analysis of the decay curve is judged poor (and if instrumental distortions have been eliminated or adequately corrected for) an additional level of complexity will be added to the proposed decay function. An example would be the decay of short-lived species quenched irreversibly in viscous solvents, which involves an exponential dependence on the square root of time (Ware and Nemzek, 1975). The new more complex decay function is then tested in the data analysis routine.

Occasionally, though not frequently, a simpler decay law may be recovered from data analysis than is consistent with a proposed hypothesis. Before discarding the hypothesis the limitations of the deconvolution procedure should be checked. Double exponential decays with closely spaced decay times may not be resolvable upon deconvolution. Too great a reliance on the so-called Occam's razor ("simplest is correct") might cause a limitation of the technique to be misinterpreted. In this regard it is helpful to have some knowledge about the decay times under investigation and to carry out preliminary tests of the analysis technique on data synthesized with, for example, the routine CRACK in Appendix 6.A1. Extensive tests of this nature should be made before accepting the results of triple exponential analysis. In addition to computer generated decays, triple and quadruple component decay curves should be constructed from simpler decays with known decay times. Triple exponential analysis should then be performed and the results compared with the known decay parameters. It is our opinion, based on analyses both successful and unsuccessful of triple exponential decays, that physically valid parameters are extremely difficult to attain, and that all the results of this type of fitting should be carefully scrutinized and checked against other measurements.

In the following sections we indicate some criteria by which the success of least-squares analysis in terms of an assumed decay function may be judged. If these criteria are favourable in a given experiment, sufficient confidence may be placed in the results to warrant interpretation by comparison with the properties of a model proposed on the basis of independent experiments, or by analogy with the known behaviour of similar molecular systems.

6.7.2 The reduced chi-square, χ_ν^2

This statistic may be calculated from the equation

$$\chi_\nu^2 = \frac{\chi^2}{n_2 - n_1 + 1 - p}, \tag{6.59}$$

where χ^2, n_1 and n_2 are as defined in Equation 6.58 and p is the number of variable parameters in the fitting function. χ_ν^2 should be close to 1 for Poisson distributed data (Bevington, 1969, p. 189). Values much less than 1 (0.75) are usually a symptom of data in which the number of counts is not large enough.

Actual χ_ν^2 values are usually greater than 1. It would render least-squares fitting much more clear-cut if a maximum value for χ_ν^2 could be stated with certainty. Unfortunately, definite statements of this kind cannot be made. Although χ_ν^2 values in excess of c. 1.02 probably invalidate the fit on the assumption of strict Poisson statistics (Bevington, 1969; Table C-4), the exact form of the noise in SPC experiments is thought to have a slight non-Poissonian contribution (Grinvald and Steinberg, 1974) and "good" χ_ν^2 values range from 0.8 to 1.2. Higher values may have to be accepted in some instances. The author upon whose work we have relied almost entirely for our formulation of the least-squares routine would grant that χ_ν^2 values of 1.5 are "reasonably close" to 1 (Bevington, 1969, p. 190). We feel that 1.5 is somewhat large when undistorted data are analysed, but would be reluctant to be more definite about an upper limit other than remarking that in our experience analyses yielding χ_ν^2 values greater than 1.3 should be scrutinized very closely.

As a criterion for unsuccessful fitting the χ_ν^2 value alone is inadequate since low values are occasionally obtained when the fit proves to have been a

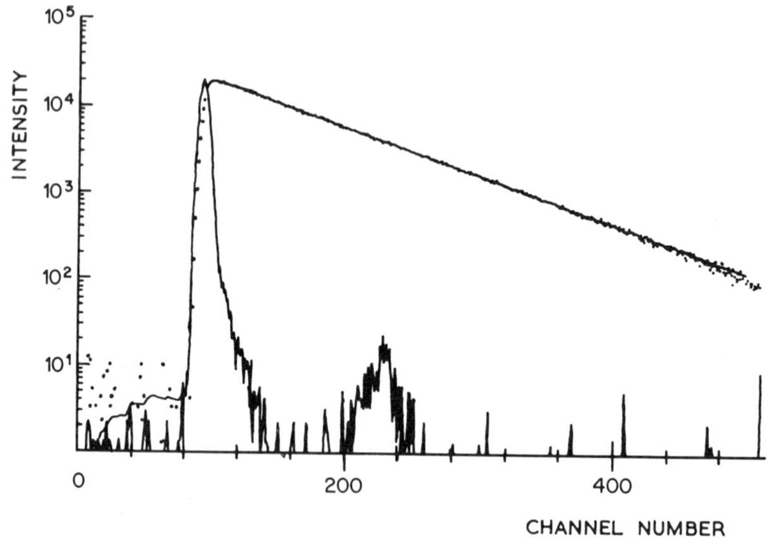

Figure 6.2 Example of a poor fit that yields an acceptable value of χ_ν^2. Decay, following 302 nm excitation, of 1-methylindole in cyclohexane. Points, observed decay data; line, best fitting single exponential (the instrument response function is also represented by a line). $\chi_\nu^2 = 1.20$. See Fig. 6.4(a) for evidence of poor fit.

failure. This is illustrated in Fig. 6.2, where the fitting, as judged by the autocorrelation function of the weighted residuals (see Section 6.7.6) was poor, yet the χ_ν^2 value was quite acceptable.

6.7.3 The standard normal variate, Z

Since acceptable values of χ_ν^2 depend critically upon ν, it is useful to calculate from χ_ν^2 a statistic called the standard normal variate, which is independent of ν. The calculation is

$$Z = \frac{\nu(\chi_\nu^2 - 1)}{\sqrt{(2\nu)}}. \tag{6.60}$$

Catterall and Duddell (1982) discuss the application of Z in detail and give equations for estimating the likelihood of obtaining a particular value. They recommend that acceptable values of Z should lie between -3 and $+3$, which corresponds, for 256 data points, to $0.735 < \chi_\nu^2 < 1.265$. As with χ_ν^2 acceptable values of Z may be obtained even for poor fits.

6.7.4 Graphical comparison of observed and calculated curves

Such a comparison, although crude, may be of some value when the fit is poor. A rather obvious example can be seen in Fig. 6.1 where the oscillations of the data points about the fitted curve are a well-known manifestation of r.f. interference. If flash lamp electrodes are badly corroded an intense secondary peak in the pump pulse profile may produce a shoulder in the decay curve that does not fit upon deconvolution; the region of poor fitting should be obvious when the two curves are plotted together. Therefore, this type of comparison is an effective diagnostic of instrumental artefacts. A word of warning should be given, however. If the two curves are not superimposable in the long-time tail region an extra long-lived component might be inferred. In fact, since least-squares fitting gives greater weight to the more precise data at shorter times, the analysis, by forcing the fit at the maximum in the decay, may induce a purely artificial divergence at long times.

6.7.5 Plot of the weighted residuals, $r(t_i)$

The weighted residual in channel i is calculated from the equation

$$r(t_i) = \sqrt{W_i}[I_0(t_i) - Y(t_i)]. \qquad (6.61)$$

Various weighting factors have been suggested but that given here is the most usual. Following the notation of Equation 6.42 we can write

$$r(t_i) = \frac{I_0(t_i) - Y(t_i)}{\sqrt{I(t_i)}}, \qquad (6.62)$$

where $Y(t_i)$ is the curve calculated with the final values of the decay parameters (see footnote p. 172). Residuals from successful fits when plotted against channel number should be randomly distributed about zero. Since the parameters are estimated from the data, in theory the residuals cannot be completely uncorrelated (McKinnon *et al.*, 1977), but small correlations usually go unobserved because of the large number of degrees of freedom. In Fig. 6.3(a) are plotted the residuals resulting from the fit of a double exponential function to a triple component decay curve. The plot in Fig. 6.3(b), resulting from the correct triple exponential analysis, is clearly an improvement.

It can be seen that this plot is a very clear test when gross non-randomness is present. However it can be quite difficult to distinguish small non-random deviations resulting from a poor fit from those resulting from the unavoidable correlations referred to in the preceding paragraph. In addition examination of the plot is a subjective visual test and, as is usual with such tests, must be interpreted with great care.

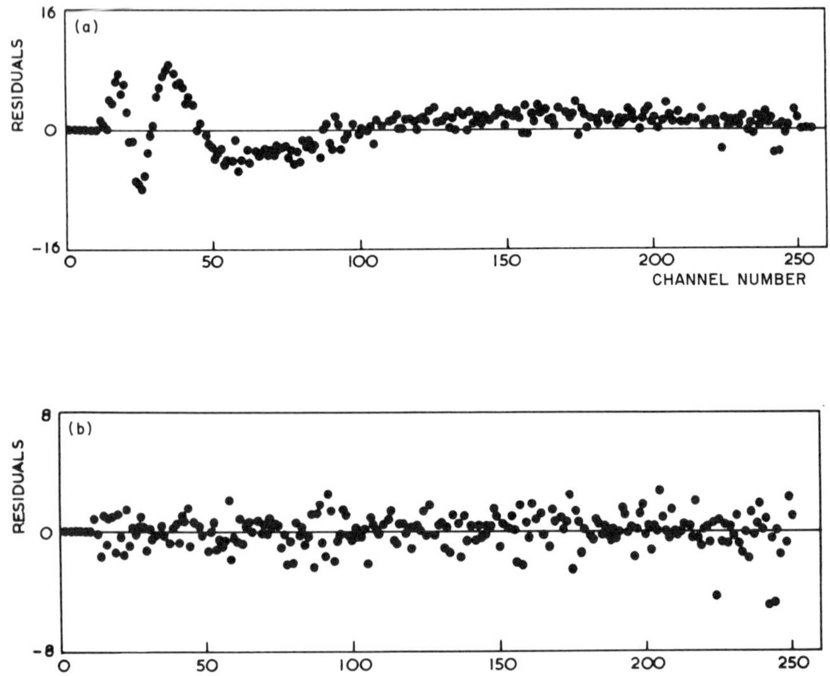

Figure 6.3 Plots of weighted residuals resulting from fits to simulated three-component data. Real instrument response function convolved with the function, $G(t) = 0.25\ \exp(-t/2.5) + 0.07\ \exp(-t/10.0) + 0.025\ \exp(-t/40)$. (a) Weighted residuals resulting from two component fit. (b) Weighted residuals resulting from three component fit.

6.7.6 Autocorrelation function of the weighted residuals, $cr(i)$

This function is calculated with Equation 6.63 (Grinvald and Steinberg, 1974):

$$Cr(j) = \frac{\dfrac{1}{m} \displaystyle\sum_{i=n_1}^{n_1+m-1} r(t_i) r(t_{i+j})}{\dfrac{1}{n_3} \displaystyle\sum_{i=n_1}^{n_2} [r(t_i)]^2}, \tag{6.63}$$

where

$$n_3 = n_2 - n_1 + 1.$$

$cr(j)$ is the correlation between the residual in channel i and the residual in channel $i+j$ summed over a selected number of i channels. An upper limit for j is set at $n_3/2$ so that the number of terms in the numerator may be sufficient for proper averaging [if there are 100 fitting channels (n_3) and j is allowed to extend to 90, say, i can run only from 1 to 10]. m is defined as $n_3 - j$ and varies for each sum.

By definition $cr(0) = 1$, that is, each residual is perfectly correlated with itself. If the observed function $I_0(t_i)$ is randomly scattered about the calculated curve $Y(t_i)$, and if n tends to infinity, $cr(j)$ will tend to zero for $j \neq 0$. But since m is finite, for a good fit the autocorrelation function shows high frequency low amplitude oscillations about zero when plotted against j.

Although this plot requires again a subjective evaluation, it is a more sensitive indication of poor fitting than the plot of residuals. However, the low frequency oscillations observed in good fits may be difficult to distinguish from fluctuations symptomatic of a poor choice of fitting function or of distorted data. A clear indication of a poor fit (owing to an incorrect fitting function) is illustrated by the plot in Fig. 6.4(a). This autocorrelation function resulted from the fit presented in Fig. 6.2. The function in Fig. 6.4(b) indicates a good fit and shows the low amplitude oscillations referred to. It should be emphasized that these oscillations should closely resemble a regular waveform. Published autocorrelation plots (e.g., Grinvald *et al.*, 1975; Badea and Brand, 1979) with irregular structure, frequently obtained with flash lamp

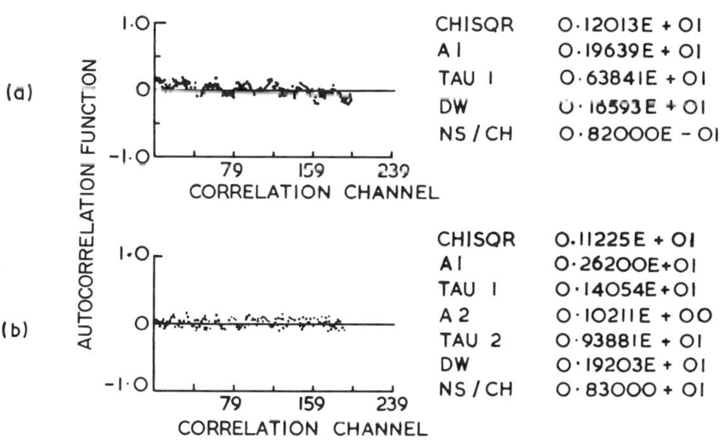

Figure 6.4 Plots of autocorrelation function of weighted residuals resulting from poor (a) and acceptable (b) fits. (a) This function results from the fit described in Fig. 6.2. (b) Double exponential fit to the decay of 1-cyanonaphthalene quenched by trimethylamine in the gas phase. The biexponentially results from reversible exciplex formation.

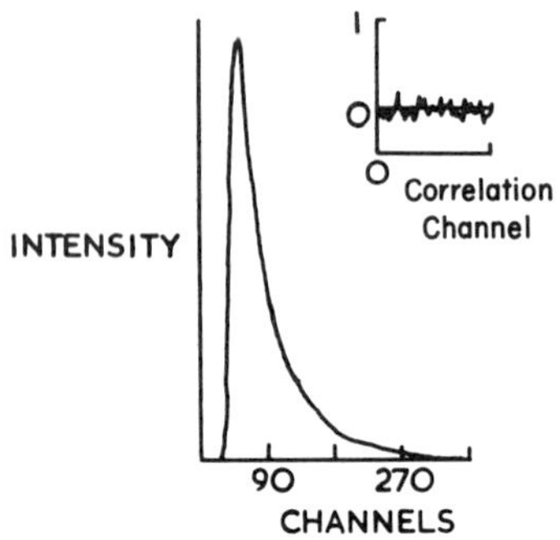

Figure 6.5 Decay of an equimolar mixture of anthracene and 9-cyanoanthracene in ethanol. The (acceptable) autocorrelation function of the residuals resulting from a double exponential fit is shown in the insert. Flash lamp excitation (adapted from Badea and Brand, 1979).

excitation sources, are symptomatic of instrumental distortions such as r.f. noise interference (see Section 3.3.6). An example of an autocorrelation function which was acceptable for a flash lamp excitation source is illustrated in Fig. 6.5 and may be compared to the plot in Fig. 6.4(b) which was obtained with a laser excitation source.

6.7.7 Variation of the fitting range

One of the advantages of curve fitting in deconvolution is that the region of fitting may be varied. As has been stated already, rising edge distortions may be avoided by this aspect of the technique. Furthermore, an extra short- or long-lived component can frequently be detected by varying the fitting range. Clearly the only effect such a variation should have on the recovered parameters is an increase or decrease in the uncertainty associated with them. For greater confidence in the results one fits over the full range of channels. However, if significant changes in a decay time occur when the fitting range is varied, the implication is that another lifetime is contributing to the decay (provided that instrumental distortions can be discounted). Pre-exponential factors are usually determined with less accuracy than decay times (Isenberg, 1973a); therefore moderate changes in these parameters upon change of fitting range may be accepted as long as the τ values do not vary.

6.7.8 Variation of initial guesses

The values finally determined for the decay parameters should not depend on the choice of initial estimates. Such a dependence is symptomatic of instrumental distortions or impurity fluorescence rather than an incorrect choice of fitting functions.

6.7.9 The Durbin–Watson parameter, *DW*

This parameter was introduced by Durbin and Watson (1950, 1951) to test for serial correlation in linear regression analysis of econometric data. The defining equation is

$$DW = \frac{\sum_{i=n_1+1}^{n_2} [r(t_i) - r(t_{i-1})]^2}{\sum_{i=n_1}^{n_2} [r(t_i)]^2}. \tag{6.64}$$

The above equation indicates the Durbin–Watson parameter has some resemblance to the autocorrelation function of the residuals for an autocorrelaction channel $j = 1$. Since DW is a numerical value its use is not open to the objection frequently levelled at plots of residuals, that is, that they involve subjective decisions based on visual inspection. Application of the Durbin–Watson parameter to SPC data analysis was rejected by McKinnon *et al.* (1977), but we now use it regularly and find it frequently more sensitive than χ_ν^2. Durbin and Watson (1951) tabulated upper and lower limits, DU and DL, for DW for different numbers of data points and for different numbers of variable parameters. As their paper states, "if the observed DW is less than DL the value is significant, while if it is greater than DU the value is not significant. If it lies between DL and DU the test is inconclusive." Accordingly it should be possible to establish a lower limit for DW. If the observed value falls below this limit an incorrect fitting function or distorted data are probable. In the tables in the paper by Durbin and Watson (1951) DL values are given for up to 100 data points and five fitting parameters. Simple extrapolation of these values combined with numerous calculations of DW lead us to conclude that values less than 1.7, 1.75 and 1.8 are indicative, respectively, of poor fits over 256 or 512 data points to single, double and triple exponential functions. For example the DW value for the poor fit illustrated by the decay in Fig. 6.2 and the autocorrelation function of the residuals in Fig. 6.4(a) was 1.66, slightly too low for a single exponential fit over 400 channels. A double exponential fit to these data yielded an acceptable value of 1.89. (It should be remarked however, that the second component merely fits some instrumental distortion since this decay is actually a single exponential). It is interesting to note that these values seem to apply even when the data are not Poisson distributed (Roberts, 1980). A more accurate calculation of limits confirms the results of the simple extrapolation (Skelton, 1983).

6.7.10 Mean residual $\bar{r}$ and standard deviation of residuals, σ_r

Since the residuals are presumed to belong to a standard normal distribution they should have a mean, $\bar{r}$, of zero and a standard deviation, σ_r of 1. These quantities are given by the equations

$$\bar{r} = \sum_{i=n_1}^{n_2} r(t_i)/n_3 \tag{6.65}$$

$$\sigma_r = \left(\sum_{i=n_1}^{n_2} [r(t_i) - \bar{r}]^2 \right)^{1/2} \Big/ (n_3 - 1)^{1/2}. \tag{6.66}$$

They are not particularly sensitive and seem to be of little value compared to the other tests.

6.7.11 Measure of skewness, *SK* (Irvin and Livingston, 1974)

A particularly skewed fit occurs when negative values of the residuals lie close to the mean while positive values show large deviations. Negative skewness on the other hand occurs when the negative values show large deviations. A measure of skewness, *SK*, can be estimated with the expression

$$SK = \left[\frac{n_3}{\left(\sum_{i=n_1}^{n_2} [r(t_i) - \bar{r}]^2 \right)^3} \right]^{1/2} \times \sum_{i=n_1}^{n_2} [r(t_i) - \bar{r}]^3. \tag{6.67}$$

For a normal distribution *SK* is approximately normally distributed with a mean of zero and a standard deviation given by $(6/n_3)^{1/2}$. Absolute values of *SK* greater than $(6/n_3)^{1/2}$ are therefore indicative of a skewed fit. The sensitivity of this parameter has not been widely tested in SPC data analysis. Our experience with it is that skewness is difficult to interpret but in general good fits, as judged by other criteria, are rarely skewed.

6.7.12 Measure of Kurtosis, *K* (Pearson, 1905)

A further type of departure from normality is called kurtosis (cf. Greek *κύρτωσις*, a convexity). A measure of this departure can be calculated from the expression (Snedecor and Cochran, 1968):

$$K = \frac{n_3 \sum_{i=n_1}^{n_2} [r(t_i) - \bar{r}]^4}{\left(\sum_{i=n_1}^{n_2} [r(t_i) - \bar{r}]^2 \right)^2}. \tag{6.68}$$

For the normal distribution this ratio has the value 3. If the value is greater than 3 there is an excess of residual values near the mean and far from it (leptokurtosis). Ratios less than 3 result from distributions with a flatter top than the normal (platykurtosis). An approximation of the standard deviation of K is given by $(24/n_3)^{1/2}$ when the sample size is over 1000. Because sample sizes in SPC analysis are usually smaller than this, it is difficult to estimate the expected value of K for normally distributed residuals.

6.7.13 The root mean square deviation of the residuals

This quantity, sometimes symbolized as RMS or RMSR, is the square root of χ_ν^2 (Knight and Selinger, 1971). As such it should be close to 1 for a good fit but, as in the case of χ_ν^2, a definite upper limit cannot be set.

6.7.14 Other tests

Other suggested fitting criteria include the parameter correlation matrix and support plane confidence intervals (McKinnon *et al.*, 1977). The former is in theory very useful but it is extremely difficult to estimate an acceptable degree of parameter correlation (Beddard, 1981). McKinnon *et al.* (1977) have calculated the non-linear support plane confidence interval for a particular parameter, but give few details of the results. Another little used criterion is the generalized statistical test of Irwin *et al.* (1981). It possesses the advantage of taking explicit account of errors in the instrument response function, but at the time of writing has been tested only on simulated data. The reader is referred to Marquardt (1963) for a full discussion of these and other parameter statistics.

6.8 Synthetic Data

As already stated, the effectiveness of any deconvolution technique must be evaluated by an analysis of real experimental data. Multiexponential decays, however, pose a problem for testing purposes, since if two or more fluorophores, the lifetimes of which are well known, are mixed in the same solution interactions arising from energy or charge transfer may complicate the resulting decay. Consequently multiexponential decays should be constructed by measuring known single exponential decays under identical conditions and adding the resulting curves in the MCA or computer. The excitation wavelength and instrument response function must be the same for all curves.

Frequently, preliminary testing of a routine is carried out with synthetic data, about the decay times of which there is no uncertainty. For these data the instrument response function can be real or calculated with some reasonable generating equation; forms such as a Gaussian of chosen FWHM (Ware *et al.*, 1973), Equation 6.69 (McKinnon *et al.*, 1977) and Equation 6.70 (Greer *et al.*, 1981) are satisfactory.

$$P_0(t) \propto 5.802t^2 e^{-0.4t} \qquad (t \text{ in channels}) \tag{6.69}$$

$$P_0(t) \propto e^{-t/0.66} - e^{-t/0.55} \tag{6.70}$$

The instrument response function (if calculated it is not usual to add noise) is then convolved with the decay function of interest. To the resultant curve, $C(t)$, Poisson noise is added. In fact it is usual to add Gaussian noise according to the equation

$$C_0(t) = C(t) + N[C(t)]^{1/2}, \tag{6.71}$$

where $C_0(t)$ represents the noisy curve and N is a random, normally distributed number with a mean of zero and a standard deviation of 1. For values of $C(t) > 20$ addition of Gaussian noise is equivalent to addition of Poisson noise (Greer *et al.*, 1981).

A routine for calculation of synthetic data, which requires input of a real instrument response function, is given in Appendix 6.A1.

References

Abramson, A. S., Spears, K. G. and Rice, S. A. (1972). *J. Chem. Phys.* **56**, 2291–2308.

Almgren, M. (1973). *Chem. Scripta* **3**, 145–153.

Ameloot, M. and Hendrickx, H. (1982). In "Deconvolution and Reconvolution of Analytical Signals" (Bouchy, M., ed.), pp. 277–299. E.N.S.I.C.–I.N.P.L., Nancy.

Andre, J. C., Vincent, L. M., O'Connor, D. and Ware, W. R. (1979). *J. Phys. Chem.* **83**, 2285–2294.

Andre, J. C., Viouy, J. L., Bouchy, M., Vincent, L. M. and Valeur, B. (1982). In "Deconvolution and Reconvolution of Analytical Signals" (Bouchy, M., ed.), pp. 223–246. E.N.S.I.C.–I.N.P.L., Nancy.

Badea, M. G. and Brand, L. (1979). *Methods Enzymol.* **61**, 378–425.

Beddard, G. S. (1981). Personal communication.

Behlen, J. M. and Rice, S. A. (1981). *J. Chem. Phys.* **75**, 5672–5684.

Bevington, P. B. (1969). "Data Reduction and Error Analysis for the Physical Sciences", Chapter II. McGraw-Hill, New York.

Bouchy, M. (1982). In "Deconvolution and Reconvolution of Analytical Signals" (Bouchy, M., ed.), pp. 211–222. E.N.S.I.C.–I.N.P.L., Nancy.

Catterall, T. and Duddell, J. (1982). In "Deconvolution and Reconvolution of Analytical Signals" (Bouchy, M., ed.), pp. 445–459. E.N.S.I.C.I.N.P.L., Nancy.

Coates, P. B. (1968). *J. Phys. E.* **1**, 878–879.

Coates, P. B. (1972a). *J. Phys. E.* **5**, 148–150.
Coates, P. B. (1972b). *Rev. Sci. Instrum.* **43**, 1355–1356.
Cooley, J. W. and Tukey, J. W. (1965). *Math. Comput.* **19**, 297–315.
Davis, C. C. and King, T. A. (1970). *J. Phys. A* **3**, 101–109.
Demas, J. N. and Adamson, A. W. (1971). *J. Phys. Chem.* **75**, 2463–2466.
Dromey, R. G. and Morrison, J. D. (1970). *Int. J. Mass Spect. and Ion Phys.* **4**, 475–482.
Durbin, J. and Watson, G. S. (1950). *Biometrika* **37**, 409–428.
Durbin, J. and Watson, G. S. (1951). *Biometrika* **38**, 159–178.
Easter, J. H., De Toma, R. P. and Brand, L. (1976). *Biophys. J.* **16**, 571–583.
Gafni, A., Modlin, R. L. and Brand, L. (1975). *Biophys. J.* **15**, 263–280.
Greer, J. M., Reed, T. W. and Demas, J. N. (1981). *Anal. Chem.* **53**, 710–714.
Gregory, T. A. and Helman, W. P. (1972). *J. Chem. Phys.* **56**, 377–385.
Grinvald, A. and Steinberg, I. Z. (1974). *Anal. Biochem.* **59**, 583–598.
Grinvald, A., Schlessinger, J., Pecht, J. and Steinberg, I. Z. (1975). *Biochem.* **14**, 1921–1929.
Grinvald, A. (1976). *Anal Biochem.* **75**, 260–280.
Hall, P. and Selinger, B. (1981). *J. Phys. Chem.* **85**, 2941–2946.
Harris, C. H. and Selinger, B. K. (1979). *Aust. J. Chem.* **32**, 2111–2119.
Helman, W. P. (1971). *Int. J. Radiat. Phys. Chem.* **3**, 283–290.
Holzapfel, C. (1974). *Rev. Sci. Instrum.* **45**, 984–896.
Hui, M. H. and Ware, W. R. (1976). *J. Amer. Chem. Soc.* **98**, 4712–4717.
Irwin, J. A. Quickenden, T. I. and Sangster, D. T. (1981). *Rev. Sci. Instrum.* **52**, 191–194.
Irvin, D. J. G. and Livingston, A. E. (1974). *Comp. Phys. Commun.* **7**, 95–113.
Isenberg, I. and Dyson, R. D. (1969). *Biophys. J.* **9**, 1337–1350.
Isenberg, I. (1973a). *J. Chem. Phys.* **59**, 5696–5707.
Isenberg, I. (1973b). *J. Chem. Phys.* **59**, 5708–5713.
Isenberg, I., Dyson, R. D. and Hanson, R. (1973). *Biophys. J.* **13**, 1090–1115.
Jezequel, J. Y., Bouchy, M. and Andre, J. C. (1982). *Anal. Chem.* **54**, 2199–2209.
Knight, A. E. and Selinger, B. K. (1971). *Spectrochem. Acta* **27A**, 1223–1234.
Knutson, J. R., Beechem, J. M. and Brand, L. (1983). *Chem. Phys. Lett.* **102**.
Marquardt, D. W. (1963). *J. Soc. Ind. Appl. Math. (SIAM J.)* **11**, 431–441.
McKinnon, A. E., Szabo, A. G. and Miller, D. R. (1977). *J. Phys. Chem.* **81**, 1564–1570.
Meech, S. R., O'Connor, D. V., Roberts, A. J. and Phillips, D. (1981). *Photochem. Photobiol.* **33**, 159–172.
Meech, S. R. (1983). Thesis. University of Southampton.
Nemzek, T. L. (1975). Thesis. University of Minnesota.
O'Connor, D. V., Ware, W. R. and Andre, J. C. (1979). *J. Phys. Chem.* **83**, 1333–1343.
Pearson, K. (1905). *Biometrika* **IV**, 181–185.
Phillips, D. and O'Connor, D. V. (1982). Personal communication.
Powell, D. R. and McDonald, J. R. (1972). *Comp. J.* **15**, 148–155.
Provencher, S. W. (1976a). *J. Chem. Phys.* **64**, 2772–2777.
Provencher, S. W. (1976b). *Biophys. J.* **16**, 27–41.
Robbins, R. J., Fleming, G. R., Beddard, G. S., Robinson, G. W., Thistlethwaite, P. J. and Wolfe, G. J. (1980). *J. Amer. Chem. Soc.* **102**, 6271–6279.
Roberts, A. J. (1980). Personal communication.
Skelton, P. (1983). Personal communication.
Small, E. W. and Isenberg, I. (1977). *J. Chem. Phys.* **66**, 3347–3351.
Snedecor, G. W. and Cochran, W. G. (1968). "Statistical Methods", 6th edn., p. 183. The Iowa State University Press, Ames, Iowa.

Striker, G. (1982). In "Deconvolution and Reconvolution of Analytical Signals" (Bouchy, M., ed.), pp. 329–354. E.N.S.I.C.–I.N.P.L., Nancy.
Valeur, B. and Moriez, J. (1973). *J. Chem. Phys.* **70**, 500–506.
Valeur, B. (1978). *Chem. Phys.* **30**, 85–93.
Ware, W. R., Doemeny, L. J. and Nemzek, T. L. (1973). *J. Phys. Chem.* **77**, 2038–2048.
Ware, W. R. and Nemzek, T. L. (1975). *J. Chem. Phys.* **62**, 477–489.
Wild, U. P., Holzwarth, A. R. and Good, H. P. (1977). *Rev. Sci. Instrum.* **38**, 1621–1627.

Appendix 6.A1

In this appendix is listed a Fortran computer routine that generates synthetic decay curves with added Gaussian noise. The routine is designed for calculations with a real instrument response function. In addition, it was written for use with a Perkin Elmer Interdata 7/32 computer.

```
$BATCH
C  PROGRAM CRACK.FTN
C  CALCULATES A DECAY CURVE USING AN INPUT PUMP PULSE
C  AND A DECAY FUNCTION CONSISTING OF A SUM OF EXPONENTIALS.
C  THE CONVOLUTION INTEGRAL,I(T)=P(T)@G(T),IS ASSUMED
C  ALSO ADDS GAUSSIAN RANDOM NOISE WITH A MEAN OF ZERO
C  AND A STANDARD DEVIATION OF 1.
C  DIMENSIONED FOR UP TO 30 EXPONENTIALS AND 512 DATA POINTS
C
C  THE PROGRAM READS FROM THE TERMINAL(CARDS) ON LOGICAL
C  DEVICE 5, AND WRITES TO IT ON LOGICAL DEVICE 4
       DIMENSION A(60),P(512),C1(512),C(512),ITITLE(17)
       INTEGER*2 IFF(9)
C  A(60) : PRE-EXPONENTIAL FACTORS AND TAU VALUES IN G(T)
C  P(512) : ARRAY OF PUMP PULSE DATA POINTS
C  C(512) : ARRAY OF CALCULATED CURVE DATA POINTS
C  C1(512) : FOR INTERMEDIATE RESULTS
C  IFF(9) : ARRAY OF CHARACTERS IN FILE NAMES
C  NCHAN : NO. OF DATA POINTS(CHANNELS)
C  FACT : NANOSECONDS/CHANNEL
C  ITITLE(17) : ARRAY CONTAINING A TITLE FOR UNFORMATTED
C                DATA FILES
1      WRITE(4,2)
2      FORMAT(1X,' HOW MANY CHANNELS ?')
       READ(5,3) NCHAN
3      FORMAT(I3)
       NCH=NCHAN-7
201    FORMAT(9A2)
203    FORMAT(1X,' ISTAT FOR OPENW = ',I5)
C  RECORD 1 ON FILES OF UNFORMATTED DATA IS ASSUMED BY
C  ANALYSIS ROUTINES(EXCEPT TRES ROUTINES) TO CONTAIN
C  AN IDENTIFYING TITLE.BLANKS ARE ACCEPTABLE
       WRITE(4,4)
4      FORMAT(1X,' TYPE A TITLE:(17A4)')
       READ(5,5) ITITLE
```

```
5     FORMAT(17A4)
      WRITE(4,204)
204   FORMAT(1X,' TYPE FILENAME OF FLASH CURVE')
      READ(5,201) IFF
C  FILE CONTAINING PUMP PULSE IS OPENED ON LOGICAL DEVICE 1
      CALL OPENW(1,IFF,7,0,0,I)
      IF(I.NE.0) WRITE(4,203) I
C  DATA FROM MCA ARE UNFORMATTED.RECORD 1 OF FILE IS A TITLE
C  DATA READ IN TO A FILE MANUALLY ARE ASSUMED TO BE IN
C  FORMAT(F6.0,7(1X,F6.0)) AND TO HAVE NO TITLE
      WRITE(4,213)
213   FORMAT(1X,' FORMATTED(1) OR UNFORMATTED(2) DATA ?')
      READ(5,19) KFOR
19    FORMAT(I1)
      IF(KFOR.EQ.2) GO TO 214
C  FORMATTED DATA ARE READ IN
      DO 81 I=1,NCH,8
      M=I+7
      READ(1,8) (P(J),J=I,M)
81    CONTINUE
      GO TO 215
8     FORMAT(F6.0,7(1X,F6.0))
C  UNFORMATTED DATA ARE READ IN
214   IK=(NCHAN/64)+1
      DO 212 I=2,IK
      READ(1,REC=I) (P(J),J=((I-2)*64)+1,(I-1)*64)
212   CONTINUE
C  THE INPUT FILE IS CLOSED
215   CALL CLOSE(1,I)
C  PUMP DATA MAY HAVE BEEN COLLECTED BACKWARDS IN MCA AS
C  E.G. FOR MHZ EXCITATION RATE
      WRITE(4,91)
91    FORMAT(1X,' IF CURVF IS BACKWARDS TYPE 1')
      READ(5,19) IAB
      IF(IAB.NE.1) GO TO 9
C  IF CURVE IS DACKWARDS INVFRT IT
      NP=NCHAN/2
      DO 92 I=1,NP
      K=NCHAN+1-I
      Z=P(K)
      P(K)=P(I)
      P(I)=Z
92    CONTINUE
C  THE BACKGROUND NOISE(COUNTS/CHANNEL) IS NOW READ IN
9     WRITE(4,10)
10    FORMAT(1X,' TYPE BACKGROUND:(F7.3)')
      READ(5,11) BKG
11    FORMAT(F7.3)
C  THE BACKGROUND IS SUBTRACTED
      P(1)=0.0
      DO 12 I=1,NCHAN
      P(I)=P(I)-BKG
      IF(P(I).LT.0.0) P(I)=0.0
```

Appendix 6.A1 (*continued*)

```
12    CONTINUE
C  THE CALCULATION IS CARRIED OUT IN TERMS OF NANOSECS
C  RATHER THAN CHANNELS
      WRITE(4,14)
14    FORMAT(1X,' TYPE NANOSECONDS/CHANNEL:(F7.3)')
      READ(5,11) FACT
C  THE NO. OF EXPONENTIAL COMPONENTS IS CHOSEN
17    WRITE(4,18)
18    FORMAT(1X,' HOW MANY COMPONENTS:(I2) ?')
      READ(5,191) IC
191   FORMAT(I2)
C  READ IN THE A FACTORS AND TAU VALUES
      DO 20 J=1,IC
      J1=2*J-1
      J2=J1+1
      WRITE(4,815) J
815   FORMAT(1X,' COMPONENT ',I3,': TYPE A AND TAU:(2F7.3)')
      READ(5,816) A(J1),A(J2)
20    CONTINUE
816   FORMAT(2F7.3)
C  THE CONVOLUTION INTEGRAL IS CALCULATED USING THE
C  TRAPEZOIDAL RULE
      DO 820 I=1,NCHAN
820   C(I)=0.0
      C1(1)=0.0
      DO 819 J=1,IC
      J1=2*J-1
      J2=J1+1
      DO 818 I=2,NCHAN
      C1(I)=EXP(-FACT/A(J2))*C1(I-1)
     1        +A(J1)*FACT*0.5*(P(I-1)*EXP(-FACT/A(J2))+P(I))
      C(I)=C(I)+C1(I)
818   CONTINUE
819   CONTINUE
C  ANALYSIS ROUTINES ASSUME THAT THE PUMP PULSE AND
C  THE DECAY CURVE HAVE THE SAME FORMAT.HENCE IF
C  THE PUMP PULSE WAS BACKWARDS THE DECAY CURVE IS
C  INVERTED
      IF(IAB.NE.1) GO TO 209
      DO 210 I=1,NP
      K=NCHAN+1-I
      Z=C(K)
      C(K)=C(I)
      C(I)=Z
210   CONTINUE
C  CHECK THAT DATA POINTS ARE NOT <0
209   DO 233 I=1,NCHAN
      IF(C(I).LT.0.0) C(I)=0.0
233   CONTINUE
C  THE CALCULATED DATA ARE LISTED
C  IREP IS A FLAG. =1 SKIP THE LISTING
      IREP =0
```

```
207   DO 206 I=1,NCH,8
      M=I+7
      WRITE(4,811) (C(J),J=I,M)
206   CONTINUE
811   FORMAT(1X,8(F6.0,1X))
      IF(IREP.EQ.1) GO TO 88
      IREP=1
C  IF DESIRED,NORMALISE(BY HEIGHT) TO A CHOSEN NO. OF COUNTS
      WRITE(4,82)
82    FORMAT(1X,' TYPE 1 TO NORMALISE')
      READ(5,19) INQ
      IF(INQ.NE.1) GO TO 88
      WRITE(4,83)
83    FORMAT(1X,' TYPE NUMBER OF COUNTS IN THE MAXIMUM(F8.1)')
      READ(5,84) CNOR
84    FORMAT(F8.1)
C  THE MAXIMUM NO. OF COUNTS IN THE CURVE IS FOUND
      CMAX=C(1)
      DO 85 I=1,NCHAN
      CMAX=AMAX1(C(I),CMAX)
85    CONTINUE
C  HEIGHT NORMALISATION
      DO 86 I=1,NCHAN
86    C(I)=C(I)*CNOR/CMAX
C  THE NORMALISED DATA ARE LISTED
      GO TO 207
C  GAUSSIAN NOISE MAY BE ADDED
88    WRITE(4,90)
90    FORMAT(1X,' TYPE 1 TO ADD NOISE')
      READ(5,19) INW
      IF(INW.NE.1) GO TO 901
C  THE NOISE ADDING PROCEDURE IS BASED ON THE
C  ROUTINES GAUSS AND RANDU IN THE IBM SYSTEM 360
C  SCIENTIFIC SUBROUTINE PACKAGE,VERSION 3
      WRITE(4,101)
101   FORMAT(1X,' TYPE A NUMBER:(I2)')
      READ(5,111) IX
111   FORMAT(I2)
      NMOD=MOD(IX,2)
      IF(NMOD.EQ.0) IX=IX+1
      DO 121 I=1,NCHAN
      CALL GLUSS(IX,1.0,0.0,V)
      C(I)=C(I)+V*SQRT(C(I))
121   CONTINUE
C  THE CONVOLVED CURVE IS WRITTEN TO A FILE
901   WRITE(4,902)
902   FORMAT(1X,' TYPE 1 TO SAVE THE CURVE')
      READ(5,19) INE
      IF(INE.NE.1) GO TO 30
      WRITE(4,205)
205   FORMAT(1X,' TYPE FILENAME FOR CALCULATED CURVE')
      READ(5,201) IFF
C  THE OUTPUT FILE IS GENERATED AND OPENED ON LOG. DEV. 2
```

Appendix 6.A1 (*continued*)

```
        CALL CFILW(IFF,2,256,1,1,0,0,I)
        IF(I.NE.0) WRITE(4,202) I
202     FORMAT(1X,' ISTAT FOR CFILW =',I5)
        CALL OPENW(2,IFF,7,0,0,I)
        IF(I.NE.0) WRITE(4,203) I
C   DATA ARE WRITTEN IN THE SAME FORMAT AS THE GENERATING
C   PUMP PULSE
        IF(KFOR.EQ.2) GO TO 217
        DO 821 I=1,NCH,8
        M=I+7
        WRITE(2,8) (C(J),J=I,M)
821     CONTINUE
        GO TO 218
217     WRITE(2) ITITLE
        DO 211 J=2,IK
        WRITE(2,REC=J) (C(I),I=(J-2)*64+1,(J-1)*64)
211     CONTINUE
C   THE OUTPUT FILE IS CLOSED
218     CALL CLOSE(2,I)
30      WRITE(4,31)
31      FORMAT(1X,' TYPE 1 FOR ANOTHER FLASH PROFILE',/,
       1 1X,'       2 FOR ANOTHER DECAY FUNCTION',/,
       2 1X,'       3 TO EXIT')
        READ(5,19) ITEST
        IF(ITEST-2) 1,17,32
32      STOP
        END
C
C   SUBROUTINE GLUSS
C   SAME AS IBM SUBROUTINE GAUSS
        SUBROUTINE GLUSS(IX,S,AM,V)
        A=0.0
        DO 50 I=1,12
        CALL RANDU(IX,IY,Y)
        IX=IY
50      A=A+Y
        V=(A-6.0)*S+AM
        RETURN
        END
C
C   IBM SUBROUTINE RANDU
        SUBROUTINE RANDU(IX,IY,YFL)
        IY=IX*65539
        IF(IY)5,6,6
5       IY=IY+2147483647+1
6       YFL=IY
        YFL=YFL*0.4656613E-9
        RETURN
        END
$BEND
```

Appendix 6.A2

This Fortran routine deconvolves a set of instrument response function and decay curve data to a single exponential or a sum or difference of two exponentials. A zero-time shift (Equation 6.13) option is included. Analysis for scattered light based on Equation 6.10 requires some simple modifications in the subroutine FUNCT1.

```
LIST FX:CREWS.FTN

$BATCH
C  PROGRAMME CREWS
C  DECONVOLUTION OF DECAY DATA TO A SINGLE
C  EXPONENTIAL OR A SUM OR DIFFERENCE OF TWO
C  EXPONENTIALS.
C  METHOD OF NON-LINEAR LEAST SQUARES.
C  IFP - STORE FOR PUMP PULSE FILENAME
C  IFY - STORE FOR DECAY CURVE FILENAME
C  NCHAN - NO. OF EXPERIMENTAL DATA POINTS
C  FACT - NANOSECONDS/CHANNEL
C  BKGDP,BKGDY - CONSTANT BACKGROUNDS
C  IH - FLAG =1 REPEAT ANALYSIS ON SAME DATA
C            =2 ANALYSE NEW DATA
C            =3 FINISH
C
      INTEGER*2 IFP,IFY
      COMMON/AREA7/IFP(9),IFY(9)
      IH=2
1001  WRITE(4,3059)
3059  FORMAT(1X,' TYPE PUMP FILENAME')
      READ(5,307)IFP
      WRITE(4,306)
306   FORMAT(1X,' TYPE DECAY FILENAME')
      READ(5,307) IFY
307   FORMAT(9A2)
C
C  DATA IN MCA MAY BE INVERTED(LASER) OR NORMAL(LAMP)
C
      WRITE(4,453)
453   FORMAT(1X,' LAMP(1) OR LASER(2) DATA ? ')
      READ(5,11) IFLL
C
C  THE BACKGROUNDS ARE ESTIMATED BEFORE DATA ANALYSIS
C
11    FORMAT(I1)
      WRITE(4,5)
5     FORMAT(1X,' NO. OF CHANNELS,NS/CH,FLASH BKG,DECAY BKG',
     1 ':(I3,3F7.3)')
      READ(5,6) NCHAN,FACT,BKGDP,BKGDY
6     FORMAT(I3,3F7.3)
```

Appendix 6.A2 *(continued)*

```
C
C  THE CURVES FOR THE ANALYSIS ARE SET UP
C
       CALL SXSOMX(NCHAN,IFLL,BKGDP,BKGDY,FACT,IH)
       IF(IH.EQ.3) GO TO 426
       GO TO 1001
426    STOP
       END
C  SUBROUTINE SXSOMX
C  SETTING UP CURVES AND PARAMETRES
C  AG - ARRAY FOR FITTING PARAMETERS
C  P - ARRAY FOR PUMP PULSE DATA
C  Y - ARRAY FOR RAW AND BACKGROUND-CORRECTED DECAY DATA
C  C - ARRAY FOR CALCULATED DECAY CURVE
C  RES - ARRAY FOR WEIGHTED RESIDUALS
C  CRES - ARRAY FOR AUTOCORRELATION FN OF RESIDUALS
C
       SUBROUTINE SXSOMX(NCHAN,IFLL,BKGDP,BKGDY,FACT,IH)
        DIMENSION AG(5)
       INTEGER*2 IFP,IFY
       INTEGER*2 IFD(9),IOV1(4)
       COMMON/AREA1/P(512)
       COMMON/AREA2/Y(512),C(512)
       COMMON/AREA7/IFP(9),IFY(9)
       COMMON/AREA8/RES(512)
       COMMON/AREA9/CRES(258)
       DATA IFD/'FX:SXOS1.OVL/G    '/,IOV1/'SXOS1    '/
C
C  FILES CONTAINING PUMP PULSE AND DECAY DATA ARE
C  OPENED ON LOGICAL UNITS 1 AND 2 AND DATA READ IN
       CALL OPENW(1,IFP,7,0,0,I)
       IF(I.NE.0) WRITE(4,309) I
       CALL OPENW(2,IFY,7,0,0,I)
       IF(I.NE.0) WRITE(4,309) I
309    FORMAT(1X,' ISTAT FOR OPENW = ',I5)
11     FORMAT(I1)
       IK=NCHAN/64+1
       DO 1000 I=2,IK
       READ(1,REC=I)(P(J),J=(I-2)*64+1,(I-1)*64)
       READ(2,REC=I)(Y(J),J=(I-2)*64+1,(I-1)*64)
1000   CONTINUE
C
C  LOGICAL UNITS 1 AND 2 ARE CLOSED
C
       CALL CLOSE(1,I)
       CALL CLOSE(2,I)
       IF(IFLL.EQ.1) GO TO 13
C
C  THE CURVES ARE ARRANGED TO HAVE TIME INCREASING
C  WITH CHANNEL NUMBER.
C
       NP=NCHAN/2
```

```
        DO 12 I=1,NP
        K=NCHAN+1-I
        R=P(K)
        S=Y(K)
        P(K)=P(I)
        Y(K)=Y(I)
        P(I)=R
        Y(I)=S
 12     CONTINUE
C
C  THE CURVES ARE CORRECTED FOR CONSTANT BACKGROUND
C
 13     Y(1)=BKGDY
        P(1)=BKGDP
        DO 25 I=1,NCHAN
        P(I)=P(I)-BKGDP
        Y(I)=Y(I)-BKGDY
 25     CONTINUE
C
C  SET UP STARTING PARAMETERS AND CHOOSE FITTING FUNCTION
C
 26     CALL SXCALC(FACT,NCHAN,N1,N2,CHISQR,BKGDY,AG,LALY,NALY)
        CALL OPENW(1,IFD,0,0,0,ISTAT)
        IF(ISTAT.NE.0) STOP
        CALL IFETCH(IOV1,1,4,ISTAT)
        IF(ISTAT.NE.0) STOP
C
C  THE SUCCESS OF THE FIT IS JUDGED BY CERTAIN CRITERIA
C
        CALL RESID3(NCHAN,N1,N2,BKGDY,DB)
C
C  THE CURVES CAN BE PLOTTED
C
C
C  SUBROUTINE CPLOT IS NOT LISTED
C
        WRITE(4,442)
 442    FORMAT(1X,' PLOT ?')
        READ(5,11) IPLOT
        IF(IPLOT.EQ.0) GO TO 420
        CALL CPLOT(CHISQR,DB,NCHAN,N1,N2,BKGDP,BKGDY,FACT,IFLL,
       1AG,NALY)
C
C  THE PLOTTING FILE WAS OPENED ON LOGICAL UNIT 1
C  IT IS NOW CLOSED
C
 420    CALL CLOSE(1,ISTAT)
        WRITE(4,421)
 421    FORMAT(1X,' TYPE 1 TO RE-ANALYSE THE SAME DATA',/,X,
       1 '      2 TO SUBMIT NEW DATA',/,X,'      3 TO EXIT')
        READ(5,11) IH
        IF(IH.NE.1) GO TO 261
        LALY=NALY
        GOTO 26
```

Appendix 6.A2 (*continued*)

```
261   RETURN
      END
C
C  SUBROUTINE SXCALC
C  IN THIS THE STARTING PARAMETERS ARE CHOSEN BY THE USER
C
C  A - =AG - ARRAY OF FITTING PARAMETERS
C  SIGMAA - STANDARD DEVIATIONS ON FITTING PARAMETERS
C  D - ARRAY OF PARTIAL DERIVATIVES
C  KSH - FLAG =1 IF ZERO-TIME SHIFT IS TO BE FITTED
C  N1,N2 - FIRST AND LAST CHANNELS OF THE ANALYSIS REGION
C
      SUBROUTINE SXCALC(FACT,NCHAN,N1,N2,CHISQR,BKGDY,A,LALY,NALY)
      INTEGER*2 JFD(9),IOV2(4)
      DIMENSION A(5)
      COMMON/AREA1/P(512)
      COMMON/AREA2/Y(512),C(512)
      COMMON/AREA4/D(512,5)
      COMMON/AREA6/SIGMAA(5)
      COMMON/AREA8/RES(512)
      COMMON/AREA9/CRES(258)
      DATA JFD/'FX:SXOXS.OVL/G    '/,IOV2/'SXOXS   '/
C
C  CHOOSE THE CHANNEL REGION FOR ANALYSIS
C
      WRITE(4,27)
27    FORMAT(1X,' READ IN CHANNEL FITTING RANGE:(2I3)')
      READ(5,28) N1,N2
28    FORMAT(2I3)
C
C  CHOOSE THE FITTING FUNCTION
C
      WRITE(4,291)
291   FORMAT(1X,' TYPE 1 FOR SINGLE EXPONENTIAL ANALYSIS,
     1 2 FOR DOUBLE',$)
      READ(5,1)NALY
C
C  CAN CHOOSE TO FIT A ZERO-TIME SHIFT
C
      WRITE(4,292)
292   FORMAT(/,1X,' TYPE 1 TO FIT A SHIFT, 0 OTHERWISE',$)
      READ(5,1)KSH
1     FORMAT(I1)
      IF(IH.EQ.1.AND.LALY.EQ.NALY)GOTO 1341
      IF(NALY.EQ.2)GOTO 5555
C
C  INITIAL ESTIMATES FOR THE FITTING PARAMETERS
C
      WRITE(4,29)
29    FORMAT(1X,' ESTIMATE A AND TAU:(2F5.2)')
      READ(5,30) A(1),A(2)
30    FORMAT(2F5.2)
```

```
1341  IF(IH.EQ.1.AND.KSH.EQ.1.AND.LALY.EQ.NALY)GOTO 5557
      A(3)=0.0
      A(4)=0.0
      A(5)=0.0
      GOTO 5557
5555  IF(IH.EQ.1.AND.LALY.EQ.NALY)GOTO 5558
      WRITE(4,293)
293   FORMAT(/,1X,' ESTIMATE A1,TAU1,A2,TAU2:(4F5.2)')
      READ(5,301)A(1),A(2),A(3),A(4)
301   FORMAT(4F5.2)
5558  IF(IH.EQ.1.AND.KSH.EQ.1)GOTO 5557
      A(5)=0.0
5557  CALL OPENW(2,JFD,0,0,0,ISTAT)
      IF(ISTAT.NE.0) STOP
      CALL IFETCH(IOV2,2,4,ISTAT)
      IF(ISTAT.NE.0) STOP
C
C  CHANGE TAU VALUES TO UNITS OF CHANNELS FOR CALCULATION
C
      CALL PRUE(FACT,A,2,1,NALY)
C
C  CURVE FITTING
C
      CALL CALC1(N1,N2,FACT,A,CHISQR,BKGDY,KSH,NALY)
      CALL CLOSE(2,ISTAT)
C
C  CHANGE TAU VALUES TO UNITS OF NANOSECONDS FOR DISPLAY
C
      CALL PRUE(FACT,A,1,1,NALY)
      WRITE(4,54)A(1),SIGMAA(1),A(2),SIGMAA(2)
54    FORMAT(1X,' A1   =',E14.5,10X,' STD. DEV. =',E14.5,/,X,
     1 ' TAU1 =',E14.5,10X,' STD. DEV. =',E14.5)
      IF(NALY.EQ.1)GOTO 5559
      WRITE(4,541)A(3),SIGMAA(3),A(4),SIGMAA(4)
541   FORMAT(1X' A2   =',E14.5,10X,' STD. DEV. =',E14.5,/,X,
     1 ' TAU2 =',E14.5,10X,' STD. DEV. =',E14.5)
5559  WRITE(4,542)A(NALY*2+1)
542   FORMAT(1X,' SHIFT =',E14.5)
      RETURN
      END
C
C  SUBROUTINE PRUE
C  CONVERTS QUANTITIES FROM CHANNEL UNITS TO UNITS OF
C  NANOSECONDS BY MULTIPLYING BY NS/CH(FACT)
C  AND VICE VERSA
C
      SUBROUTINE PRUE(FACT,A,M1,M2,M3)
      DIMENSION A(5),SIGMAA(5)
      COMMON/AREA6/SIGMAA
      IF(M1.EQ.2)GOTO 10
      A(2)=A(2)*FACT
      A(1)=A(1)/FACT
      IF(M3.EQ.1)GOTO 1000
```

Appendix 6.A2 *(continued)*

```
        A(4)=A(4)*FACT
        A(3)=A(3)/FACT
  1000  IF(M2.EQ.2)GOTO 100
        SIGMAA(2)=SIGMAA(2)*FACT
        SIGMAA(1)=SIGMAA(1)/FACT
        IF(M3.EQ.1)GOTO 100
        SIGMAA(4)=SIGMAA(4)*FACT
        SIGMAA(3)=SIGMAA(3)/FACT
        GOTO 100
  10    A(2)=A(2)/FACT
        A(1)=A(1)*FACT
        IF(M3.EQ.1)GOTO 1001
        A(4)=A(4)/FACT
        A(3)=A(3)*FACT
  1001  IF(M2.EQ.2)GOTO 100
        SIGMAA(2)=SIGMAA(2)/FACT
        SIGMAA(1)=SIGMAA(1)*FACT
        IF(M3.EQ.1)RETURN
        SIGMAA(4)=SIGMAA(4)/FACT
        SIGMAA(3)=SIGMAA(3)*FACT
  100   RETURN
        END
  $BEND


 FJ      -END OF TASK CODE=       0


 LIST FX:SXOXS.FTN

  $BATCH
  C  SUBROUTINE CALC1
  C  THE CURVE FITTING ROUTINE
  C  IT IS BASED DIRECTLY ON BEVINGTON(1969),P237
  C
  C  B - ARRAY FOR STORING PREVIOUSLY CALCULATED A
  C  ALPHA - CURVATURE MATRIX
  C  BETA - SEE TEXT EQUATION 6.49
  C  ARRAY - ALPHA MODIFIED FOR EASE OF INVERSION
  C        - ALSO USED TO STORE THE INVERTED MATRIX
  C  FLAMDA - PROPORTION OF GRADIENT SEARCH INCLUDED
  C  JCOUNT - FLAG ON NO. OF ITERATIONS(LIMIT=30)
  C  ICOUNT - FLAG ON NO. OF MATRIX INVERSIONS WITHIN
  C           A GIVEN ITERATION(LIMIT=5)
  C
        SUBROUTINE CALC1(N1,N2,FACT,A,CHISQR,BKGDY,KSH,NALY)
        DIMENSION A(5),B(5),BETA(5),ALPHA(5,5),ARRAY(5,5)
        COMMON/AREA1/P(512)
        COMMON/AREA2/Y(512),C(512)
        COMMON/AREA4/D(512,5)
```

```
      COMMON/AREA6/SIGMAA(5)
C
C  A(L3) IS THE SHIFT PARAMETER
C
      L3=NALY*2+KSH
C
C  INITIAL VALUE OF CHI-SQUARE IS CALCULATED
C
67    CALL FUNCT1(N1,N2,A,CHISQR,BKGDY,L3)
      DO 570 J=1,5
570   SIGMAA(J)=0.0
      JCOUNT=0
      FLAMDA=0.01
80    JCOUNT=JCOUNT+1
C
C  INITIALISE ARRAYS
C
      DO 91 J=1,5
      BETA(J)=0.0
      DO 91 K=1,J
91    ALPHA(J,K)=0.0
      DO 110 I=N1,N2
C
C  THE FIT IS WEIGHTED BY THE RECIPROCALS OF THE RAW DATA
C
      W=Y(I)+BKGDY
      IF(W.LT.1.0) W=1.0
      W=1.0/W
C
C  EVALUATE ALPHA AND BETA MATRICES
C
      DO 100 J=1,L3
      BETA(J)-BETA(J)+W*(Y(I)-C(I))*D(I,J)
      DO 100 K=1,J
100   ALPHA(J,K)=ALPHA(J,K)+W*D(I,J)*D(I,K)
110   CONTINUE
C
C  THE CURVATURE MATRIX IS SYMMETRICAL
C
      DO 150 J=1,L3
      DO 150 K=1,J
150   ALPHA(K,J)=ALPHA(J,K)
C
C  STORE CHI-SQUARE
C
      CHISQ1=CHISQR
      ICOUNT=0
160   DO 174 J=1,L3
      DO 173 K=1,L3
      SROOT=SQRT(ALPHA(J,J)*ALPHA(K,K))
      IF(SROOT.EQ.0.0) GO TO 561
C
C  MODIFY CURVATURE MATRIX FOR EASE OF INVERSION
```

Appendix 6.A2 (*continued*)

```
C
173   ARRAY(J,K)=ALPHA(J,K)/SROOT
174   ARRAY(J,J)=1.0+FLAMDA
      GO TO 562
561   WRITE(4,563)
563   FORMAT(1X,' DIVIDING BY 0')
      GO TO 560
C
C  INVERT CURVATURE MATRIX
C
562   CALL MATINV(ARRAY,L3,DET)
      ICOUNT=ICOUNT+1
C
C  CALCULATE NEW PARAMETERS BY ADDITION OF CALCULATED
C  INCREMENTS
C
      DO 201 J=1,L3
      B(J)=A(J)
      DO 201 K=1,L3
      B(J)=B(J)+BETA(K)*ARRAY(J,K)/SQRT(ALPHA(J,J)*ALPHA(K,K))
201   CONTINUE
C
C  THE DECAY TIMES ARE CONSTRAINED TO BE +VE
C
      IF(B(2).LE.0.0) B2=-B2
      IF(NALY.EQ.1)GOTO 3021
      IF(B(4).LE.0.0)B4=-B4
C
C  NEW CONVOLVED CURVE AND CHI-SQUARE CALCULATED
C
3021  CALL FUNCT1(N1,N2,B,CHISQR,BKGDY,L3)
C
C  NEW AND OLD VALUES OF CHI-SQUARE ARE COMPARED
C
      IF(CHISQ1-CHISQR)302,303,303
C
C  NEW IS SMALLER.ACCEPT NEW PARAMETERS
C
303   DO 304 J=1,L3
      A(J)=B(J)
C
C  CALCULATE STANDARD DEVIATIONS USING INVERTED
C  CURVATURE MATRIX(ERROR MATRIX)
C
      SIGMAA(J)=SQRT(ARRAY(J,J)/ALPHA(J,J))
304   CONTINUE
C
C  STOP IF NO CONVERGENCE AFTER 30 ITERATIONS
C
      IF(JCOUNT.EQ.30) GO TO 560
C
C  TEST FOR CONVERGENCE
```

```
C
      IF((CHISQ1-CHISQR).LT.0.000001) GO TO 560
C
C  SLIDING DOWN THE GREEN HILL
C
      FLAMDA=FLAMDA/10.0
      GO TO 80
C
C  CLIMBING UP THE GRADIENT
C
302   FLAMDA=FLAMDA*10.0
C
C  CANNOT FIND A BETTER MINIMUM.RE-CALCULATE
C  OPTIMUM PARAMETERS
C
      IF(ICOUNT.EQ.5) GO TO 305
      IF(ICOUNT.EQ.6) GO TO 330
      GO TO 160
305   FLAMDA=FLAMDA/10**6
      GO TO 160
330   CHISQR=CHISQ1
560   WRITE(4,53)JCOUNT,CHISQR
53    FORMAT(1X,I3,' ITERATIONS',5X,' CHISQR = ',E14.5)
      CALL FUNCT1(N1,N2,A,CHISQR,BKGDY,L3)
      RETURN
      END
C
C  SUBROUTINE FUNCT1
C  IN WHICH THE CONVOLUTION INTEGRAL AND PARTIAL
C  DERIVATIVES ARE EVALUATED USING RECURSION RELATIONS
C
      SUBROUTINE FUNCT1(N1,N2,A,CHISQR,BKGDY,L3)
      DIMENSION A(5)
      COMMON/AREA1/P(512)
      COMMON/AREA2/Y(512),C(512)
      COMMON/AREA4/D(512,5)
      LL3=5
      IF(L3.LT.4)LL3=3
      DO 1 J=1,5
      DO 1 I=1,N2
      D(I,J)=0.0
1     CONTINUE
      IBIG=N2-1
C
C  TAKE ACOUNT OF ZERO-TIME SHIFT
C
      ISH=INT(A(LL3))
      IF(A(LL3).LT.0.0) ISH=ISH-1
      FR=A(LL3)-FLOAT(ISH)
      IST=ISH+1
      IF(IST.LT.1)IST=1
      G1=P(IST)+FR*(P(IST+1)-P(IST))
      ETAS=EXP(-1.0/A(2))
```

Appendix 6.A2 (*continued*)

```
      IF(L3.LT.4)GOTO 101
      G2=G1
      ETAT=EXP(-1.0/A(4))
101   C(1)=G1*A(1)
      IF(L3.LT.4)GOTO 102
      C(1)=C(1)+G2*A(3)
102   IF(L3.EQ.4.OR.L3.EQ.2)GOTO 11
      GS1=P(IST+1)-P(IST)
      IF(L3.LT.4)GOTO 103
      GS2=GS1
103   D(1,L3)=GS1*A(1)
      IF(L3.LT.4)GOTO 11
      D(1,L3)=D(1,L3)+GS2*A(3)
11    D(1,1)=G1
      BB2A=A(1)/(A(2)*A(2))
      G1T=0.0
      IF(L3.LT.4)GOTO 104
      DD2A=A(3)/(A(4)*A(4))
      G2T=0.0
      D(1,3)=G2
104   DO 12 I=1,IBIG
      IST=ISH+I
      IF(IST.LE.0)IST=1
      IF(IST.GT.N2)IST=N2
      SIST=P(IST)*0.5
      SJST=P(IST+1)*0.5
      SJJST=P(IST+2)*0.5
      SK=SIST+FR*(SJST-SIST)
      SSK=SJST+FR*(SJJST-SJST)
      G1T=ETAS*(G1T+SK+G1)
      G1=ETAS*(SK+G1)+SSK
      C(I+1)=G1*A(1)
      D(I+1,1)=G1
      D(I+1,2)=G1T*BB2A
      IF(L3.LT.4)GOTO 105
      G2T=ETAT*(G2T+G2+SK)
      G2=ETAT*(G2+SK)+SSK
      C(I+1)=C(I+1)+A(3)*G2
      D(I+1,3)=G2
      D(I+1,4)=G2T*DD2A
C
C  THE PARTIAL DERIVATIVE W.R.T. THE SHIFT ENTAILS
C  KNOWLEDGE OF DP(I)/DT. THIS IS CALCULATED BY
C  LINEAR INTERPOLATION.
C
105   IF(L3.EQ.4.OR.L3.EQ.2)GOTO 12
      GS1=ETAS*(GS1+SJST-SIST)+SJJST-SJST
      D(I+1,L3)=A(1)*GS1
      IF(L3.LT.4)GOTO 12
      GS2=ETAT*(GS2+SJST-SIST)+SJJST-SJST
      D(I+1,L3)=D(I+1,L3)+A(3)*GS2
12    CONTINUE
```

```
C
C  CALCULATE NO. OF DEGREES OF FREEDOM
C
      NPTS=N2-N1+1
      NFREE=NPTS-L3
      FREE=NFREE
      CHISQ=0.0
      IF(NFREE.LE.0) GO TO 404
C
C  EVALUATE CHI-SQUARE
C
      DO 3 I=N1,N2
      W=Y(I)+BKGDY
      IF(W.LT.1.0) W=1.0
      W=1.0/W
      CHISQ=CHISQ+W*(Y(I)-C(I))*(Y(I)-C(I))
3     CONTINUE
404   CHISQR=CHISQ/FREE
      RETURN
      END
C
C  SUBROUTINE MATINV
C  MATRIX INVERSION ROUTINE
C  FROM BEVINGTON(1969),P302
C
      SUBROUTINE MATINV(ARRAY,NORDER,DET)
      DIMENSION ARRAY(5,5)
      DIMENSION IK(5),JK(5)
      DET=1.0
      DO 100 K=1,NORDER
      AMAX=0.0
21    DO 30 I=K,NORDER
      DO 30 J=K,NORDER
      IF(ABS(AMAX)-ABS(ARRAY(I,J)))24,24,30
24    AMAX=ARRAY(I,J)
      IK(K)=I
      JK(K)=J
30    CONTINUE
      IF(AMAX) 41,32,41
32    DET=0.0
      GO TO 140
41    I=IK(K)
      IF(I-K)21,51,43
43    DO 50 J=1,NORDER
      SAVE=ARRAY(K,J)
      ARRAY(K,J)=ARRAY(I,J)
50    ARRAY(I,J)=-SAVE
51    J=JK(K)
      IF(J-K)21,61,53
53    DO 60 I=1,NORDER
      SAVE=ARRAY(I,K)
      ARRAY(I,K)=ARRAY(I,J)
60    ARRAY(I,J)=-SAVE
```

Appendix 6.A2 (*continued*)

```
61     DO 70 I=1,NORDER
       IF(I-K) 63,70,63
63     ARRAY(I,K)=-ARRAY(I,K)/AMAX
70     CONTINUE
       DO 80 I=1,NORDER
       DO 80 J=1,NORDER
       IF(I-K) 74,80,74
74     IF(J-K)75,80,75
75     ARRAY(I,J)=ARRAY(I,J)+ARRAY(I,K)*ARRAY(K,J)
80     CONTINUE
       DO 90 J=1,NORDER
       IF(J-K) 83,90,83
83     ARRAY(K,J)=ARRAY(K,J)/AMAX
90     CONTINUE
       ARRAY(K,K)=1.0/AMAX
100    DET=DET*AMAX
       DO 130 L=1,NORDER
       K=NORDER-L+1
       J=IK(K)
       IF(J-K) 111,111,105
105    DO 110 I=1,NORDER
       SAVE=ARRAY(I,K)
       ARRAY(I,K)=-ARRAY(I,J)
110    ARRAY(I,J)=SAVE
111    I=JK(K)
       IF(I-K) 130,130,113
113    DO 120 J=1,NORDER
       SAVE=ARRAY(K,J)
       ARRAY(K,J)=-ARRAY(I,J)
120    ARRAY(I,J)=SAVE
130    CONTINUE
140    RETURN
       END
$BEND

FJ      -END OF TASK CODE=       0
*

LIST FX:SXOS1.FTN

$BATCH
C  SUBROUTINE RESID3
C  CALCULATES PARAMETERS WITH WHICH TO JUDGE THE FIT
C  BY: DECAY CURVE BACKGROUND
C  DB: DURBIN-WATSON PARAMETER
C  FZ: MEAN RESIDUAL
C  FZS: STANDARD DEVIATION OF RESIDUALS
C  F3MU: SKEWNESS FACTOR
C  S3MU: STANDARD DEVIATION OF SKEWNESS FACTOR
C  F4MU: KURTOSIS FACTOR
C  S4MU - STANDARD DEVIATION ON KURTOSIS FACTOR
C
```

```
      SUBROUTINE RESID3(NCHAN,N1,N2,BY,DB)
      COMMON/AREA2/Y(512),C(512)
      COMMON/AREA8/RES(512)
      COMMON/AREA9/CRES(258)
      FZ=0.0
      SS=0.0
      PS=0.0
C  WEIGHTED RESIDUALS ARE CALCULATED
      DO 444 I=N1,N2
      W=Y(I)+BY
      IF(W.LT.1.0) W=1.0
      W=1.0/W
      DIFF=Y(I)-C(I)
      RES(I)=DIFF*SQRT(W)
      IF(I.EQ.N1) GO TO 447
C  SUMS FOR DURBIN-WATSON PARAMETER CALCULATED
      SS=SS+((RES(I)-RES(I-1))*(RES(I)-RES(I-1)))
447   PS=PS+RES(I)*RES(I)
444   FZ=FZ+RES(I)
      NPTS=N2-N1+1
      FN=NPTS
C  MEAN RESIDUAL
      FZ=FZ/FN
      F2MU=0.0
      F3MU=0.0
      F4MU=0.0
C  SUMS FOR OTHER PARAMETERS
      DO 451 I=N1,N2
      PX=RES(I)-FZ
      F2MU=F2MU+PX*PX
      F3MU=F3MU+PX*PX*PX
      F4MU-F4MU+PX*PX*PX*PX
451   CONTINUE
      FZS=SQRT(F2MU/(FN-1.0))
      F3MU=F3MU*SQRT(FN)/(F2MU*SQRT(F2MU))
      F4MU=(F4MU*FN/(F2MU*F2MU))-3.0
      S3MU=SQRT(6.0/FN)
      S4MU=SQRT(24./FN)
      DB=SS/PS
      WRITE(4,446)FZ,FZS,F3MU,S3MU,F4MU,S4MU,DB
446   FORMAT(X,' MEAN RESIDUAL =',10X,E14.5,/,X,' STD. ',
     1 'DEV. OF RESIDUALS =',X,E14.5,/,X,' SKEWNESS FACTOR =',
     2 8X,E14.5,/,X,' SHOULD BE LESS THAN :',4X,E14.5,/,X,
     3 ' KURTOSIS FACTOR =',8X,E14.5,/,X,' SHOULD BE LESS',
     4 ' THAN :',4X,E14.5,/,X,' DURBIN-WATSON PARAMETER =',E14.5)
C
C  CALCULATION OF AUTOCORRELATION FUNCTION OF WEIGHTED RESIDUALS
C
C  THIS CALCULATION WOULD NORMALLY BE PERFORMED IN THE
C  PLOTTING ROUTINE
C
      FNPH=FN*0.5
      NPH=INT(FNPH)
```

Appendix 6.A2 (*continued*)

```
          PS=PS/FN
          DO 747 J=0,NPH
          SUMIJ=0.0
          M=NPTS-J
          FM=M
          M2=N2-J
          DO 748 I=N1,M2
          SUMIJ=SUMIJ+RES(I)*RES(I+J)
    748   CONTINUE
          SUMIJ=SUMIJ/FM
          CRES(J)=SUMIJ/PS
    747   CONTINUE
          RETURN
          END
```

7 Nanosecond Time-resolved Emission Spectroscopy

7.1 Introduction

In the block diagram of a simple SPC instrument, shown previously in Fig. 2.1, F1 and F2 denote filters. If the intensity in the light source is high enough these filters may be replaced by scanning monochromators, at which stage the instrument, since it incorporates a light source, detector and recorder (usually an MCA), can function as a standard spectrofluorimeter. The simplest configuration is one where the TAC is bypassed and the signals from the discriminator on the emission channel (CFTD) are routed, possibly after amplification, direct to the MCA. With the MCA operated in multichannel scaling mode and the channel advance synchronized with the wavelength drive of one of the monochromators the resulting histogram of counts in the MCA channels represents the fluorescence excitation or fluorescence spectrum of the sample. This type of instrument is a simple photon counting fluorimeter (Arecchi *et al.*, 1966), and is rather insensitive since the intensity in the pulsed flash lamp is orders of magnitude less than in a conventional arc lamp (see Section 3.3.4). An elegant instrument of this type, employing a high intensity source, has been described by Tuan and Wild (1973, 1974).

If the pulsed source is retained but the TAC reintroduced to correlate the fluorescence photons with the excitation events (that is, the electronic components and settings are the same as for decay curve measurements except for the mode of operation of the MCA), a fluorescence spectrum considerably more revealing than a normal spectrum can be collected. This arises from the fact that the amplitudes of the TAC output pulses are proportional to the times of occurrence of the corresponding fluorescence photons. Selection of TAC pulses according to voltage, which is easy to accomplish, is a selection of fluorescence events according to time of occurrence after excitation. Fluorescence spectra recorded, or constructed, by resolving, with respect to wavelength or energy, single photon pulses that have been selected from a limited time interval, are referred to as time-resolved emission spectra or simply TRES. We give TRES the symbol $S(\lambda, t_k)$.

The relationship between the normal fluorescence spectrum, $F(\lambda)$, the decay functions at a fixed wavelength, $G(\lambda_e,t)$, and the time-resolved spectra can be deduced from the diagram in Fig. 7.1, which is an effort to represent the three-dimensional surface of the intensity at all wavelengths and times during the fluorescence decay. For the moment we specify any point on this surface by $I_n(\lambda_j,t_k)$. It can be seen that any plane parallel to the I_n plane represents the emission spectrum at a particular time (the TRES). Therefore the digital decay function at the wavelength λ_j is made up of the points

$$I_n(\lambda_j,t_1), I_n(\lambda_j,t_2), \ldots I_n(\lambda_j,t_k)\ldots \equiv G(\lambda_j,t) \tag{7.1}$$

the TRES (in digital form) at time t_k is made up of

$$I_n(\lambda_1,t_k), I_n(\lambda_2,t_k), \ldots I_n(\lambda_j,t_k)\ldots \equiv S(\lambda,t_k) \tag{7.2}$$

and the total emission spectrum (in digital form) is made up of the points

$$\int_0^\infty I_n(\lambda_1,t)\,\mathrm{d}t, \int_0^\infty I_n(\lambda_2,t)\,\mathrm{d}t, \ldots \int_0^\infty I_n(\lambda_j,t)\,\mathrm{d}t, \ldots \equiv F(\lambda) \tag{7.3}$$

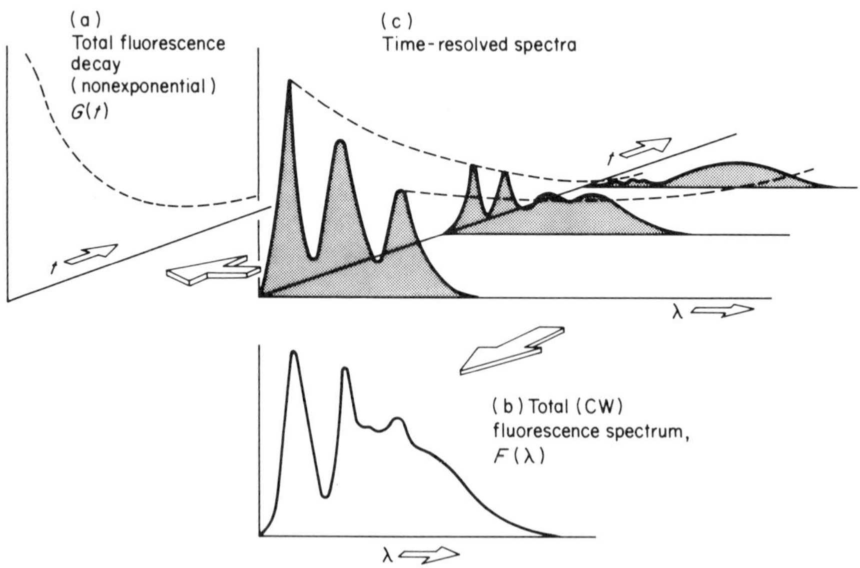

Figure 7.1 Representation of emission intensity distribution as a function of time and wavelength (after Beavan *et al.*, 1979).

It will be seen that the intensity function $I_n(\lambda,t)$ is merely a dummy form of $G(\lambda,t)$ and that

$$F(\lambda) \sim \int_0^\infty G(\lambda_1,t)\,dt, \int_0^\infty G(\lambda_2,t)\,dt, \ldots$$

$$\int_0^\infty G(\lambda_j,t)\,dt \tag{7.4}$$

and

$$S(\lambda,t_k) \sim G(\lambda_1,t_k), G(\lambda_2,t_k), \ldots G(\lambda_j,t_k) \tag{7.5}$$

In the time-resolved spectral measurement we have outlined, the TRES are made up of points not on the deconvolved functions $G(\lambda,t)$ but on the convolved decay curves, $I(\lambda,t)$. The question of convolution is an important one in the consideration of TRES and we shall return to it in the following sections.

7.2 Applications and Limitations

Time-resolved emission spectroscopy can yield both qualitative and quantitative information about many photophysical processes that cannot be gleaned from normal fluorescence spectra of even extremely high optical resolution. One such process is a simple exciplex equilibrium, in which an excited precursor is quenched reversibly by a ground state molecule to form an excited state complex, precursor and complex having different decay times. If photons emitted only at long times are wavelength-resolved the resulting spectrum will have a much reduced and perhaps negligible contribution from the fluorescence of the short-lived species. Similarly, excited state molecules having two fluorescing configurations of differing decay times are ideal candidates for time-resolved spectral studies. However, if the interconversion between the different emitting species in both of these molecular systems is uncomplicated, it can be studied easily by means of conventional decay time measurements. Time-resolved spectra may be helpful, and even necessary, in characterizing the process under investigation, but perhaps most of the quantitative information will be deduced from decay curve analysis. It is in their application to excited state phenomena that admit of no simple kinetic analysis or that yield decay curves the lifetimes of which can be associated with no simple process or group of processes that time-resolved spectra have most value.

One such phenomenon is the much studied solvent relaxation round an excited state fluorophore. Time-dependent shifts in the fluorescence spectrum may accompany the interactions between solvent and excited solute, the decay of which will not be interpretable in terms of a simple exponential model if the timescale on which the solvent relaxation occurs is comparable to the normal lifetime of the solute (Bakshiev, 1965). An understanding of the interactions occurring can be attained by means of time-resolved spectra and can serve as a basis for study of, for instance, the effect of the local environment in biological molecules on the spectral characteristics of fluorescent probes (Cockle and Szabo, 1981).

For example, 8-anilino-1-naphthalene sulphonate (ANS) has been much used as a fluorescent probe in biological systems since it exhibits different fluorescence characteristics when bound to proteins, lipids or cellular membranes. This is illustrated by the fluorescence spectra of ANS shown in Fig. 7.2 which reveal a clear shift in emission maximum for the probe bound to a lipid and to the model protein, bovine serum albumin (BSA) (Lee *et al.*, 1975).

It would be useful if in a real membrane with both protein and lipid components, the spectral differences seen in Fig. 7.2 could be distinguished, thus allowing protein sites and lipid sites to be discriminated. However, the spectral differences are small and the actual fluorescence of ANS under continuous illumination in a real membrane, such as the cilia of a small organism *Tetrahymena pyroformis*, is a single irresolvable peak (Fig. 7.3).

If recourse is made to time-resolved fluorescence spectroscopy, however, the fact that the decay time of lipid-bound ANS is relatively short (*c.* 4 ns), whereas that of protein bound ANS is long (*c.* 16 ns) permits the desired spectral resolution, as shown in Fig. 7.4. Here the early "gated" spectrum resembles the spectrum of ANS bound to the model lipid in Fig. 7.2, and the "late-gated" spectrum is very similar to the ANS bound to BSA. The added dimension of time-resolution thus permits the two classes of site to be

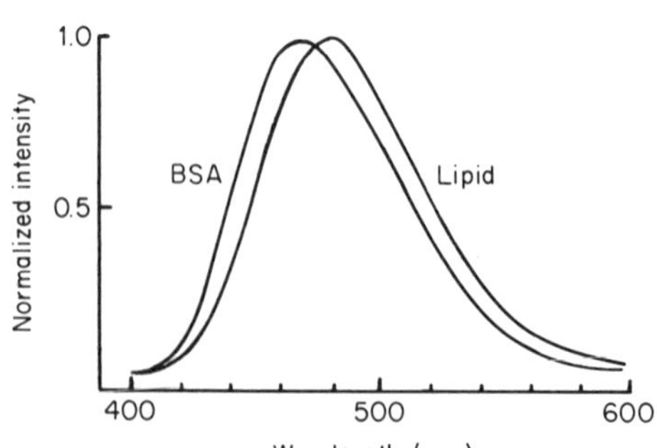

Figure 7.2 Total fluorescence emission spectra for ANS bound to liposomes of egg yolk phostatidylcholine (LIPID) and bovine serum albumin (BSA) (after Lee *et al.*, 1975).

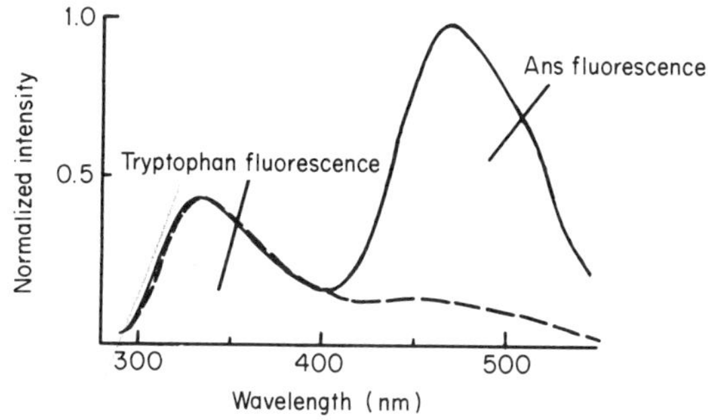

Figure 7.3 Total fluorescence emission spectra for ANS bound to microsomal membranes, - - - - -. Fluorescence from blank microsome sample, ——. Fluorescence from microsomes plus ANS (1.7×10^{-5} M) (after Lee *et al.*, 1975).

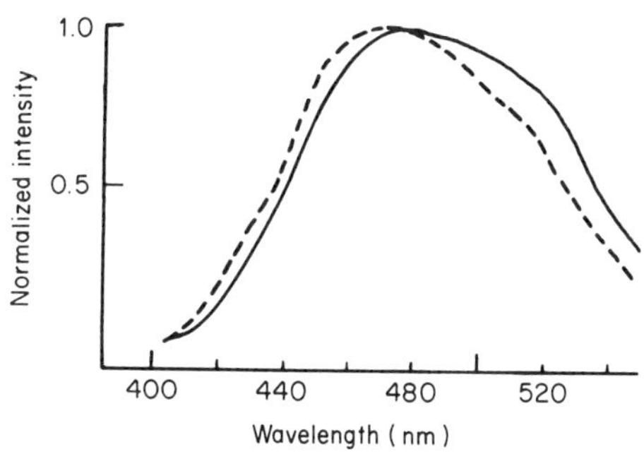

Figure 7.4 Time-resolved fluorescence emission spectra for ANS bound to microsomal membranes. ——, Early gated spectrum; - - - - -, late gated spectrum (after Lee *et al.*, 1975).

distinguished. Use can then be made of the spectral discrimination in biochemical studies on membranes.

As a further example of the use of TRES, we turn to the field of synthetic polymers. There is now increasing evidence that, owing to the variety of excimer conformations that pendant aromatic groups in polymer chains can assume, fluorescence decays in such systems must be described by multi-exponential decay laws (Roberts *et al.*, 1981; Hoyle and Guillet, 1979). Although models involving three exponential components have gained some acceptance it is conceivable that fluorescence decays in polymeric systems are really much more complex. Again, here, time-resolved spectra are invaluable.

For example in the synthetic copolymer poly(vinylnaphthalene)-poly(methylmethacrylate) the decay of the fluorescence at any wavelength requires a minimum of three exponential components for successful fitting. However, the time-resolved spectra shown in Fig. 7.5 (Roberts *et al.*, 1981) can all be synthesized by summing only two components, suitably weighted. This reveals that the origin of the complexity in fluorescence decay lies in the kinetics of interconversion of these two fluorescent species and their ability for participation in energy migration, rather than the existence of multiple overlapping emitting species.

Even in one-component molecular systems departures from simple kinetics may be observed. Isolated aromatic molecules in the gas phase excited to truly single vibronic levels may undergo intramolecular vibrational redistribution in competition with short-lived fluorescence (Lambert *et al.*, 1981; Cureton *et al.*, 1981) and furnish intriguing examples of multiexponential decay behaviour. There has long been considerable experimental and theoretical interest in this topic, which is now being enhanced with the advent of picosecond time resolution in single photon counting apparatus. There is no doubt that TRES resolved on a picosecond time scale can provide a long-sought experimental tool in the investigation of intra molecular vibrational redistribution (Moore *et al.*, 1983).

As a final example of an area where TRES could provide vital insight into excited state processes we would cite the transient effect associated with concentration gradients in excited state diffusion-controlling reactions. Because of this effect the quenching rate "constant" is time dependent and the kinetic equations cannot be solved analytically without assumptions that may not be valid in all time regions (Nemzek and Ware, 1975). When reversible chemical reaction, as in exciplex systems, occurs on the same timescale as diffusion the kinetics take on a further level of complexity (Andre *et al.*, 1979). Conventional decay curves from such systems have decay functions, $G(t)$, that cannot be described by a simple functional form and that produce highly "distorted" decay data especially at short times (Hui and Ware, 1976). Time and wavelength resolution of the fluorescence emitted from such systems offers an effective means of monitoring the behaviour of the molecules step by step during the time evolution of the reaction.

Clearly time-resolved spectroscopy can be of great value in the understanding of many photophysical processes that attract much current interest. In addition it can serve as a background (scattered light) suppressor in analytical fluorescence techniques (Haugen and Lytle, 1981). It may therefore be wondered why the technique has not to date been implemented more widely. One of the reasons lies in the inherent intractability of the processes referred to, involving as they do fluorescence, sometimes on very short timescales, from molecules reacting in a little understood way to stimuli that may themselves be difficult to characterize properly. Of more importance, however, are the intensity problems posed by the measurement, having their

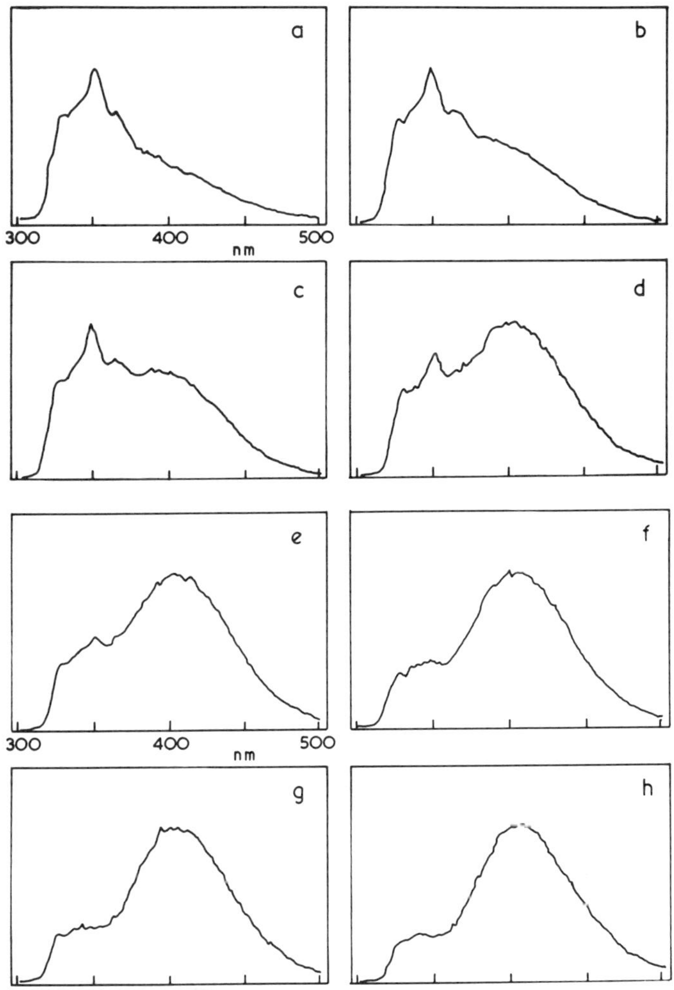

Figure 7.5 Time-resolved fluorescence of poly(vinylnaphthalene)-poly(methyl-methacrylate) copolymer. Delays Δt (a) 0 ns, (b) 3.2 ns, (c) 6.4 ns, (d) 12.8 ns, (e) 19.2 ns, (f) 25.6 ns, (g) 32.0 ns and (h) 38.0 ns (after Roberts *et al.*, 1981).

source in thc necessarily weak output of conventional nanosecond flash lamps especially if exciting radiation is isolated with a monochromator. In many applications highly monochromatic exciting light is a strict requirement of the investigation, in which case the intensity losses in the two monochromators may be so great as to preclude the measurement of accurate time-resolved spectra. Finally, it will be realized that the spectra recorded in the manner we have outlined are distorted by convolution with the exciting pulse and the detector response and may, depending on the timescale of interest, be of limited quantitative value. Nevertheless the ability of TRES to provide a qualitative picture of a process can often indicate the most effective means of proceeding with further investigation. Moreover, with the advent of picosecond lasers and detectors convolution effects in time-correlated measurements are being confined to ever shortening times with the consequence that TRES undistorted on a nanosecond timescale can now be applied to an increasing number of short-time phenomena. In addition, monochromatic laser lines of relatively high intensity now render the recording of TRES a matter of some ease. We describe the details of the measurement in Section 7.4.

If the time resolution of the SPC instrument is not high enough or if time-resolved spectra are required for kinetic analysis of picosecond events, a procedure has been developed whereby TRES can be constructed from decay functions from which distortions due to the finite width of the pumping pulse and detector response have been removed. This procedure is described in Section 7.5.

7.3 Other Methods of Spectral Time Resolution

Before discussing spectral time resolution on present day SPC instruments we will briefly mention some other techniques by which spectra resulting from short time fluorescence have been measured. In one of these methods large amounts of a non-specific quencher (such as oxygen) are added to the sample under investigation. If both long-lived and short-lived species are present, proportionately more of the fluorescence from the long-lived species will be quenched and the measured spectrum will result from short-lived unquenched fluorescence. Although the introduction of a high concentration of fluorescence quencher might be totally inappropriate in many studies, this method has some advantages and has been implemented, for example, in a series of elegant experiments concerned with vibrational redistribution in the gas phase (Dolson *et al.*, 1981). In some spectral measurements the dispersed light from a grating is allowed to impinge on a diode array operated by an optical multichannel analyser (OMA). This method is usually employed in conjunction with a pulsed light source operated at low repetition rate. A suitable gating pulse can be delivered to the OMA at any time after the light pulse, in which case the data collection will take place only in a time interval determined by the gating pulse (O'Connor *et al.*, 1983). An extremely sophisticated time gating technique has been described by Moore *et al.*, (1983), and entails down conversion of fluorescence by non-linear mixing with a picosecond laser pulse, the gating time being set by the width of this pulse. Finally, we mention a technique which was the direct precursor of the method to be described in the following section and employed gating of the fluorescence photomultiplier with a travelling wave gate circuit (Ware *et al.*, 1968; Chakrabarti and Ware, 1971; Egawa *et al.*, 1971). With the advent of TAC based lifetime instruments TRES measured by the travelling wave technique have become less common.

7.4 Direct Recording of TRES

Time-resolved emission spectroscopy with the single photon technique was apparently first performed by Ware and co-workers (1971). The principle of the method is depicted in Fig. 7.6. Output pulses from the TAC, of maximum amplitude, say, 10 V, are routed to a single channel analyser which processes pulses falling within a certain voltage range. A single channel analyser (SCA)

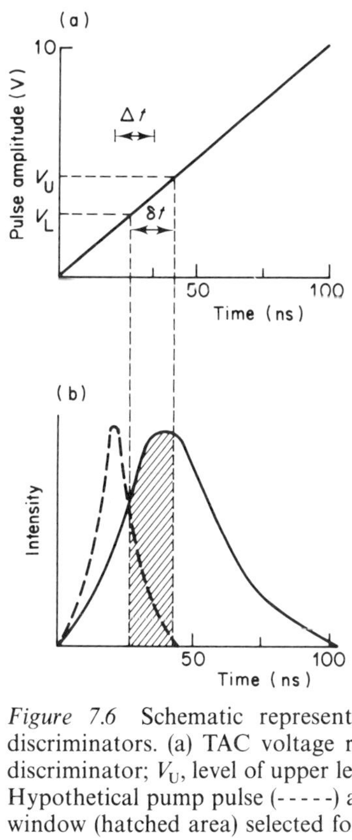

Figure 7.6 Schematic representation of selection of time window by voltage discriminators. (a) TAC voltage ramp (TAC range 100 ns). V_L, level of lower level discriminator; V_U, level of upper level discriminator; Δt, delay time; δt, gate width. (b) Hypothetical pump pulse (-----) and convolved decay curve (——). Photons in time window (hatched area) selected for spectral analysis.

is essentially a timing discriminator with two variable levels V_L and V_U. Input pulses amplitudes lower than V_L and higher than V_U are rejected. The SCA may be a separate electronic component but is usually incorporated in the multichannel analyser. The output pulses from the SCA are routed to the analogue-to-digital converter and thence to the storage channels. In Fig. 7.6 the voltage selection of TAC conversions is shown as time selection in the decay curve. We shall refer to the time, Δt, corresponding to the TAC pulse amplitude $[V_L+(V_U-V_L)/2]$ as the delay time and the time window, δt, corresponding to the voltage range $V_U - V_L$ as the gate width. Usually an accuracy in Δt and δt corresponding to the width of one MCA channel can be attained.

In order to choose the discriminator levels the decay curve is first collected in the normal way with the MCA in pulse height analysis mode. During this collection the lower and upper level discriminators are adjusted until the decay is building up only in channels corresponding to the time window of interest. The MCA channels are then cleared, the MODE switch turned to multichannel scaling and MCA collection started synchronously with the monochromator wavelength drive. Spectral resolution is governed not only by the normal slit width and scanning speed effects but also by the dwell time per channel in the MCA. Most MCAs have an internal clock with which the channel advance can be controlled. If not, or if dwell times unattainable with the MCA clock are desired, clocks designed specifically for control of multichannel scaling are available as NIM standard modules (see Section 5.2.10). If channel advance is controlled by an external clock this must normally be started at the same time as the data collection. Automatic operation of the data collection is of course possible by means of a simple electronic circuit. Since MCA channels now correspond to increments in wavelength the accumulated histogram of counts is a wavelength-resolved spectrum of the photons emitted during the interval δt. It should be remembered that these time-resolved spectra are not corrected for the wavelength response of the detecting system.

For instrument response functions of finite FWHM the delay time Δt must be defined arbitrarily since a unique time of excitation cannot be established, as reference to Fig. 7.6 demonstrates. It is therefore usual to specify Δt as the time elapsed between the time the instrument response function reaches its maximum and the midpoint of the time interval δt. In order to determine this time the MCA is returned to pulse height analysis mode and counts collected (and stored) without alteration of the upper and lower level discriminators. Then these levels are adjusted to the normal settings for decay curve measurements and an instrument response function collected. There is now a record of the position of Δt with respect to the instrument response function as well as of the gate width δt.

Constant intensity in the pump pulse is of crucial importance when TRES are measured in this simple way. However, since this stability is necessary only during one wavelength scan a simple stratagem can be adopted to correct for steady drifts in intensity. The same spectrum is scanned through fairly rapidly a number of times and alternate scans are driven in a different wavelength direction. Since the channel can usually be advanced only in the forward direction the alternate scans must be added to different memory groups of the MCA. Finally the spectrum in the second memory group is altered to run in the same wavelength direction as the first spectrum and added to it (in the computer processing). A more satisfactory procedure is to advance the wavelength and channel number in accordance with a predetermined photon flux from the exciting light, which could be monitored with an additional photomultiplier or photodiode (Ware *et al.*, 1971). A circuit designed for this purpose has been described by Rughooputh *et al.* (1983). Background counts could also be averaged out by means of a rotating shutter.

As we have already stated, optical resolution in the spectrum is dependent on the slit width and scanning speed of the monochromator. In addition, many MCAs have a maximum of 1024 channels, so that, if it is desired to scan a wide wavelength range, the number of available channels may also impose a limitation on the optical resolution. This limitation could obviously be overcome by a rapid transfer of data from the MCA into a secondary store (e.g., a computer) and an automatic control over the wavelength drive.

Time resolution in this type of TRES can be quite high since counts accumulating in a single MCA channel could be selected for spectral analysis. Unfortunately, as was stated previously an absolute zero-time is sometimes impossible to define; the width of the instrument response function relative to the delay time, Δt, is crucial in the determination of time resolution. For example, a spectrum of photons emitted at late times, when the excitation has long ceased, will not be badly distorted since convolution at very late times is negligible. It is the spectra of photons emitted in a time window occurring shortly after the maximum in the instrument response function that have low, convolution affected, time resolution.

In the attainment of good time resolution intensities may have to be drastically curtailed, especially if the delay time is set at very short or very long times. A flash lamp monochromator excitation source will probably require very careful optical alignment if the intensities necessary for the recording of accurate (the accuracy depending on pulse stability and therefore short collection times) and well resolved spectra are to be achieved. Some recent semi-quantitative studies into flash lamp generated TRES demonstrate the power of the technique (Hara *et al.*, 1980; Hoyle and Guillet, 1979). The disadvantages, related to intensity and time resolution, associated with

TRES measurement are greatly reduced if a pulsed laser serves as the excitation source. High photon fluxes in a narrow energy band, and sometimes in a pulse of a few picoseconds duration, render measurement of TRES from photons even in a very narrow gate width relatively facile.

TRES generated by three techniques are illustrated in Fig. 7.7. The phenomenon under investigation was exciplex formation, in cyclohexane solution, between 1,4-dicyanonaphthalene in its S_1 state and ground state 2,5-dimethyl-2,4-hexadiene. The total fluorescence spectrum, in which the structureless red-shifted exciplex band is clearly visible, is illustrated in Fig. 7.8. A CW dye laser (pumped by an Ar^+ ion laser) with cavity-dumped and frequency doubled output provided excitation pulses of 7 ns duration at 5 MHz. In Fig. 7.7(a) are shown the separated monomer and exciplex spectra, the latter obtained by subtracting the fluorescence spectrum of the sample to which oxygen was added (see Section 7.3) from the spectrum of the degassed sample (Johnson, 1975). The spectra in Fig. 7.7(b) are TRES measured as outlined in this section. Although the late gated spectrum was measured at a delay time of 58 ns, well removed from the maximum in the instrument response function, it is obviously contaminated with some monomer emission. This spectrum would tend to confirm the view of Easter *et al.* (1976) that TRES distorted by convolution are unsuitable for quantitative kinetic analysis, for which it is preferable to have spectra of the type illustrated in Fig. 7.7(c). Generation of these spectra is described in the following section.

7.5 Construction of TRES from Deconvolved Decay Functions

TRES measured directly are composed not of the true decay functions as given by Equation 7.5, but of convolved decay functions,

$$S(\lambda, t_k) \sim I(\lambda_1, t_k), I(\lambda_2, t_k), \ldots, I(\lambda_j, t_k). \qquad (7.6)$$

In order to construct TRES free of convolution effects it is necessary to measure decay curves at a number of wavelengths, extract true decay functions from them by deconvolution and normalize and rearrange these $G(\lambda, t)$s (Easter *et al.*, 1976; Rayner and Szabo, 1978). In Appendix 7.A1 is listed a computer routine with which these steps may be carried out. A detailed description of the procedure will now be given.

First, decay curves are collected at various wavelengths spanning the emission spectrum of the sample. Since spectral resolution is determined by the number of decays that go to make up the TRES (see Equation 7.5), a large number of measurements will be necessary for even moderate resolution. Time profiles of the pump pulse (instrument response functions) are collected

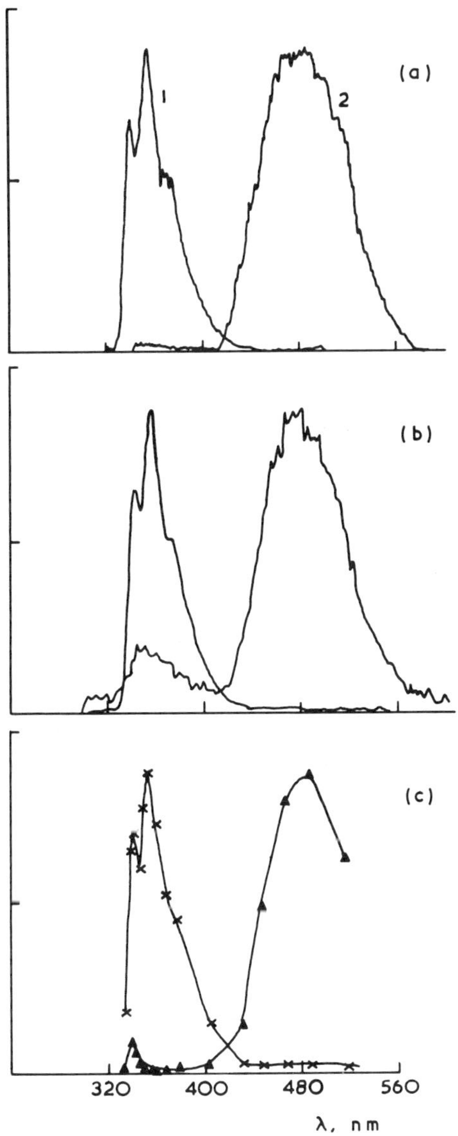

Figure 7.7 Spectra for the 1,4-dicyanonaphthalene (DCNN) 2,5-dimethyl-2,4-hexadiene (HD) exciplex system in degassed cyclohexane. (a) Spectra separated using the differential quenching technique: (1) monomer; (2) exciplex. (b) Directly measured TRES: (1) EGS $\Delta t = 2.56$ ns, $\delta t = 8.32$ ns; (2) LGS $\Delta t = 57.6$ ns, $\delta t = 19.84$ ns. (c) TRES constructed from deconvolved decays: (1) EGS $\Delta t = 1.28$ ns, $\delta t = 0.64$ ns; (2) LGS $\Delta t = 54.4$ ns, $\delta t = 0.64$ ns. Spectra are normalized to maximum intensity (cf. Fig. 7.8).

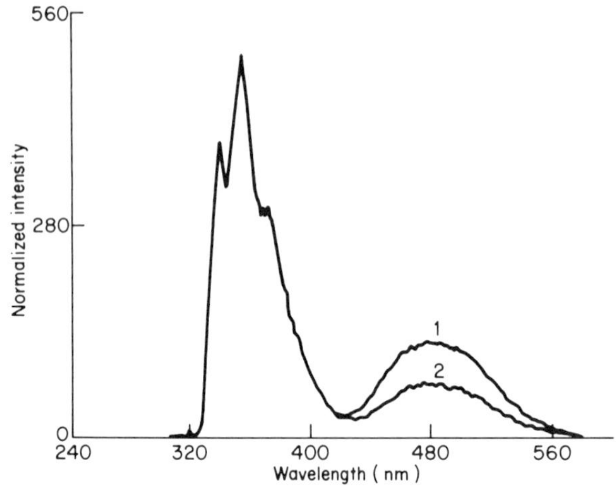

Figure 7.8 Fluorescence spectra of DCNN–HD in (1) degassed cyclohexane; (2) aerated cyclohexane.

at regular intervals, although with a stable excitation source it should not be necessary to measure an instrument response function for each decay.

These decays are now deconvolved. If the system under study is one for which the form of the decay functions, $G(t)$, cannot be specified, deconvolution is perhaps best achieved with the exponential series method (Section 6.5.8), although Fourier transform techniques (Section 6.5.6) could also be used. In the deconvolution routine CEDRIC listed in Appendix 7.A1 $G(t)$ is represented by a sum of ten exponentials and, in fitting, only the pre-exponential factors are allowed to vary (Ghiggino *et al.*, 1981). Given that TRES are usually desired to isolate short time processes from those occurring at longer times, least-squares fitting should be carried out over the whole curve rather than merely over the decay. It must be emphasized that when fitting encompasses the rising edge of the decay curve there is a heightened need to eliminate experimental distortions affecting short time data, such as scattered light and photomultiplier wavelength effects. (In the listed routines the deconvolved decay functions, $G(\lambda, t)$ are transferred to an output file and can be plotted in another routine (not listed).)

It is now necessary to normalize the decay functions to the intensity in the total fluorescence spectrum, $F(\lambda)$ or $F(\nu)$, as consideration of Fig. 7.1 and Equation 7.4 indicates. It would be equally valid to normalize to constant excitation light flux intensity. Normalization to the spectrum is usually chosen since, if this spectrum has been corrected for the wavelength dependence of the detector*, corrected TRES are generated at this stage. In

*This correction must obviously be distinguished from the wavelength correction in deconvolution.

addition, if the total spectrum is known as a function of energy, $F(\bar{v})$, the TRES will also be known as a function of energy, $S(\bar{v},t)$, a form that is most suitable for spectral calculations. As Easter *et al.* (1976) point out, the total spectrum and the decays should be measured under the same conditions of spectral bandwidth. Normalized decays, $G^0(\bar{v},t)$ are calculated with the equation (written in terms of wavenumber),

$$G^0(\bar{v}_j,t) = \frac{G(\bar{v}_j,t)F(\bar{v}_j)}{\int_0^\infty G(\bar{v}_j,t)\,\mathrm{d}t}, \tag{7.7}$$

Where $F(\bar{v}_j)$ is the intensity in the total spectrum at wavenumber, $\bar{v}_j$ (normalization is carried out in the routine CLAY).

There are now available from the normalized decay functions, derived for, say, m wavenumbers, the array of intensity data, $G^0(\bar{v}_j,t_k)$, j running from 0 to m and k from 0 to n, say. Time-resolved spectra, $S(\bar{v},t_k)$ may therefore be assembled for any time, t_k, merely by taking the t_k point from the decay functions ordered according to the wavenumber at which they were measured (see Equation 7.5). (This array manipulation is carried out in the routine TUMBLE. TRES may then be listed and plotted (Routine BUMBLE).)

Most of the crucial aspects of this procedure for construction of TRES have been covered in the preceding chapter on data analysis. Some further points, however, are worthy of mention. As we have already stated, spectral resolution is determined both by the number of wavelengths at which decay curves were measured and by the more usual slit width factor. Collection of the requisite number of decay curves and instrument response functions, and their ensuing deconvolution, may be both a time consuming and a tedious operation, which might be rendered less so by computer control of data collection. Laser excitation sources can reduce collection times by perhaps an order of magnitude. Time resolution in the spectrum depends on the accuracy with which deconvolution has been achieved and is probably greater than that determined by the MCA channel width.

Choice of a ten exponential function as a representation for $G(t)$ may give rise to the difficulties with spurious oscillations referred to in Section 6.5.8, so that fitting to a sum of three or four exponentials might be preferred. However, the latter may not lead to as acceptable a fit (Meech *et al.*, 1981). If a ten or fifteen exponential series is chosen we would emphasize the need to inspect the recovered decay functions graphically for oscillations. If present, they usually affect only the tails of the decay functions, at which times TRES may not be required. Normalization to the intensity in the total fluorescence spectrum will be seriously affected by these oscillations, since the area under the decay function (see Equation 7.7), if calculated analytically, will be grossly in error. The area under the function specified previously in Equation 6.57 as:

$$G(t) = \sum_{j=1}^{l} a_j \exp(-t/\gamma_j)$$

is given by:

$$\int_0^\infty G(t)\,\mathrm{d}t = \sum_{j=1}^{l} a_j\gamma_j. \tag{7.8}$$

If oscillations occur in the function after the time, say t_n, integration to infinity is obviously unacceptable. The area under the actual curve could be calculated either numerically (Equation 7.9) or analytically (Equation 7.10):

$$\int_0^{t_n} G(t)\,\mathrm{d}t = \sum_{i=1}^{n} G(t_i) \tag{7.9}$$

$$\int_0^{t_n} G(t)\,\mathrm{d}t = \sum_{j=1}^{l} a_j\gamma_j(1 - \mathrm{e}^{-t_n/\gamma_j}) \tag{7.10}$$

A "cut-off" correction would have to be devised for functions that had not decayed almost to zero by the time t_n. For data that give rise to spurious oscillations it would be preferable to reduce the number of exponentials in the fitting function to three or four and allow the γ_j also to vary. (The routine CREWS listed in Appendix 6.A2 could easily be adapted to this purpose.)

To conclude this section we refer to the spectra presented in Fig. 7.7(c), which are TRES assembled from 15 deconvolved decay functions derived using a sum of four exponentials in the fitting routine. The deconvolved functions were normalized to the intensity in the total spectrum (Fig. 7.8). The system under study was again the exciplex between 1,4-dicyanonaphthalene and 2,5-dimethyl-2,4-hexadiene in cyclohexane solution. In Fig. 7.9 are shown the deconvolved decay functions at 354 nm and 486 nm as well as the fits to the observed data. The unquenched lifetime of 1,4-dicyanonaphthalene is 1.7 ns while the lifetime of the exciplex (as derived from a measurement at very high diene concentrations) is 35 ns. These decay times are widely enough spaced so that, if convolution effects were not present, a direct measurement of TRES at early and late times would separate the monomer from the exciplex spectrum. Convolution effects however lead to the distorted exciplex spectrum shown in Fig. 7.7(b). Given that the resolution in the TRES constructed from decay curves is clearly not satisfactory, these spectra (Fig. 7.7(c)) and the true separated spectra (Fig. 7.7(a)) are quite similar, the former being obviously superior to the directly measured TRES with respect to time resolution and convolution effects.

7.6 Decay-associated Spectra

It will be clear that in a mixture of two fluorophores, time-resolved emission spectra can reproduce the emission of the separate components only if the

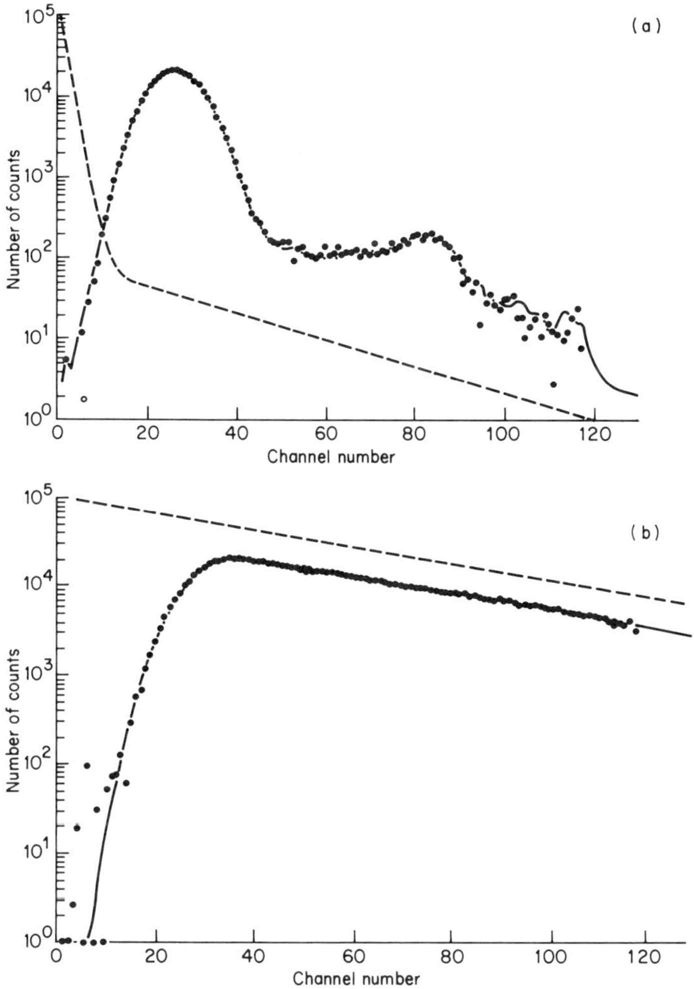

Figure 7.9 Decay data for DCNN–HD in degassed cyclohexane. ●●●, observed decays, 0.64 ns per channel; ——, reconvolved curves; - - - - -, decay functions, $G(t)$. (a) 354 nm; (b) 486 nm.

associated lifetimes are quite different. If the fluorophore lifetimes are similar a further decomposition of the TRES is required in order to obtain spectra characteristic of the individual emitting species. Such spectra have been termed decay-associated spectra and may be derived in the following way (Knutson *et al.*, 1982): suppose the observed emission results from two components, having decay times, τ_1 and τ_2, and spectral distributions $f_1(\lambda)$

and $f_2(\lambda)$. The directly measured TRES will then consist of two contributions, namely

$$S(\lambda,\Delta t_j)=f_1(\lambda)i_1(\Delta t_j)+f_2(\lambda)i_2(\Delta t_j), \tag{7.11}$$

where

$$i_i(\Delta t_j)=\int_{\Delta t_j}^{\Delta t_j+\delta t} I_i(t)\,\mathrm{d}t \tag{7.12}$$

with

$$I_i(t)=a_1\int_0^t P(t')\mathrm{e}^{-(t-t')/\tau_i}\,\mathrm{d}t' \tag{7.13}$$

Suppose now we measure TRES at two delay times Δt_1 and Δt_2. We have the equations

$$\left.\begin{aligned} S(\lambda,\Delta t_1)=f_1(\lambda)i_1(\Delta t_1)+f_2(\lambda)i_2(\Delta t_1)\\ S(\lambda,\Delta t_2)=f_1(\lambda)i_1(\Delta t_2)+f_2(\lambda)i_2(\Delta t_2)\end{aligned}\right\} \tag{7.14}$$

and we wish to determine $f_1(\lambda)$ and $f_2(\lambda)$. First, the decay at any wavelength is deconvolved and τ_1 and τ_2 determined. Then $i_1(\Delta t_1)$, $i_1(\Delta t_2)$, $i_2(\Delta t_1)$ and $i_2(\Delta t_2)$ are reconstructed using Equations 7.13 and 7.12, and Equation 7.14 is solved for $f_1(\lambda)$ and $f_2(\lambda)$. If deconvolved TRES are available Equation 7.11 becomes

$$S(\lambda,t_j)=f_1(\lambda)\mathrm{e}^{-t_j/\tau_1}+f_2(\lambda)\mathrm{e}^{-t_j/\tau_2}, \tag{7.15}$$

and the solution when τ_1 and τ_2 are known is even more straightforward. The extension to cases with three or more fluorophores and solution of a system of equations like Equation 7.14 by matrix inversion is equally simple. Figure 7.10 is presented as an example of two-component separation. In the upper part are shown early gated and late gated TRES, measured directly as outlined in Section 7.4, for a mixture of anthracene and 9-cyanoanthracene in ethanol. The decomposition of these spectra into decay-associated spectra is shown in the lower part where it can be seen that the individual spectra of anthracene and 9-cyanoanthracene are reproduced almost exactly. This method is simple but offers a means of elucidating vital information about many complex phenomena, notably heterogeneity in biological systems (Knuston *et al.*, 1982).

7.7 Conclusion

As the discussion in Sections 7.4 and 7.5 indicate, convolution effects are difficult to eliminate from directly measured time-resolved spectra. The result

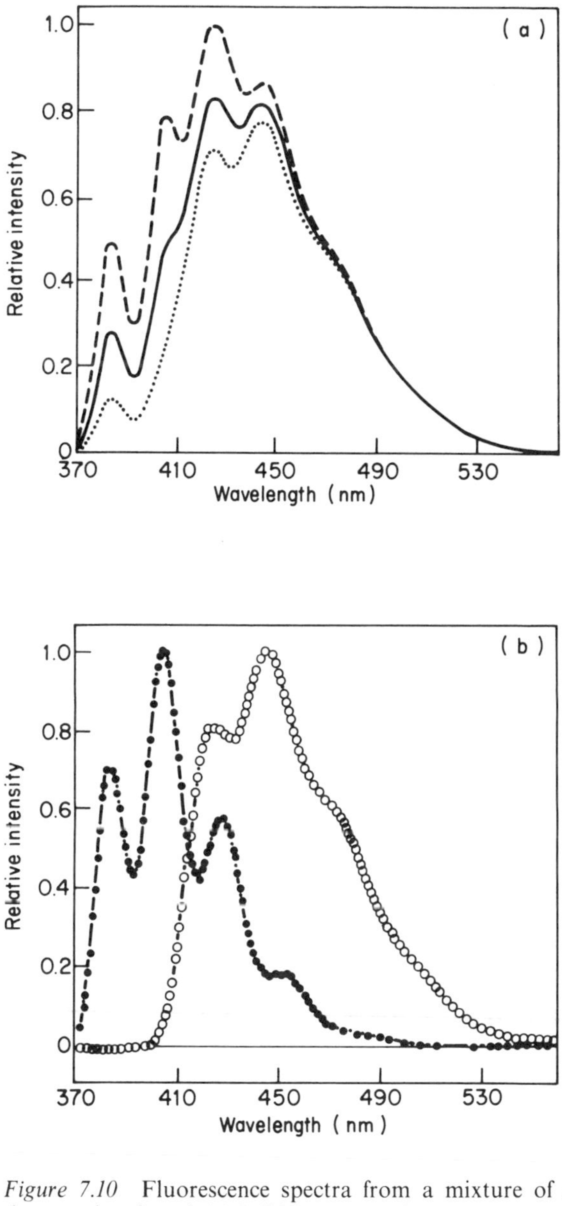

Figure 7.10 Fluorescence spectra from a mixture of anthracene and 9-cyanoanthracene in ethanol. (a) Solid curve, steady state spectrum; dashed curve, EGS, $\Delta t = 1$ channel, $\delta t = 121$ channels; dotted curve, LGS, $\Delta t = 122$ channels, $\delta t = 387$ channels. (b) Solid curves, spectra of pure anthracene and pure 9-cyanoanthracene; solid circles, normalized decay-associated spectrum of component 1; open circles, normalized decay-associated spectrum of component 2 (after Knutson *et al.*, 1982).

is that these types of TRES are, on the whole, unsuitable for quantitative analysis (Gafni *et al*., 1977). On the other hand, measurement and analysis (in terms of six or more variable parameters) of numerous decay curves is certainly time consuming and could perhaps be quite costly as regards computer time. In addition, TRES constructed from deconvolved decays will invariably have poorer optical resolution than directly measured TRES. Furthermore, while distortions other than those arising from convolution have not been discussed in this chapter, their presence, unless corrected for, could seriously interfere with the accurate construction of spectra. If kinetic analysis related, say, to study of the phenomena referred to in Section 7.2 were to be attempted through the construction of TRES, great care would have to be taken in eliminating experimental distortions and in manipulating and analysing decay curve data. For an initial qualitative investigation of a system under study, directly measured TRES are obviously to be preferred. In short, the type of TRES required depends on the information sought.

Every TRES is merely a time slice from the total spectrum and may therefore include contributions from individual components. We have illustrated a simple means by which the latter can be extracted from the TRES. In many complex systems, for instance where the exact form of the decay law is unknown, determination of decay-associated spectra may be difficult or impossible. Nevertheless, by combining time resolution with spectral resolution in the generation of TRES it should be possible to achieve a qualitative picture and to construct a model of the physics involved upon which further analysis can be based.

References

Andre, J. C., Bouchy, M. and Ware, W. R. (1979). *Chem. Phys.* **37**, 107–137.

Arecchi, J. T., Gatti, E. and Sona, A. (1966). *Rev. Sci. Instrum.* **37**, 942–948.

Beavan, S. W., Hargreaves, J. S. and Phillips, D. (1979). *Adv. Photochem.* **11**, p. 234.

Bakshiev, M. C. (1965). *Opt. Spectrosc.* **19**, 196–200.

Chakrabarti, S. K. and Ware, W. R. (1971). *J. Chem. Phys.* **55**, 5494–5498.

Cockle, S. A. and Szabo, A. G. (1981). *Photochem. Photobiol.* **34**, 23–27.

Cureton, L. G., Hara, K., O'Connor, D. V. and Phillips, D. (1981). *Chem. Phys.* **63**, 31–49.

Dolson, D. A., Parmenter, C. S. and Stone, B. M. (1981). *Chem. Phys. Lett.* **81**, 360–364.

Easter, J. H., De Toma, R. P. and Brand, L. (1976). *Biophys. J.* **16**, 571–583.

Egawa, K., Nakashima, N., Mataga, N. and Yamanaka, C. (1971). *Bull. Chem. Soc. Japan* **44**, 3287–3292.

Gafni, A., De Toma, R. P., Manrow, R. E. and Brand, L. (1977). *Biophys. J.* **17**, 155–168.

Ghiggino, K. P., Lee, A. G., Meech, S. R., O'Connor, D. V. and Phillips, D. (1981). *Biochemistry* **20**, 5381–5389.

Hara, K., De Mayo, P., Ware, W. R., Weedon, A. C., Wong, G. S. K. and Wu, K. C. (1980). *Chem. Phys. Lett.* **69**, 105–108.
Haugen, G. R. and Lytle, F. E. (1981). *Anal. Chem.* **53**, 1554–1559.
Hoyle C. E. and Guillet, J. E. (1979). *Macromolecules* **12**, 956–958.
Hui, M. H. and Ware, W. R. (1976). *J. Amer. Chem. Soc.* **98**, 4712–4717.
Johnson, G. E. (1975). *J. Chem. Phys.* **62**, 4696–4709.
Knutson, J. R., Walbridge, D. G. and Brand, L. (1982). *Biochemistry* **21**, 4671–4679.
Lambert, W. R., Walker, P. M. and Zewail, A. H. (1981). *J. Chem. Phys.* **75**, 5958–5960.
Lee, A. G., Rogers, J., Wilton, D. C., Ghiggino, K. P. and Phillips, D. (1975). *FEBS. Lett.* **94**, 171–174.
Meech, S. R., O'Connor, D. V., Roberts, A. J. and Phillips, D. (1981). *Photochem. Photobiol.* **33**, 159–172.
Moore, R., Doany, F. E., Heilweil, E. J. and Hochstrasser, R. M. (1983). *Discuss. Faraday Soc.* **75**, Paper 20.
O'Connor, D., Sumitani, M., Takagi, Y., Nakashima, N., Kamogawa, K., Udagawa, Y. and Yoshihara, K. (1983). *J. Phys. Chem.* **87**, 4848–4854.
Nemzek, T. L. and Ware, W. R. (1975). *J. Chem. Phys.* **62**, 477–489.
Rayner, D. M. and Szabo, A. G. (1978). *Canad. J. Chem.* **54**, 3246–3489.
Roberts, A. J., O'Connor, D. V. and Phillips, D. (1981). *Ann. N.Y. Acad. Sci.* **366**, 109–124.
Rumbles, G. (1984). Thesis, University of London.
Tao, T. (1969). *Biopolymers* **8**, 609–632.
Tuan, V. D. and Wild, U. P. (1973). *Appl. Optics* **12**, 1286–1292.
Tuan, V. D. and Wild, U. P. (1974). *Appl. Optics* **13**, 2899–2906.
Ware, W. R., Lee, S. K. and Chow, P. (1968). *Chem. Phys. Lett.* **2**, 356–358.
Ware, W. R., Lee, S. K., Brant, G. J. and Chow, P. (1971). *J. Chem. Phys.* **54**, 4729–4737.

Appendix 7.A1

In this appendix are listed Fortran computer routines designed for the construction of time-resolved emission spectra from decay data. Since the routines were written for use in a fairly small Perkin-Elmer Interdata 7/32 minicomputer with a Tektronix Model 43 plotter hard-wired, the code may have features puzzling to the reader about which some brief comments might be in order. It is assumed that data on disc files are stored in unformatted files with 64 data-words to a record. In order to minimize the size of routines at run time we have divided the TRES construction operations into four main sections. Dimensioned arrays are kept to a minimum so that data are usually written to holding disc files when not being manipulated. It is also assumed that raw decay data were collected with fast repetition rate laser excitation and consequently with the TAC in "inverted" mode (Section 2.2.6). Raw decay data on disc must therefore be inverted with respect to time in the software. Data for the deconvolution routine are assumed to be stored in a single file in the order $P_1(t)$, $I_1(t)$, $P_2(t)$, $I_2(t)$, etc., even if an instrument response was not measured for each decay curve. Finally, while we have not listed the two main programs that are designed primarily for plotting purposes we would emphasize that the deconvolved decay functions should always be inspected graphically for irregular behaviour. It should also be mentioned that in the main routine (not listed) for plotting the total fluorescence spectrum, this spectrum is corrected for the wavelength dependence of the detecting optics and converted from a wavelength to an energy spectrum if so desired.

The deconvolution routine CEDRIC, which carries out least-squares fitting to a 10 exponential function (Section 6.5.8) was tested in the following way. A 3-exponential decay curve was synthesized in the routine CRACK (Appendix 6.A1) from a real instrument response function $P_{or}(t)$ and the decay function

$$G(t) = 0.1e^{-t/1.0} + 1.0e^{-t/5.0} + 0.5e^{-t/10.0}.$$

Poisson noise was added (Section 6.8) and the data deconvolved in the routine CEDRIC with the set of (fixed) γ_j (see Equation 6.57) chosen as

$$\gamma_1 = 1.0, \qquad \gamma_2 = 2.0, \ldots, \qquad \gamma_{10} = 10.0.$$

After fitting ($\chi_\nu^2 = 0.97$) the values of the variable pre-exponential coefficients were, 0.17; -0.32; 0.56; -0.19; 0.45; 0.47; 0.11; -0.15; 0.005; 0.50. Next, this 10 exponential function was convolved with $P_{or}(t)$ in CRACK, noise was added and the decay curve so synthesized was deconvolved to a sum of 3 exponentials ($\chi_\nu^2 = 1.04$). The recovered function was

$$G(t) = 0.12e^{-t/0.99} + 0.97e^{-t/4.99} + 0.53e^{-t/9.75}$$

in good agreement with the original function. This testing procedure was carried out principally to check the mathematical operations (and possible typing errors) of the code. Some such procedure should be followed initially by the user adapting the listing to his or her own computer.

Except for the plotting sections, most of which are perhaps specific to the computer–plotter interface with which the routines were first run, COMMENT statements have been inserted to explain the operations performed.

```
LIST FX:CEDRIC.FTN

$BATCH
C  PROGRAM CEDRIC.FTN
C  DECONVOLVES SPC DECAY CURVES USING EXPONENTIAL SERIES METHOD
C  DIMENSIONED FOR 10 EXPONENTIALS AND 512 DATA POINTS
      INTEGER*2 IFP(9),IFR(9)
```

```
      DIMENSION A(10),T(10)
      COMMON/AREA2/Y(512),C(512)
C   IFP : FILENAME FOR INPUT DATA
C   IFR : FILENAME FOR OUTPUT DECAY FNS AND RECONVOLVED DECAYS
C   A : PRE-EXPONENTIAL FACTORS IN EXP SERIES
C   T : TAU VALUES IN EXP SERIES(FIXED)
C   C(512) : FINAL RECONVOLVED CURVE
C   Y(512) : RAW DECAY CURVE DATA AND CORRECTED DATA
C   NCU : NUMBER OF DECAY CURVES(WAVELENGTHS)
C   NC : NUMBER OF CHANNELS
C   NPAR : NUMBER OF FITTING PARAMETERS(10)
C   FACT : NANOSECONDS/CHANNEL
C
C   IN REPLY TO QUESTIONS ASKING FOR YES OR NO,
C   GIVE 1 FOR YES, 0 FOR NO
C
      WRITE(4,2)
2     FORMAT(1X,' HOW MANY CURVES:(I3) ?')
      READ(5,3) NCU
3     FORMAT(I3)
      WRITE(4,4)
4     FORMAT(1X,' HOW MANY CHANNELS:(I3) ?')
      READ(5,3) NC
      WRITE(4,5)
5     FORMAT(1X,' HOW MANY NANOSECONDS/CHANNEL:(F7.3) ?')
      READ(5,6) FACT
6     FORMAT(F7.3)
71    WRITE(4,7)
7     FORMAT(1X,' WHAT IS THE FILENAME FOR THE DECAY CURVES ?')
      READ(5,8) IFP
8     FORMAT(9A2)
C   INPUT FILE IS OPENED ON LOGICAL DEVICE 1
      CALL OPENW(1,IFP,7,0,0,I)
      IF(I.NE.0) GO TO 71
14    WRITE(4,13)
13    FORMAT(1X,' WHAT IS THE FILENAME FOR THE RESULTS ?')
      READ(5,8) IFR
C   OUTPUT FILE IS MADE AND THEN OPENED ON LOGICAL DEVICE 2
      CALL CFILW(IFR,2,256,1,1,0,0,I)
      IF(I.NE.0) GO TO 14
      CALL OPENW(2,IFR,7,0,0,I)
      IF(I.NE.0) GO TO 14
C   FILE RECORDS MUST BE WRITTEN SEQUENTIALLY. HENCE
C   SINCE ALL THE INFORMATION TO BE WRITTEN ON RECORD 1
C   IS NOT YET KNOWN A DUMMY IS WRITTEN. IT WILL BE
C   OVERWRITTEN LATER
      WRITE(2,REC=1) FACT
      NP=NC/2
      NPAR=10
      IK=NC/64
      IK1=IK+1
      KJI=0
C   THE DECAYS ARE ANALYSED ONE BY ONE
```

Appendix 7.A1 *(continued)*

```
          DO 1001 JI=1,NCU
          CALL RSET(JI,IK,NC,NCU,FACT,NP,N1,N2,SB)
          NI=0
C   IF THE FLAG NI IS SET=1 IN THE ROUTINE CALC10 THE FIT
C   HAS FAILED AND THE RESULTS ARE REJECTED
C
C   FOR THE DATA AFTER THE FIRST SET THE STARTING VALUES FOR
C   THE PARAMETERS ARE TAKEN AS THE RECOVERED VALUES FROM THE
C   PREVIOUS SET
          IF(JI.GT.1) GO TO 10
111       NI=0
C   A SET OF 10 TAU VALUES IS CHOSEN.THEY ARE NOT VARIED IN
C   THE FIT OR FROM DATA SET TO ANOTHER.
C   CHOOSE TAU VALUES SUCH THAT FACT/TAU < 80
          WRITE(4,9)
9         FORMAT(1X,' CHOOSE THE 10 TAU VALUES:(10F6.2)')
          READ(5,12)(T(J),J=1,NPAR)
12        FORMAT(10F6.2)
C   FOR THE FIRST DECAY STARTING VALUES FOR THE VARIABLE A FACTORS
C   ARE CHOSEN. IT IS RECOMMENDED TO SET THEM ALL AT THE SAME
C   VALUE(E.G. 0.1) BUT ALTERNATELY +VE AND -VE
          WRITE(4,11)
11        FORMAT(1X,' GIVE STARTING VALUES FOR THE A FACTORS:(10F6.2)')
          READ(5,12)(A(J),J=1,NPAR)
C
C   CURVE FITTING
10        CALL CALC10(FACT,NC,N1,N2,NI,NPAR,A,T,SB)
C   IF THE FITTING HAS FAILED TRY NEW STARTING VALUES
          IF(NI.EQ.1) GO TO 111
C   THE QUALITY OF THE FIT IS JUDGED BY THE VALUE OF THE
C   REDUCED CHI-SQUARE AND THE DATA ACCEPTED,REJECTED OR
C   RE-ANALYSED
          WRITE(4,15)
15        FORMAT(1X,' DO YOU WANT TO INCLUDE THIS RESULT ?')
          READ(5,16) INC
16        FORMAT(I1)
          IF(INC.EQ.1) GO TO 20
          WRITE(4,17)
17        FORMAT(1X,' DO YOU WANT TO RE-ANALYSE THIS CURVE ?')
          READ(5,16) IRE
          IF(IRE.EQ.1) GO TO 111
          GO TO 1001
C   THE ACCEPTED DECAY FNS AND RECONVOLVED CURVES ARE WRITTEN
C   TO THE OUTPUT FILE TOGETHER WITH SOME INFORMATION TO BE
C   USED BY SUBSEQUENT ROUTINES
20        WRITE(4,21)
21        FORMAT(1X,' DO YOU WANT A PLOT ?')
          READ(5,16) IP
          WRITE(4,22)
22        FORMAT(1X,' WHAT IS THE WAVELENGTH:(F7.3) ?')
          READ(5,6) Z
          M=(KJI*IK1)+2
```

```
      KJI=KJI+1
      WRITE(2,REC=M)SB,JI,IP,Z,N1,N2,(A(J),J=1,NPAR),(T(J),J=1,NPAR)
      DO 18 J=1,IK
      KJ3=M+J
18    WRITE(2,REC=KJ3)(C(I),I=(J-1)*64+1,J*64)
1001  CONTINUE
C  THE OUTPUT FILE IS NOW REWOUND AND DATA CHARACTERISING
C  THE COMPLETE DATA SET IS WRITTEN ON RECORD 1
      REWIND 2
      WRITE(2,REC=1)KJI,NC,FACT,NPAR
C  KJI IS A COUNT OF THE NUMBER OF CURVES ACCEPTED
C
C  INPUT AND OUTPUT FILES ARE CLOSED
      CALL CLOSE(1,I)
      CALL CLOSE(2,I)
      STOP
      END
C  SUBROUTINE RSET
C  READS IN THE RAW DATA,INVERTS THE CURVES,CALCULATES AND
C  SUBTRACTS BACKGROUNDS AND DETERMINES CHANNEL FITTING RANGE
C  JI : DECAY CURVE NUMBER
C  IK : NUMBER OF RECORDS ON THE FILE OCCUPIED BY ONE CURVE
C  NP : NC/2 : NEEDED FOR INVERTING CURVES
C  N1,N2 : LIMITS OF CHANNEL FITTING RANGE
C  SB : BACKGROUND NOISE(COUNTS/CHANNEL) IN DECAY CURVE
C  P(512) : PUMP PULSE PROFILE
C
      SUBROUTINE RSET(JI,IK,NC,NCU,FACT,NP,N1,N2,SB)
      COMMON/AREA1/P(512)
      COMMON/AREA2/Y(512),C(512)
C  ON THE INPUT FILE THE DATA ARE ARRANGED IN RECORDS OF
C  LENGTH 64(UNFORMATTED DATA). THE CURVES FOLLOW EACH OTHER
C  IN SEQUENCE P(1,NC),Y(1,NC),P(2,NC),Y(2,NC)..P(NCU,NC),Y(NCU,NC)
      KI=1+(JI-1)*IK*2
      KJ=KI+IK-1
      MI=KJ+1
      MJ=MI+IK-1
      IM=IK*2
      DO 911 J=KI,KJ
      JM=MOD(J,IM)
      READ(1,REC=J)(P(I),I=(JM-1)*64+1,JM*64)
911   CONTINUE
      DO 912 J=MI,MJ
      IK2=J-IK
      JM=MOD(IK2,IM)
      READ(1,REC=J)(Y(I),I=(JM-1)*64+1,JM*64)
912   CONTINUE
C  THE DATA ARE ASSUMED TO BE STORED IN THE MCA WITH
C  TIME INCREASING FROM RIGHT TO LEFT,AS THEY WOULD BE
C  E.G. FOR MHZ EXCITATION RATE
      DO 3 I=1,NP
      K=NC+1-I
      R=P(K)
```

Appendix 7.A1 *(continued)*

```
          S=Y(K)
          P(K)=P(I)
          Y(K)=Y(I)
          P(I)=R
          Y(I)=S
  3       CONTINUE
  C   THE BACKGROUNDS ARE CALCULATED ON THE ASSUMPTION THAT
  C   THE FIRST 21 CHANNELS IN EACH CURVE CONTAIN ONLY
  C   DARK COUNTS
          RB=0.0
          SB=0.0
          DO 91 I=2,21
          RB=RB+P(I)
          SB=SB+Y(I)
  91      CONTINUE
          RB=RB/20.0
          SB=SB/20.0
  C   THE CURVES ARE CORRECTED FOR THE DARK BACKGROUND
          P(1)=RB
          Y(1)=SB
          DO 5 I=1,NC
          P(I)=P(I)-RB
          IF(P(I).LT.0.0) P(I)=0.0
          Y(I)=Y(I)-SB
          IF(Y(I).LT.0.0) Y(I)=0.0
  5       CONTINUE
  C   THE CHANNEL FITTING RANGE IS CHOSEN FROM Y(N1)=1000
  C   TO Y(N2)=75(OR Y(NC-12) IF Y(NC)>75)
          DO 6 I=1,NC
          IF(Y(I).LT.1000.) GO TO 6
          N1=I
          GO TO 7
  6       CONTINUE
  7       DO 8 I=N1,NC
          IF(Y(I).GT.75.0) GO TO 8
          N2=I
          GO TO 9
  8       CONTINUE
          N2=NC-12
  9       RETURN
          END
  C
  C   SUBROUTINE CALC10
  C   CURVE FITTING TO A 10 EXPONENTIAL SERIES WITH VARIABLE
  C   A FACTORS
  C   FOLLOWS CLOSELY BEVINGTON P.237
          SUBROUTINE CALC10(FACT,NC,N1,N2,NI,NPAR,A,T,SB)
          DIMENSION ALPHA(10,10),BETA(10),B(10),ARRAY(10,10),A(10),T(10)
          COMMON/AREA1/P(512)
          COMMON/AREA2/Y(512),C(512)
          COMMON/AREA6/D(10,512)
  C   ALPHA(10,10) : MATRIX OF PRODUCTS OF PARTIAL DERIVATIVES
```

```
C                    ALSO CALLED THE ERROR MATRIX
C   ARRAY(10,10) : ERROR MATRIX ADJUSTED FOR EASE OF INVERSION
C   ALPHA AND ARRAY ALSO ARE USED FOR THE INVERTED MATRIX
C   BETA(10) : ARRAY OF THE PRODUCTS (OBS-CALC)*PART.DERIV.
C   D(10,512) : ARRAY OF PARTIAL DERIVATIVES
C   B(10) : STORE FOR A VALUES
C
C   INITIAL CHI-SQUARE AND PARTIAL DERIVATIVES ARE CALCULATED
          CALL FUNC10(N1,N2,FACT,CHISQR,NC,A,T,SB,1,NPAR)
          WRITE(4,671) CHISQR
  671     FORMAT(1X,' CHISQR = ',E14.6)
          JCOUNT=0
C   JCOUNT IS THE NUMBER OF RECONVOLUTIONS THAT PRODUCE A
C   LOWER VALUE OF CHISQR. IT IS NOT ALLOWED TO EXCEED 30
          FLAMDA=0.05
C   FOR THE MEANING OF FLAMDA SEE TEXT (SECTION 6.5.7)
  80      JCOUNT=JCOUNT+1
C   CALCULATE THE ALPHA AND BETA ARRAYS
          DO 91 J=1,NPAR
          BETA(J)=0.0
          DO 91 K=1,J
  91      ALPHA(J,K)=0.0
          DO 110 I=N1,N2
C   THE FIT IS POISSON WEIGHTED WITH THE RECIPROCALS OF
C   THE RAW DATA COUNTS
          W=Y(I)+SB
          IF(W.LT.1.0) W=1.0
          W=1.0/W
          DO 100 J=1,NPAR
          BETA(J)=BETA(J)+W*(Y(I)-C(I))*D(J,I)
          DO 100 K=1,J
  100     ALPHA(J,K)=ALPHA(J,K)+W*D(J,I)*D(K,I)
  110     CONTINUE
C   THE ALPHA MATRIX IS SYMMETRICAL
          DO 150 J=1,NPAR
          DO 150 K=1,J
  150     ALPHA(K,J)=ALPHA(J,K)
          CHISQ1=CHISQR
          ICOUNT=0
C   ICOUNT IS THE NUMBER OF TIMES THE ERROR MATRIX(WITH
C   DIFFERENT VALUES OF FLAMDA EACH TIME) IS INVERTED
C   IT NOT ALLOWED TO EXCEED 6
C
C   THE ERROR MATRIX IS ADJUSTED FOR EASE OF INVERSION
  160     DO 174 J=1,NPAR
          DO 173 K=1,NPAR
          SROOT=SQRT(ALPHA(J,J)*ALPHA(K,K))
          IF(SROOT.EQ.0.0) GO TO 561
  173     ARRAY(J,K)=ALPHA(J,K)/SROOT
C   APPLICATION OF THE MARQUARDT METHOD
  174     ARRAY(J,J)=1.0+FLAMDA
          GO TO 562
C   HERE THE FITTING HAS FAILED. SET NI=1 AND TRY AGAIN
```

Appendix 7.A1 (*continued*)

```
561   WRITE(4,563)
563   FORMAT(1X,' DIVISION BY 0')
      NI=1
      GO TO 901
C  THE ADJUSTED ERROR MATRIX IS INVERTED
562   CALL MATINV(ARRAY,NPAR,DET)
      ICOUNT=ICOUNT+1
C  INCREMENTS IN THE FITTING PARAMETERS ARE CALCULATED AND
C  ADDED TO THE CURRENT ESTIMATES OF THE PARAMETERS
C  THE CALCULATION TAKES ACCOUNT OF THE ADJUSTMENT PERFORMED
C  IN THE DO LOOP 173
      DO 201 J=1,NPAR
      B(J)=A(J)
      DO 201 K=1,NPAR
      B(J)=B(J)+BETA(K)*ARRAY(J,K)/SQRT(ALPHA(J,J)*ALPHA(K,K))
201   CONTINUE
C  WITH THE RECALCULATED PARAMETERS A NEW VALUE OF
C  CHI-SQUARE AND PARTIAL DERIVATIVES ARE CALCULATED
      CALL FUNC10(N1,N2,FACT,CHISQR,NC,B,T,SB,1,NPAR)
C  IF NEW CHI-SQUARE IS < OLD KEEP CURRENT PARAMETER VALUES
C  IF OLD < NEW, RE-INVERT ERROR MATRIX ACCORDING TO
C  MARQUARDT METHOD
      IF(CHISQ1-CHISQR) 302,303,303
303   DO 304 J=1,NPAR
304   A(J)=B(J)
      IF(JCOUNT.EQ.30) GO TO 560
C  CHECK FOR CONVERGENCE
      IF((CHISQ1-CHISQR).LT.0.000001) GO TO 560
C  THE SEARCH HAS NOW REACHED A POINT ON THE HYPERSURFACE
C  CLOSER TO THE MINIMUM. SO REDUCE FLAMDA AND RE-CALCULATE
C  PARAMETER INCREMENTS
      FLAMDA=FLAMDA/10.0
      GO TO 80
C
302   FLAMDA=FLAMDA*10.0
      IF(ICOUNT.EQ.5) GO TO 305
      IF(ICOUNT.EQ.6.AND.JCOUNT.LE.2) GO TO 361
      IF(ICOUNT.EQ.6.AND.JCOUNT.GT.2) GO TO 330
      GO TO 160
C  IF NO MORE THAN 2 RECONVOLUTIONS HAVE BEEN PERFORMED
C  THE FIT IS CONSIDERED UNSUCCESSFUL
361   NI=1
      GO TO 901
305   FLAMDA=FLAMDA/10**6
      GO TO 160
330   CHISQR=CHISQ1
560   WRITE(4,622)JCOUNT,N1,N2,CHISQR
622   FORMAT(1X,I3,' CONVOLUTIONS',/,1X,' FITTING BETWEEN',
     1 ' CHANNELS ',I3,' AND ',I3,2X,' CHISQR = ',E14.5,/,
     1 7X,' A FACTORS',11X,'TAU')
      DO 623 J=1,NPAR
      WRITE(4,624) A(J),T(J)
```

```
623    CONTINUE
624    FORMAT(4X,E14.6,5X,F10.4)
C  RECALCULATE FITTED CURVE FOR BEST PARAMETERS
       CALL FUNC10(N1,N2,FACT,CHISQR,NC,A,T,SB,0,NPAR)
901    RETURN
       END
C
C  SUBROUTINE FUNC10
C  CALCULATES THE CONVOLUTION PRODUCT BETWEEN P(NC) AND
C  THE 10 EXPONENTIAL SERIES. ALSO CALCULATES THE PARTIAL
C  DERIVATIVES.
       SUBROUTINE FUNC10(N1,N2,FACT,CHISQR,NC,A,T,SB,ID,NP)
       DIMENSION A(10),T(10),CC(10)
       COMMON/AREA1/P(512)
       COMMON/AREA2/Y(512),C(512)
       COMMON/AREA6/D(10,512)
C
C  ID IS A FLAG:=1 CALCULATE DERIVS,=0 DO NOT
       IF(ID.EQ.0) GO TO 303
       DO 1 J=1,NP
       DO 1 I=N1,N2
       D(J,I)=0.0
1      CONTINUE
C  THE INTEGRATION IS PERFORMED USING THE TRAPEZOIDAL RULE
303    CMULT=FACT*0.5
       C(1)=0.0
       DO 304 J=I,NP
       CC(J)=CMULT*A(J)*P(1)
304    C(1)=C(1)+CC(J)
       DO 2 I=2,N2
       C(I)=0.0
       DO 21 J=1,NP
       CC(J)=EXP(-FACT/T(J))*CC(J)+A(J)*CMULT*(P(I-1)
      1 *EXP(-FACT/T(J))+P(I))
       IF(ID.EQ.0) GO TO 21
       D(J,I)=CC(J)/A(J)
21     C(I)=C(I)+CC(J)
2      CONTINUE
C  THE REDUCED CHI-SQUARE IS CALCULATED
       NPTS=N2-N1+1
       NFREE=NPTS-10
       FREE=FLOAT(NFREE)
       CHISQ=0.0
       DO 3 I=N1,N2
C  THE FITTING IS POISSON WEIGHTED WITH THE RECIPROCALS OF
C  THE RAW DATA POINTS
       W=Y(I)+SB
       IF(W.LT.1.0) W=1.0
       W=1.0/W
       CHISQ=CHISQ+W*(Y(I)-C(I))*(Y(I)-C(I))
3      CONTINUE
       CHISQR=CHISQ/FREE
       RETURN
```

Appendix 7.A1 *(continued)*

```
      END
C
C  SUBROUTINE MATINV
C  MATRIX INVERSION ROUTINE TAKEN DIRECT FROM BEVINGTON
C  FOR COMMENTS SEE BEVINGTON P.301
      SUBROUTINE MATINV(ARRAY,NORDER,DET)
      DIMENSION ARRAY(NORDER,NORDER)
      DIMENSION IK(10),JK(10)
      DET=1.0
      DO 100 K=1,NORDER
      AMAX=0.0
21    DO 30 I=K,NORDER
      DO 30 J=K,NORDER
      IF(ABS(AMAX)-ABS(ARRAY(I,J))) 24,24,30
24    AMAX=ARRAY(I,J)
      IK(K)=I
      JK(K)=J
30    CONTINUE
      IF(AMAX) 41,32,41
32    DET=0.0
      GO TO 140
41    I=IK(K)
      IF(I-K) 21,51,43
43    DO 50 J=1,NORDER
      SAVE=ARRAY(K,J)
      ARRAY(K,J)=ARRAY(I,J)
50    ARRAY(I,J)=-SAVE
51    J=JK(K)
      IF(J-K) 21,61,53
53    DO 60 I=1,NORDER
      SAVE=ARRAY(I,K)
      ARRAY(I,K)=ARRAY(I,J)
60    ARRAY(I,J)=-SAVE
61    DO 70 I=1,NORDER
      IF(I-K) 63,70,63
63    ARRAY(I,K)=-ARRAY(I,K)/AMAX
70    CONTINUE
      DO 80 I=1,NORDER
      DO 80 J=1,NORDER
      IF(I-K) 74,80,74
74    IF(J-K) 75,80,75
75    ARRAY(I,J)=ARRAY(I,J)+ARRAY(I,K)*ARRAY(K,J)
80    CONTINUE
      DO 90 J=1,NORDER
      IF(J-K) 83,90,83
83    ARRAY(K,J)=ARRAY(K,J)/AMAX
90    CONTINUE
      ARRAY(K,K)=1.0/AMAX
100   DET=DET*AMAX
      DO 130 L=1,NORDER
      K=NORDER-L+1
      J=IK(K)
```

```
      IF(J-K) 111,111,105
105   DO 110 I=1,NORDER
      SAVE=ARRAY(I,K)
      ARRAY(I,K)=-ARRAY(I,J)
110   ARRAY(I,J)=SAVE
111   I=JK(K)
      IF(I-K) 130,130,113
113   DO 120 J=1,NORDER
      SAVE=ARRAY(K,J)
      ARRAY(K,J)=-ARRAY(I,J)
120   ARRAY(I,J)=SAVE
130   CONTINUE
140   RETURN
      END
$BEND
```

```
LIST FX:CLAY.FTN
```

```
$BATCH
C  PROGRAM CLAY.FTN
C  CONSTRUCTS TIME-RESOLVED SPECTRA FROM DECONVOLVED
C  DECAY FNS AND A TOTAL FLUORESCENCE SPECTRUM
      INTEGER*2 IFR(9),IFC(9),IGRAF(3)
      DIMENSION MIXER(64),S(512)
      COMMON/AREA7/A(20)
      COMMON/AREA8/G(512)
      COMMON/AREA1/IFC
      COMMON/AREA2/TRES(64),TLAM(64)
      DATA IGRAF/'GRAF: '/
C  IFR : NAME OF FILE CONTAINING THE DECAY FNS
C        I.E. OUTPUT FILE FROM CEDRIC
C  IFC : NAME OF FILE CONTAINING (CORRECTED) SPECTRUM
C        ALSO USED FOR HOLDING FILES
C  IGRAF : NAME USED TO IDENTIFY THE PLOTTER
C  MIXER(64) : ARRAY CONTAINING CHANNEL NOS. OF TRES TO BE
C              PLOTTED
C  S(512) : (CORRECTED) SPECTRUM DATA POINTS
C  A(10),T(10) : A FACTORS AND TAU VALUES OF DECAY FNS
C  G(512) : ARRAY CONTAINING DECONVOLVED DECAY CURVE
C  TRES(64) : ARRAY CONTAINING THE TIME-RESOLVED SPECTRM
C  TLAM(64) : WAVELENGTHS OF THE POINTS IN TRES
C
C  TOTAL SPECTRUM READ IN FROM FILE ON LOGICAL DEVICE 2
7     WRITE(4,6)
6     FORMAT(1X,' TYPE CORRECTED SPECTRUM FILENAME')
      READ(5,3) IFC
      CALL OPENW(2,IFC,7,0,0,I)
      IF(I.NE.0) GO TO 7
C  ON RECORD 1 IN SPECTRUM FILE IS THE INFORMATION:
C  NS : NUMBER OF POINTS
C  FSLAM : WAVELENGTH OF FIRST POINT
C  CHIPPY : INCREASE IN WAVELENGTH PER DATA POINT
```

Appendix 7.A1 (*continued*)

```
C  LUD : FLAG. IF =1 TRES IN WAVELENGTH. =2 IN ENERGY
      READ(2,REC=1) NS,FSLAM,CHIPPY,LUD
      IM=NS/64+1
      DO 8 J=2,IM
      READ(2,REC=J)(S(I),I=((J-2)*64+1),(J-1)*64)
8     CONTINUE
      CALL CLOSE(2,I)
C  OPEN DECAY FNS FILE ON LOGICAL DEVICE 1
1     WRITE(4,2)
2     FORMAT(1X,' TYPE DECONVOLUTION RESULTS FILENAME')
      READ(5,3) IFR
3     FORMAT(9A2)
      CALL OPENW(1,IFR,7,0,0,I)
      IF(I.NE.0) GO TO 1
      READ(1,REC=1) KJI,NC,FACT,NPAR
      NP2=NPAR*2
C  IK : NUMBER OF RECORDS ON THE OUTPUT FILE THAT WILL BE
C        OCCUPIED BY A DECAY CURVE
      IK=NC/64
      IK1=IK+1
C  TO REDUCE THE NUMBER OF ARRAYS,THE NORMALISED DECAY CURVES
C  ARE STORED IN A HOLDING FILE. IT IS NOW MADE AND OPENED ON
C  LOGICAL DEVICE 3
9     WRITE(4,10)
10    FORMAT(1X,' TYPE FILENAME FOR DECAYS NORMALISED TO THE TFS')
      READ(5,3) IFC
      CALL CFILW(IFC,2,256,1,1,0,0,I)
      IF(I.NE.0) GO TO 9
      CALL OPENW(3,IFC,7,0,0,I)
      IF(I.NE.0) GO TO 9
C  THE DECAY FNS ARE READ IN ONE AT A TIME
      DO 1002 JN=1,KJI
      KJ2=(JN-1)*(IK+1)+2
C  FOR THE INFORMATION IN THIS FILE SEE CEDRIC
      READ(1,REC=KJ2)SB,JI,IPL,FLAM,NQ,NW,(A(J),J=1,NP2)
      TLAM(JN)=FLAM
      IF(LUD.EQ.1) GO TO 501
C  IF LUD=2 CONVERT WAVELENGTHS TO ENERGY
      TLAM(JN)=100000./FLAM
C  CALCULATE THE DECAY CURVES
501   CALL DECFUN(NC,FACT,NPAR)
C  THE THEORETICAL AREA UNDER THE DECAY CURVE IS CALCULATED
C  IF G(T) SHOWS OSCILLATIONS AND PROVIDED IT HAS DECAYED
C  TO 0 AT LAST CHANNEL REPLACE THIS CALCULATION BY A
C  CALCULATION OF THE ACTUAL AREA(SUM OF POINTS)
      ARG=0.0
      DO 4 LA=1,NPAR
      LA2=LA+NPAR
4     ARG=ARG+A(LA)*A(LA2)
C  NORMALISE THE DECAY TO THE INTENSITY IN THE SPECTRUM
      DO 5 I=1,NC
5     G(I)=G(I)/ARG
```

```
C  INTERPOLATE THE SPECTRUM TO FIND THE INTENSITY AT THE
C  WAVELENGTH AT WHICH THE DECAY CURVE WAS MEASURED
      DO 68 J=1,NS
      FJ=FLOAT(J)
      SLAM=FSLAM+(FJ-1.)*CHIPPY
      R=FLAM-SLAM
      IF(R.LT.0.0) GO TO 69
68    CONTINUE
69    P1F=(S(J)-S(J-1))/CHIPPY
      PF=(P1F*(FLAM-SLAM+CHIPPY))+S(J-1)
      DO 71 L=1,NC
71    G(L)=G(L)*PF
C  WRITE THE NORMALISED DECAY CURVES TO THE HOLDING FILE
      DO 11 J=1,IK
      KJ3=(JN-1)*IK+J
11    WRITE(3,REC=KJ3)(G(I),I=(J-1)*64+1,J*64)
1002  CONTINUE
C  WHEN ALL THE NORMALISED DECAY CURVES HAVE BEEN WRITTEN
C  TO THE FILE CLOSE THE HOLDING FILE AND THE DECAY FNS FILE
      CALL CLOSE(1,I)
      CALL CLOSE(3,I)
C  THE TRES ARE CALCULATED
C  KIK : FLAG. IF =0 NORMALISED DECAY CURVES FILE COULD NOT
C          BE OPENED IN SUBROUTINE TUMBLE
      CALL TUMBLE(KIK,IK,KJI,LUD)
      IF(KIK.NE.0) GO TO 1003
C  TRES CAN BE LISTED
      WRITE(4,12)
12    FORMAT(1X,' DO YOU WANT TO LIST SOME TRES ?')
      READ(5,13) IL
13    FORMAT(I1)
      IF(IL.EQ.0) GO TO 101
      WRITE(4,14)
14    FORMAT(1X,' HOW MANY:(I2) ?')
      READ(5,15) NTR
15    FORMAT(I2)
C  WAVELENGTHS OR WAVENUMBERS ARRANGED IN GROUPS OF 8
      CALL RANCH(KJI,8,L1,L2)
      IF(LUD.EQ.1) GO TO 502
C  LUD INDICATES WAVELENGTHS OR WAVENUMBERS
      WRITE(4,161)
161   FORMAT(20X,' WAVENUMBERS * 0.01')
      GO TO 503
502   WRITE(4,16)
16    FORMAT(30X,' WAVELENGTHS')
503   IF(L1.EQ.0) GO TO 20
      DO 17 J=1,L1
      J1=8*J
      J2=(J-1)*8+1
      WRITE(4,18)(TLAM(K),K=J2,J1)
17    CONTINUE
20    L3=L1*8
      DO 19 J=1,L2
```

Appendix 7.A1 (*continued*)

```
          L4=L3+J
          WRITE(4,24) TLAM(L4)
 19       CONTINUE
 18       FORMAT(1X,8(F6.2,2X))
 24       FORMAT(1X,F6.2)
 35       FORMAT(I3)
 C  SPECTRUM POINTS ARRANGED IN GROUPS OF 5
          CALL RANCH(KJI,5,L1,L2)
          L3=5*L1
          DO 25 JL=1,NTR
          WRITE(4,34) JL
 34       FORMAT(1X,' TYPE CHANNEL NUMBER OF TRES',I2,1X,':(I3)')
          READ(5,35)LJ
 C  TRES ARE READ IN FROM HOLDING FILE
          READ(2,REC=LJ)(TRES(K),K=1,KJI)
          IF(L1.EQ.0) GO TO 27
          DO 28 JK=1,L1
          J1=5*JK
          J2=(JK-1)*5+1
          WRITE(4,29)(TRES(K1),K1=J2,J1)
 28       CONTINUE
 27       DO 30 J=1,L2
          L4=L3+J
          WRITE(4,31) TRES(L4)
 30       CONTINUE
 29       FORMAT(1X,5(E12.5,1X))
 31       FORMAT(1X,E12.5)
 25       CONTINUE
 C  TRES CAN BE PLOTTED
 101      WRITE(4,32)
 32       FORMAT(1X,' DO YOU WANT TO PLOT SOME TRES ?')
          READ(5,13) IL
          IF(IL.EQ.0) GO TO 1003
          WRITE(4,14)
          READ(5,15) NUM
          WRITE(4,26)
 26       FORMAT(1X,' TYPE CHANNEL NO.S OF TRES TO BE PLOTTED:(I3)')
          DO 33 JL=1,NUM
 33       READ(5,35) MIXER(JL)
 C  THE PLOTTER IS OPENED FOR PLOTTING
 36       CALL OPENW(3,IGRAF,7,0,0,I)
          IF(I.EQ.0) GO TO 37
          WRITE(4,38)
 38       FORMAT(1X,' THE PLOTTER IS NOT READY.WHEN IT IS TYPE 1')
          READ(5,13) IME
          GO TO 36
 C  PLOTTING
 37       CALL BUMBLE(NUM,MIXER,KJI,NS,FSLAM,CHIPPY,LUD)
 C  CLOSE PLOTTER AND HOLDING FILE
          CALL CLOSE(3,I)
          GO TO 101
 1003     CALL CLOSE(2,I)
```

```
      STOP
      END
C
C  SUBROUTINE DECFUN
C  CALCULATES A SUM OF EXPONENTIALS
C  DIMENSIONED FOR UP TO TEN EXPONENTIALS
      SUBROUTINE DECFUN(N,F,NPAR)
      COMMON/AREA7/A(20)
      COMMON/AREA8/G(512)
      DO 1 I=1,N
      G(I)=0.0
      FI=FLOAT(I)
      DO 2 K=1,NPAR
      K2=K+NPAR
      R=FI*F/A(K2)
      IF(R.GT.85.0) R=85.0
      P1=A(K)*EXP(-R)
      G(I)=G(I)+P1
2     CONTINUE
1     CONTINUE
      DO 3 I=1,N
      IF(G(I).LT.0.0) G(I)=0.0
3     CONTINUE
C  NORMALISE THE CURVE TO 100000 COUNTS IN THE
C  CHANNEL OF MAXIMUM COUNTS
      GMAX=G(1)
      DO 4 I=2,N
4     GMAX=AMAX1(G(I),GMAX)
      DO 5 I=1,N
5     G(I)=G(I)*100000./GMAX
      RETURN
      END
C
C  SUBROUTINE RANCH
C  CALCULATES HOW MANY TIMES(L1) L GOES INTO KJ1 AND
C  THE REMAINDER(L2)
      SUBROUTINE RANCH(KJ1,L,L1,L2)
      RL=FLOAT(KJ1)/FLOAT(L)
      L2=MOD(KJ1,L)
      IF(L2)1,1,2
1     L1=KJ1/L
      GO TO 3
2     L1=INT(RL)
3     RETURN
      END
C
C  SUBROUTINE TUMBLE
C  REARRANGES A FILE CONTAINING NORMALISED DECAY CURVES
C  TO GIVE TIME-RESOLVED SPECTRA.THE SPECTRA ARE ASSUMED
C  TO HAVE <64 POINTS AND ARE WRITTEN TO AN OUTPUT
C  FILE, ONE TRES PER RECORD
      SUBROUTINE TUMBLE(KIK,IK,NLAM,LUD)
      INTEGER*2 IFC(9),IFT(9)
```

Appendix 7.A1 (*continued*)

```
      DIMENSION XFQ(512)
      COMMON/AREA1/IFC
      COMMON/AREA8/G(512)
      COMMON/AREA2/TRES(64),TLAM(64)
C  NLAM : NUMBER OF CURVES ON THE INPUT FILE
C         = NO. OF POINTS IN THE TRES
C  IFC : INPUT FILE CONTAINING NORMALISED DECAY CURVES
C  IFT : OUTPUT FILE CONTAINING TRES
C  XFQ(512) : ARRAY OF MEAN FREQUENCIES OF EACH TRES
      KIK=0
      CALL OPENW(1,IFC,7,0,0,I)
      IF(I.NE.0) GO TO 1000
1     WRITE(4,2)
2     FORMAT(1X,' TYPE FILENAME FOR TRES')
      READ(5,3) IFT
3     FORMAT(9A2)
      CALL CFILW(IFT,2,256,1,1,0,0,I)
      IF(I.NE.0) GO TO 1
      CALL OPENW(2,IFT,7,0,0,I)
      IF(I.NE.0) GO TO 1
C  READ IN THE DECAY CURVES AND CONSTRUCT THE TRES
      DO 23 K=1,IK
      K1=64*(K-1)+1
      K2=64*K
      DO 22 I=K1,K2
      DO 21 J=1,NLAM
      IK4=IK*(J-1)+K
      READ(1,REC=IK4)(G(L),L=K1,K2)
      TRES(J)=G(I)
21    CONTINUE
C  CORRECT THE TRES IF INTENSITY VS. ENERGY IS REQUIRED
C  AND CALCULATE THE MEAN FREQUENCY OF EACH TRES
      IF(LUD.EQ.1) GO TO 11
      XF=0.0
      XM=0.0
      DO 12 L=1,NLAM
      TRES(L)=TRES(L)*TLAM(L)*TLAM(L)*0.0001
      XF=XF+TRES(L)*TLAM(L)
      XM=XM+TRES(L)
12    CONTINUE
C  WRITE TRES TO OUTPUT FILE
11    WRITE(2)(TRES(J),J=1,NLAM)
C  LIST THE MEAN FREQUENCIES
      IF(LUD.EQ.1) GO TO 22
      XFQ(I)=XF/XM
22    CONTINUE
      IF(LUD.EQ.1) GO TO 23
      WRITE(4,2105)K1,K2
2105  FORMAT(1X,'MEAN FREQ. TRES  ',I3,' TO ',I3)
      DO 2106 I=K1,K2,8
      M=I+7
      WRITE(4,2107)(XFQ(J),J=I,M)
```

```
2106  CONTINUE
2107  FORMAT(1X,8(F7.1,1X))
      WRITE(4,2108)
2108  FORMAT(1X,'TYPE 1 TO CONTINUE')
      READ(5,2109)MILL
2109  FORMAT(I1)
23    CONTINUE
C  LEAVE OUTPUT FILE OPEN. IS READ FROM IN MAIN PROGRAM
      GO TO 1001
1000  KIK=1
1001  RETURN
      END
C
C  SUBROUTINE BUMBLE
C  NO COMMENTS ARE GIVEN FOR THIS ROUTINE AS IT
C  IS FOR THE MOST PART RELEVANT TO A SPECIFIC PLOTTER
      SUBROUTINE BUMBLE(NUM,M,KJ1,NS,FSLAM,CHIPPY,LUD)
      INTEGER*2 IUS
      COMMON/AREA2/TRES(64),TLAM(64)
      DIMENSION M(64)
      DATA IUS/31/
      RANGY=2390.0/10000.0
      FNS=FLOAT(NS)
      SLAMF=FSLAM+(FNS-1.0)*CHIPPY
      FS=FSLAM
      SL=SLAMF
      IF(LUD.EQ.1) GO TO 501
      FS=100000./SLAMF
      SL=100000./FSLAM
501   ICE=0
101   WRITE(4,1)
1     FORMAT(1X,' TYPE 0 TO FINISH PLOTTING')
      READ(5,2) IPLF
2     FORMAT(I1)
      IF(IPLF.EQ.0) GO TO 401
      WRITE(4,3)
3     FORMAT(1X,' TYPE 1 FOR UNNORMALISED,SAME PLOT',/,
     1 1X,'       2 FOR UNNORMALISED,DIFFERENT PLOTS',/,
     2 1X,'       3 FOR NORMALISED,SAME PLOT',/,
     3 1X,'       4 FOR NORMALISED,DIFFERENT PLOTS')
      READ(5,2) ICAT
      IF(ICAT.GE.3) GO TO 6
      IF(ICE.EQ.1) GO TO 6
      TMAX=0.0
      ICE=1
      A=0.
      DO 1000 MI=1,NUM
      READ(2,REC=M(MI))(TRES(K),K=1,KJ1)
      TMAX=TRES(1)
      DO 4 J=2,KJ1
4     TMAX=AMAX1(TRES(J),TMAX)
      IF(TMAX-A)1000,1000,5
5     A=TMAX
```

Appendix 7.A1 (*continued*)

```
1000  CONTINUE
      REWIND 2
6     IF(ICAT.EQ.2.OR.ICAT.EQ.4) GO TO 61
      WRITE(4,10)
      READ(5,2) MAB
      CALL SIDLE
      CALL APLOT(56.0,2944.0,-1)
      IF(ICAT.EQ.3) GO TO 63
      WRITE(3,64)IUS
64    FORMAT(A2,'UNNORMALISED TIME RESOLVED SPECTRA')
      GO TO 611
63    WRITE(3,65)IUS
65    FORMAT(A2,'NORMALISED TIME RESOLVED SPECTRA')
611   CALL PAXIS(FS,SL,FIX1,FIX2,LUD)
61    DO 1001 MI=1,NUM
      READ(2,REC=M(MI))(TRES(K),K=1,KJ1)
      IF(ICAT.EQ.2.OR.ICAT.EQ.4) GO TO 7
      IF(MI.EQ.1) GO TO 8
      WRITE(4,66)
66    FORMAT(1X,' CHOOSE A PEN. READY ?')
      READ(5,2)MAB
      GO TO 8
7     WRITE(4,10)
10    FORMAT(1X,' SET UP PLOTTER.   READY ?')
      READ(5,2) MAB
      CALL SIDLE
      CALL APLOT(56.0,3030.0,-1)
      WRITE(3,67)IUS,M(MI)
67    FORMAT(A2,'FOR CHANNEL ',I3)
      CALL APLOT(56.0,2944.0,-1)
      IF(ICAT.EQ.2) GO TO 68
      WRITE(3,69)IUS
69    FORMAT(A2,'NORMALISED TIME RESOLVED SPECTRUM')
      GO TO 70
68    WRITE(3,71)IUS
71    FORMAT(A2,'UNNORMALISED TIME RESOLVED SPECTRUM')
70    CALL PAXIS(FS,SL,FIX1,FIX2,LUD)
8     IF(ICAT.LT.3) GO TO 12
      TRENT=TRES(1)
      DO 9 MK=2,KJ1
9     TRENT=AMAX1(TRES(MK),TRENT)
12    DO 11 MK=1,KJ1
      IF(ICAT.GE.3) GO TO 14
      TRES(MK)=TRES(MK)*10000./A
      GO TO 11
14    TRES(MK)=TRES(MK)*10000./TRENT
11    CONTINUE
      DO 13 MK=1,KJ1
      Y=TRES(MK)
      PW=TLAM(MK)-FIX1
      X=460.0+PW*3480.0/(FIX2-FIX1)
      Y=410.0+RANGY*Y
```

```
      KLM=0
      IF(MK.EQ.1) KLM=-1
      CALL APLOT(X,Y,KLM)
13    CONTINUE
1001  CONTINUE
      REWIND 2
      GO TO 101
401   RETURN
      END
      SUBROUTINE APLOT(X,Y,IG)
      DIMENSION ICOORD(5)
      DATA IGS/486539264/,ISYN/369098752/,ISP/'    '/
      ICOORD(1)=16777216*INT(AMOD(Y/128.0,32.0)+32.0)
      ICOORD(2)=16777216*INT(AMOD(Y,4.0)*4.0+AMOD(X,4.0)+96.0)
      ICOORD(3)=16777216*INT(AMOD(Y/4.0,32.0)+96.0)
      ICOORD(4)=16777216*INT(AMOD(X/128.0,32.0)+32.0)
      ICOORD(5)=16777216*INT(AMOD(X/4.0,32.0)+64.0)
      IF(IG)199,200,201
199   WRITE(3,12) IGS,(ICOORD(I),I=1,5)
      RETURN
200   WRITE(3,14) ISYN,ISP,(ICOORD(I),I=1,5)
      RETURN
201   WRITE(3,16) IGS,(ICOORD(I),I=1,5),(ICOORD(I),I=1,5)
      RETURN
12    FORMAT(1H ,A1,5A1)
14    FORMAT(1H ,2A1,5A1)
16    FORMAT(1H ,A1,5A1,5A1)
      END
      SUBROUTINE SIDLE
      INTEGER*2 IESC,IA,IE,IN
      DATA IESC/27/,IA/65/,IE/69/,IN/78/
      WRITE(3,1) IESC,IA,IE,IESC,IA,IN
1     FORMAT(6A2)
      CALL APLOT(0.0,3124.0,-1)
      CALL APLOT(0.0,0.0,0)
      CALL APLOT(4095.0,0.0,0)
      CALL APLOT(4095.0,3124.0,0)
      CALL APLOT(0.0,3124.0,0)
      RETURN
      END
      SUBROUTINE PAXIS(XMIN,XMAX,FIX1,FIX2,L)
      INTEGER*2 IESC,IA,IE,IN,IJ,II,ICR,IUS,ICOM,ITF,ISEV
      DATA IESC/27/,IA/65/,IE/69/,IN/78/,IJ/74/,ISEV/70/
      DATA ICR/13/,IUS/31/,ICOM/44/,II/73/,ITF/35/
      CALL APLOT(460.0,2800.0,-1)
      CALL APLOT(460.0,410.0,0)
      CALL APLOT(3940.0,410.0,0)
      CALL APLOT(100.0,1213.0,-1)
      WRITE(3,1) IESC,IA,IJ,ICR
1     FORMAT(3A2,'90',A2)
      WRITE(3,2) IUS,IESC,IA,IN
2     FORMAT(A2,'INTENSITY',3A2)
      IXMIN=INT(XMIN)
```

Appendix 7.A1 (*continued*)

```
        IXM=MOD(IXMIN,10)
        IXMIN=IXMIN-IXM
        FIX1=FLOAT(IXMIN)
        IXMAX=INT(XMAX)
        IXM=MOD(IXMAX,10)
        IXMAX=IXMAX-IXM+10
        FIX2=FLOAT(IXMAX)
        JOB=IXMAX-IXMIN
        DO 3 I=0,10
        FI=FLOAT(I)
        X=460.0+FI*348.0
        Y=410.0
        IF(I.EQ.0) GO TO 11
        CALL APLOT(X,Y,-1)
11      Y=Y+17.0
        IF(I.EQ.0) GO TO 12
        CALL APLOT(X,Y,0)
12      X=X-56.0
        Y=Y-95.0
        CALL APLOT(X,Y,-1)
        IQ=IXMIN+(JOB*I/10)
        WRITE(3,4)IUS,IQ
4       FORMAT(A2,I3)
3       CONTINUE
        IF(L.EQ.1) GO TO 51
        Y=120.0
        X=1584.0
        CALL APLOT(X,Y,-1)
        WRITE(3,52)IUS
52      FORMAT(A2,'WAVENUMBER X 10')
        X=2424.0
        Y=208.0
        CALL APLOT(X,Y,-1)
        WRITE(3,53)IUS
53      FORMAT(A2,'-2')
        X=2536.0
        Y=120.0
        CALL APLOT(X,Y,-1)
        WRITE(3,54)IUS
54      FORMAT(A2,',CM')
        X=2704.0
        Y=208.0
        CALL APLOT(X,Y,-1)
        WRITE(3,55)IUS
55      FORMAT(A2,'-1')
        GO TO 56
51      Y=146.0
        X=1808.0
        CALL APLOT(X,Y,-1)
        WRITE(3,5)IUS
5       FORMAT(A2,'WAVELENGTH,NM')
56      CALL WAIT(30,2,I)
```

```
         DO 7 I=0,10
         FI=FLOAT(I)
         X=460.0
         Y=410.0+239.0*FI
         IF(I.EQ.0) GO TO 14
         CALL APLOT(X,Y,-1)
   14    X=485.0
         IF(I.EQ.0) GO TO 13
         CALL APLOT(X,Y,0)
         X=150.0
         GO TO 15
   13    X=110.0
   15    Z=Y-20
         IQ=I*1000
         CALL APLOT(X,Z,-1)
         WRITE(3,8)IUS,IQ
    8    FORMAT(A2,I5)
    7    CONTINUE
         CALL WAIT(15,2,I)
         WRITE(3,9)IESC,IA,IN
    9    FORMAT(3A2)
         RETURN
         END
$BEND
```

8 Time Dependence of Fluorescence Anisotropy

8.1 Introduction

Time-resolved fluorescence depolarization studies have, since the early 1970s, provided an interesting method for monitoring molecular reorientational motions in solution. Until recently the relatively poor quality of much of the data available from these measurements has precluded detailed quantitative interpretations of the results. With the advent of improved experimental techniques for fluorescence decay time determinations interest in the technique has been revived.

When an ensemble of fluorescent chromophores is illuminated an anisotropy will be created in the system since only those species with a component of the absorption transition moment in the same direction as the polarization vector of the incident light will be excited. When plane-polarized excitation radiation is employed, if the fluorescence emission transition moment maintains a fixed geometrical relationship with that for the absorption process, steady-state illumination leads to three distinct cases.

8.1.1 Steady-state anisotropy measurements

The Perrin equation (8.1) relates anisotropy r (Equation 8.2) to initial anisotropy r_0, the fluorescence decay time τ_F and the rotational relaxation time, τ_r.

$$\frac{1}{r}=\frac{1}{r_0}\left[1+\frac{3\tau_F}{\tau_r}\right] \tag{8.1}$$

$$r=\frac{(I_{\|}-I_{\perp})}{(I_{\|}+2I_{\perp})} \tag{8.2}$$

Here $I_{\|}$ and $I_{\perp}$ are the fluorescence intensity components parallel and perpendicular to the plane of polarization of the electric vector of the incident radiation. The three possible cases are as follows:

(A) If $\tau_F \ll \tau_r$ then $r \approx r_0$ and τ_r is difficult to measure (the fluorescence decays before the molecule moves).

(B) If $\tau_F \gg \tau_r$ then $r=0$ and the anisotropy (and thus τ_r) cannot be accurately measured.

(C) If $\tau_F \approx \tau_r$ This represents the best case for accurate determination of $\tau_r (0<r<r_0)$.

For conventional emitting species with fluorescence decay times of the order of ns, rotational motion must be restricted if it is to occur on this timescale. This can be achieved in small molecules by use of viscous solvents, or by incorporating the chromophore into a macromolecule of synthetic or biological origin. Effects resulting from the use of unpolarized exciting radiation will be discussed later.

8.1.2 Time-resolved anisotropy measurements

Spencer and Weber (1969) have derived the relevant equations for the time dependence of $I_{\parallel}$ (Equation 8.3) and $I_{\perp}$ (Equation 8.4), for single τ_r

$$I_{\parallel}(t)=e^{-t/\tau_F}(1+2r_0e^{-t/\tau_r}) \tag{8.3}$$

$$I_{\perp}(t)=e^{-t/\tau_F}(1-r_0e^{-t/\tau_r}) \tag{8.4}$$

The total fluorescence intensity $F(t)$ is given by Equation 8.5:

$$\begin{aligned} F(t) &= I_{\parallel}(t)+2I_{\perp}(t) \\ &= 3e^{-t/\tau_F} \\ &= F_0e^{-t/\tau_F} \end{aligned} \tag{8.5}$$

and the time dependent anisotropy, $r(t)$ is given by:

$$\begin{aligned} r(t) &= [I_{\parallel}(t)-I_{\perp}(t)/(I_{\parallel}(t)+2I_{\perp}(t)] \\ &- r_0c^{-t/\tau_r} \end{aligned} \tag{8.6}$$

$F(t)$ depends upon τ_F, and $r(t)$ only upon τ_r, so that these two lifetimes can be separated. This separation is not possible in steady state measurements. (The degree of polarization, P, defined as $(I_{\parallel}-I_{\perp})/(I_{\parallel}+I_{\perp})$, is not independent of τ_F and is therefore not as useful a quantity as r). Again three limiting cases can be considered.

(a) If $\tau_F<\tau_r$, the fluorescence decays before the anisotropy decays, and hence only r_0 can be measured.

(b) If $\tau_r<\tau_F$, in contrast to steady-state measurements, τ_r can be measured in principle, for as Fig. 8.1 shows, the *decay* of the parallel and perpendicular components depends only upon τ_r. The one experimental disadvantage of this case is that those photons emitted after the period of a few times τ_r cannot contribute toward the determination of τ_r, but provided the signal-to-noise ratio is favourable, this need not be of great concern. However, there *can* be conditions in which the signal-to-noise ratio is unfavourable due to the experimental limitations dictated by the need to avoid pulse pile-up. While it is always disadvantageous to use a fluorophore with $\tau_r \gg \tau_F$ for measurement of τ_r [case (a)], one of the advantages of time-resolved anisotropy measurements lies in being able to extend the time range over which a given fluorophore can be used for accurate measurement of τ_r.

(c) $\tau_r \approx \tau_F$. This is the ideal experimental situation since almost all photons are counted within the time (equal to several rotational relaxation times) in which $r(t)$ shows measurable changes.

8.2 Effects of Polarization on Intensity Measurements

8.2.1 Instrument response

Before discussing methods of measuring time-dependent anisotropy of fluorescence, we emphasize that polarization effects can be of importance in

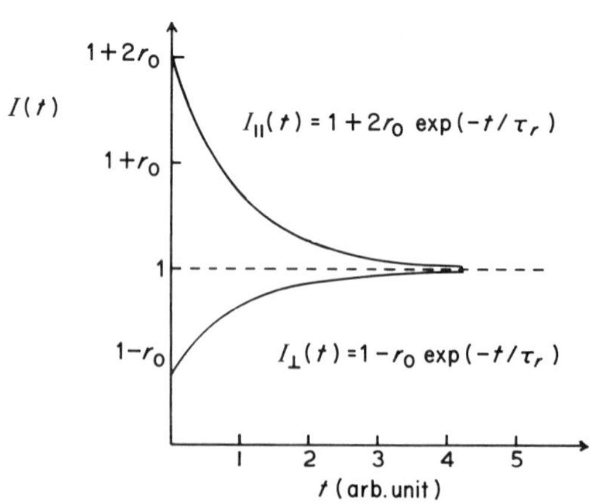

Figure 8.1 Idealized representation of the decay of $I_{\|}(t)$ and $I_{\perp}(t)$ when $\tau_r < \tau_F$. τ_r assumed to be equal to 1 time unit.

all fluorescence intensity, including conventional decay time, measurements. We must first recognize that the instrument monitoring the fluorescence, particularly the spectral dispersion element, responds differently to different polarizations of light, thus occasioning a need for a correction factor. For example, the use of diffraction gratings can yield intensities of emission which depend strongly upon orientation with respect to the plane of the grating. It is inevitably necessary when using such instruments to correct for the anisotropy in response. This instrumental anisotropy is usually termed the G factor and is defined as the ratio of the transmission efficiency for vertically polarized light to the transmission efficiency for horizontally polarized light. If in the simple instrument shown in Fig. 8.2, the two planes of exciting radiation, horizontal, H, and vertical, V, are as defined in the diagram with respect to the planes of observation of fluorescence, $I_{\|}$ and $I_{\perp}$, then for horizontally polarized exciting radiation, $I_{\|}$ and $I_{\perp}$ should be identical, or $(I_{\|}/I_{\perp})_{\mathrm{H}}=1$. If measured values of $G[=(I_{\|}/I_{\perp})_{\mathrm{H}}]$ deviate from unity, a correction factor must be applied to all measurements of $I_{\|}$ and $I_{\perp}$ excited with vertically polarized light, $(I_{\|}/I_{\perp})_{\mathrm{V}}$. Thus

$$\frac{(I_{\|})}{(I_{\perp})_{\mathrm{V\ corrected}}}=\frac{(I_{\|})}{(I_{\perp})_{\mathrm{V}}}\bigg/\frac{(I_{\|})}{(I_{\perp})_{\mathrm{H}}}. \tag{8.7}$$

Hence values of fluorescence anisotropy, r, corrected for instrumental response, would be given by:

$$r=\frac{(I_{\|}/I_{\perp})-G}{(I_{\|}/I_{\perp})+2G}. \tag{8.8}$$

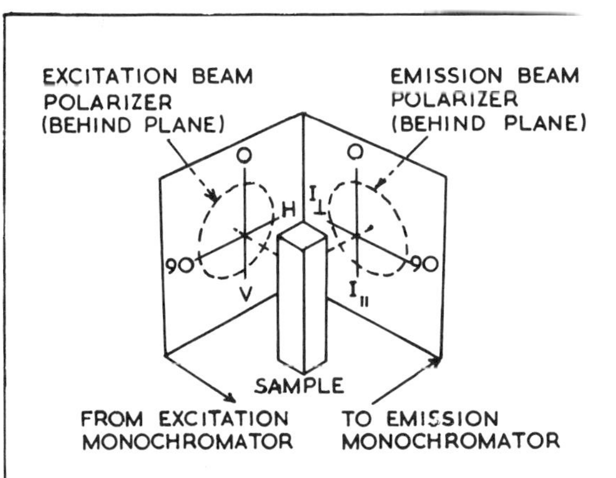

Figure 8.2 Representation of planes of polarization of incident and fluorescent light.

8.2.2 Elimination of polarization bias in intensity measurements

The removal of unwanted effects of photoselection (preferential excitation of molecules with specific orientations with respect to excitation radiation, both polarized and unpolarized) has been discussed thoroughly by Mielenz *et al.* (1976), from whom we have paraphrased the following analysis. The general case for steady-state illumination is illustrated in Fig. 8.3, in which the exciting radiation is assumed to be *partially* polarized, and is regarded as an incoherent superposition of two plane polarized components S^V and S^H having orthogonal electric field vectors, E^V and E^H. The intensity of the fluorescence emission is detected at an arbitrary viewing direction α in the horizontal plane, and is resolved into four components with electric field vectors E_V^V, E_V^H, E_H^V, and E_H^H where superscripts denote state of polarization of exciting radiation, and subscripts that of fluorescence.

The intensities of these four components are

$$I_V^V(\alpha) = (E_V^V)^2 = 1/3kS^V\tau(1+2r) \tag{8.9}$$

$$I_V^H(\alpha) = (E_V^H)^2 = 1/3kS^H\tau(1-r) \tag{8.10}$$

$$I_H^V(\alpha) = (E_H^V)^2 = 1/3kS^V\tau(1-r) \tag{8.11}$$

$$I_H^H(\alpha) = (E_H^H)^2 = 1/3kS^H\tau(1-r+3\cos^2\alpha) \tag{8.12}$$

k is a proportionality factor, τ is the overall molecular cross-section for

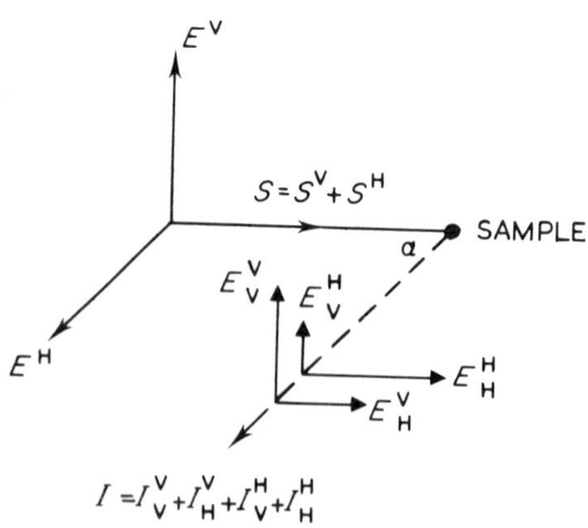

Figure 8.3 Excitation of a sample by partially polarized light composed of a superposition of two plane polarized components, S and S^H. E^V and E^H are the corresponding electric field vectors. For the other symbols see text (after Mielenz *et al.*, 1976).

fluorescence, and r is the emission anisotropy defined as in Equation 8.6, and also given by Equation 8.13 as

$$r = 1/5(3\cos^2\theta - 1), \tag{8.13}$$

where θ is the angle that the emission dipole has at the time of emission relative to the position of the absorption dipole at the moment of absorption.

There are several methods which can be used to obtain intensity measurements corrected for polarization

Method 1

The most cumbersome is to take readings, R, equivalent to Equations 8.9–8.12 using vertical and horizontal polarised excitation *and* emission.

These are then

$$R_{\rm V}^{\rm V}(\alpha) = I_{\rm V}^{\rm V}(\alpha)T_{\rm V} \tag{8.14}$$

$$R_{\rm V}^{\rm H}(\alpha) = I_{\rm V}^{\rm H}(\alpha)T_{\rm V} \tag{8.15}$$

$$R_{\rm H}^{\rm V}(\alpha) = I_{\rm H}^{\rm V}(\alpha)T_{\rm H} \tag{8.16}$$

$$R_{\rm H}^{\rm H}(\alpha) = I_{\rm H}^{\rm H}(\alpha)T_{\rm H} \tag{8.17}$$

where $T_{\rm V}$ and $T_{\rm H}$ are the response factors of the detection system (see above) for vertically and horizontally polarized light respectively. These raw data are then used to calculate the excitation (Equation 8.18) and emission (Equation 8.19) correction factors

$$F = S^{\rm V}/S^{\rm H} \tag{8.18}$$

$$G = T_{\rm V}/T_{\rm H} \tag{8.19}$$

and the emission dichroic ratio D, related to the emission anisotropy r, by Equation 8.20:

$$r = (D-1)/(D+2). \tag{8.20}$$

The information R_0, which would be obtained from an equivalent but depolarized sample is then

$$R_0 = 1/3k(S^{\rm V} + S^{\rm H})(T_{\rm V} + T_{\rm H})\tau \tag{8.21}$$

The procedure is informative, but unnecessarily time-consuming, particularly if R_0 is the only parameter required, and other alternatives are available.

Method 2 (Spencer and Weber, 1970)

This method was developed for *lifetime* measurements, $\tau_{\rm F}$, free from spurious effects of polarization, but is used for intensity measurements also. The authors used *vertically* polarized exciting radiation, and right angle ($\alpha = 90$) viewing. The emission polarizer was set at an angle ϕ, (Fig. 8.4), and the total transmitted intensity when $\phi = \cos^{-1}(1/\sqrt{3}) = 54.75^0$ or 125.25^0, is given by Equation 8.22.

$$\begin{aligned} I_{\phi}^{V}(\alpha) &= (E_{V}^{V} \cos \phi)^2 + (E_{H}^{V} \sin \phi)^2 \\ &= 1/3 I_{V}^{V}(\alpha) + 2/3 I_{H}^{V}(\alpha) \end{aligned} \tag{8.22}$$

which is *independent* of *r*, and can be written as

$$I_{\phi}^{V}(\alpha) = 1/3 k S^{V} \tau \tag{8.23}$$

The responses T_V and T_H with respect to ϕ are given by

$$\begin{aligned} T_{\phi} &= T_V \cos^2 \phi + T_H \sin^2 \phi \\ &= 1/3(T_V + 2T_H) \\ &= 1/3\, T_H(G+2). \end{aligned} \tag{8.24}$$

The detector signal,

$$\begin{aligned} R_{\phi}^{V}(\alpha) &= T_{\phi} I_{\phi}^{V}(\alpha) \\ &= 1/9 k S^{V} T_H (G+2)\tau, \end{aligned} \tag{8.25}$$

is *also* independent of *r*, and therefore represents an *unbiased* measure of the fluorescence cross-section. *Hence the use of an emission polarizer at the so-called magic angle of 54.75° (or 125.25°) with vertically polarized exciting light eliminates bias due to polarization effects for all viewing angles, not just* $\alpha = 90°$.

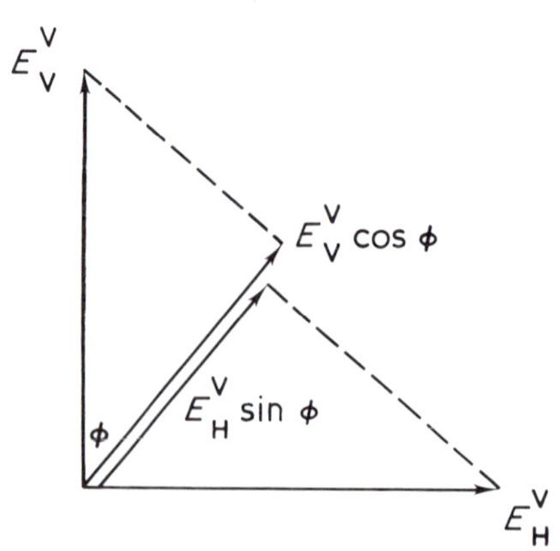

Figure 8.4 Electric field vector components of fluorescence intensity transmitted by a polarizer set at angle ϕ. Vertically polarized excitation (after Mielenz *et al.*, 1976).

Method 3

If the excitation radiation is *unpolarized* an unbiased detection signal can also be obtained if the same magic angle ϕ is used for the detector polarizer, but with α now specific. With no excitation polarizer, the emission detector signal will be

$$R_\phi(\alpha) = T_\phi T_\phi^{\mathrm{V}}(\alpha) + I_\phi^{\mathrm{H}}(\alpha), \tag{8.26}$$

where $I_\phi^{\mathrm{V}}(\alpha)$ is as in Equation 8.23, and $I_\phi^{\mathrm{H}}(\alpha)$, with reference to Fig. 8.5 is given by

$$\begin{aligned} I_\phi^{\mathrm{H}}(\alpha) &= (E_{\mathrm{V}}^{\mathrm{H}} \cos \phi)^2 + (E_{\mathrm{H}}^{\mathrm{H}} \sin \phi)^2 \\ &= 1/3 I_{\mathrm{V}}^{\mathrm{H}}(\alpha) + 2/3 I_{\mathrm{H}}^{\mathrm{H}}(\alpha) \\ &= 1/3 k S^{\mathrm{H}} \tau [1 - r(1 - 2\cos^2 \alpha)]. \end{aligned} \tag{8.27}$$

The polarization bias of I^{H} will be removed if

$$\alpha = \cos^{-1}(1/\sqrt{2}) = 45^\circ \text{ or } 135^\circ, \tag{8.28}$$

so that $I_\phi^{\mathrm{H}}(45) = 1/3 k S^{\mathrm{H}} \tau$.

Hence

$$R_\phi(45) = 1/9 k (S^{\mathrm{V}} + S^{\mathrm{H}}) T_{\mathrm{H}}(G + 2)\tau \tag{8.29}$$

and is also independent of r.

The main advantage of Method 3 is that there is no need for an excitation polarizer. Furthermore, only one measurement is required. The disadvantage is that the technique is very sensitive to angular misalignment. For $\alpha = 90^\circ$ and emission polarizer set at vertical or horizontal (Method 1), the first derivatives of the detector response with respect to α and ϕ are zero, whereas for the arrangements in Method 3, this is not so, and misalignment affects $R_\phi(45)$.

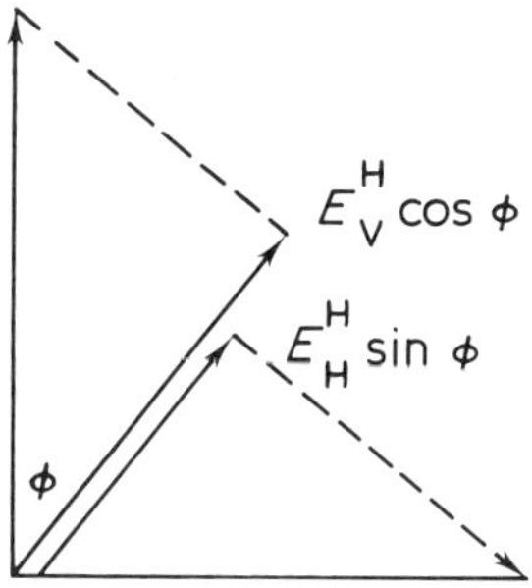

Figure 8.5 Electric field vector components of fluorescence intensity transmitted by a polarizer set at angle ϕ. Horizontally polarized excitation (after Mielenz *et al.*, 1976).

Method 4 (Spencer and Weber, 1970)

Here no excitation polarizer is used, the emission polarizer is set at $\phi = \cos^{-1}(2/3)^{1/2}$, (35.25°) and right-angle viewing ($\alpha = 90°$) is used. In this case, it can be shown (Mielenz *et al.*, 1976) that

$$R_\phi(90) = T_\phi[2/3 I^{\mathrm{V}}_{\mathrm{V}}(90) + I^{\mathrm{H}}_{\mathrm{V}}(90) + 1/3 I^{\mathrm{V}}_{\mathrm{H}}(90) + I^{\mathrm{H}}_{\mathrm{H}}(90)]$$
$$= 1/3 k S^{\mathrm{H}} T_\phi \tau(F+1) + r(F-1). \quad (8.30)$$

Hence the method is valid only if the exciting radiation is strictly unpolarized, i.e. $F = 1$.

Method 5

Here the *excitation* polarizer is set at $\phi = \cos^{-1}(2/3)^{1/2}$, and *no* emission polarizer is used, and again $\alpha = 90°$. This is the method recommended by Spencer and Weber (1970), but since in this case

$$R^\phi(90) = T_{\mathrm{V}} 2/3 I^{\mathrm{V}}_{\mathrm{V}}(90) + 1/3 I^{\mathrm{H}}_{\mathrm{V}}(90) + T_{\mathrm{H}} 2/3 I^{\mathrm{V}}_{\mathrm{H}}(90) + 1/3 I^{\mathrm{H}}_{\mathrm{H}}(90)$$
$$= 1/9 k S^{\mathrm{H}} T_{\mathrm{H}} \tau(2FG + 2F - G - 1) + r(4FG - 2F - G - 1). \quad (8.31)$$

the result will only be unbiased if $F = 1$, *and* $G = 1$.

Method 6

Here the excitation polarizer is set at $\phi = \cos^{-1}(1/3)^{1/2}$, the emission polarizer is vertical, and again $\alpha = 90°$. In this case, the response is

$$R^\phi_{\mathrm{V}}(90) = T_{\mathrm{V}}[1/3 I^{\mathrm{V}}_{\mathrm{V}}(90) + 2/3 I^{\mathrm{H}}_{\mathrm{V}}(90)]$$
$$= 1/9 k S^{\mathrm{H}} T_{\mathrm{V}} \tau(F+2) + 2r(F-1) \quad (8.32)$$

Again, the result is unbiased only if the excitation radiation is unpolarized, $F = 1$. *Use of any of the Methods 4, 5 and 6 therefore demands that scramblers be used to depolarize the exciting light and, in Method 5, the emitted light.* The emission scrambler should be placed between emission polarizers and dispersive element, if any, or between emission polarizer and detector. It would be as well at this juncture to note one further practical point concerning the use of polarizers. Glan prism polarizers are two half-prisms separated by a thin layer of air. The beam emerging from such a device is displaced with respect to the incident beam, and rotation of the polarizer causes the transmitted beam to deviate round a circle (Azumi and McGlynn, 1962). When used with an emission monochromator this causes *spectral* shifts of as much as 2 nm between vertical and horizontally polarized components which can be critical when line spectra are being observed. Such effects are not found with flat plate polarizers, which although deficient in other respects, are much easier to use.

8.3 Time-dependent Anisotropy Measurements

The time dependence of anisotropy is obtained experimentally by measurement of the time dependence of the components of fluorescence polarized parallel $I_{\parallel}(t)$, and perpendicular $I_{\perp}(t)$ to, say, vertically polarized excitation radiation, and is evaluated using Equations 8.3, 8.4 or 8.6.

Considerable care must be taken in making such apparently simple measurements. Some of the difficulties will now be discussed briefly.

8.3.1 Instrumental instability

Construction of $r(t)$ through Equation 8.6 involves the *difference* between $I_{\parallel}$ and $I_{\perp}$ and this poses the first problem. To produce good statistics, it is essential that the difference be maximized; therefore the $I_{\parallel}$ and $I_{\perp}$ curves must be collected to very high counts. Apart from the time consumed in this data collection, it increases the problems, discussed in earlier chapters, associated with excitation source instabilities, detection system drift, etc. These problems are not insurmountable, but require very careful attention to detail during the experiments; in particular, pulse profiles must be very strictly monitored before and after data collection to eliminate drift. Effects of changes in pump pulse profile can be minimized by simultaneous collection of $I_{\parallel}$ and $I_{\perp}$.

A dual detection system can be utilized for this purpose, which, however, necessitates expensive duplication of equipment. Such a system is described by Tschanz and Binkert (1976) and is shown in Fig. 8.6. P_2 is the emission analysing polarizer which is kept fixed with the transmission direction perpendicular to the plane of the figure. For measurements of anisotropy, P_1 is rotated between the horizontal and vertical positions to measure $I_{\perp}(t)$ and $I_{\parallel}(t)$, while the analyser P_3, in the reference channel, is positioned so that the direction of the transmitted component bisects the two positions of P_1, so that light detected through P_3 is independent of these positions of P_1. The alternation of *excitation* polarizer to obtain $I_{\parallel}$ and $I_{\perp}$ rather than keeping P_1 fixed and altering P_2 has some advantages, particularly if a dispersive element is used to select some fluorescence wavelengths. The detection system polarizer, P_2, since it is not altered between measurements of $I_{\parallel}$ and $I_{\perp}$, has the same response in each case, since light in the same polarization sense is always detected.

As an alternative to a polaroid sheet polarizer for P_1 with laser excitation, a half-wave plate may be used to change the plane of excitation radiation, though this has a disadvantage in that the spectral bandwidth of a half-wave plate is narrow. Half-wave plates are expensive and several will be required to span a given spectral range.

The method of determining time-dependent anisotropy with the system illustrated in Fig. 8.6 would be as follows. The excitation function, with a diffuse reflector in place of the sample, is measured. Then, with polarizer and

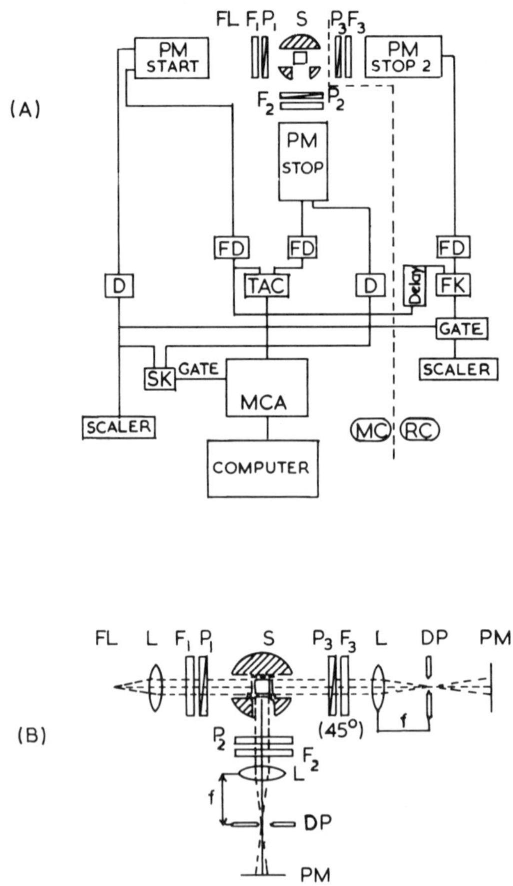

Figure 8.6 Dual detection single photon counting instrument. (A) Experimental arrangement. FL, flash lamp; S, sample; F_1, F_2, F_3, filters; P_1, P_2, P_3, polarizers; D, discriminator; FD, fast discriminator; MC, measuring channel; RC, reference channel; SK, slow coincidence; FK, fast coincidence. (B) Optical arrangement. Symbols as in (A) and L, lenses; DP, diaphragm (placed in the focal plane of the corresponding lens) (after Tschanz and Binkert, 1976).

analyser P_1 and P_2 parallel vertically, $I_{||}(t)$ is measured to a total of $m_{||}$ counts in the reference channel, 3. The polarizer P_1 is turned into the horizontal position, and a second measurement gives $I_{\perp}(t)$ and $m_{\perp}$. Before calculation of anisotropy, the intensity $I_{\perp}(t)$ must be normalized for identical excitation intensities, to $I_{||}(t)$ using the factor $m_{||}/m_{\perp}$.

As an alternative to the use of a reference channel, an experimental system can be set up in which there is a rapid alternation of observation of $I_{||}$ and $I_{\perp}$, for example, by using a two-position stepper-motor to rotate the detection polarizer between parallel and perpendicular positions on a time scale (e.g., 10 s) short compared with total data collection time. This ensures that the $I_{||}$ and $I_{\perp}$ curves are collected under identical conditions. A further non-trivial problem associated with the fact that difference curves are required involves the chemical and biological stability of samples under investigation during the long counting times.

8.3.2 Elimination of *G* factor

Even when fluorescence is not spectrally dispersed, care has to be taken to ensure that the *G* factor discussed above is unity, or if not, is taken into account in anisotropy measurements. There are several methods available to check this.

8.3.2(a) Tail-matching

Provided that $\tau_r < \tau_F$, it can be seen from Equations 8.3 and 8.4 that at long times after excitation, the curves for $I_{||}(t)$ and $I_{\perp}(t)$ as measured should become identical. If in any experiment they are not, it can usually be assumed that this is due to a non-unitary *G* factor. Hence normalizing the two decay curves on the tail of the decay eliminates the *G* factor. In Fig. 8.7 are shown uncorrected $I_{||}(t)$ and $I_{\perp}(t)$ curves for a macromolecular sample measured with laser excitation and a monochromator for dispersion of fluorescence. There is an obvious discrepancy. However, there are some serious drawbacks to tail-matching. First, the matching, or curve fitting, has to be done in a region of the decay where the statistics are poorest, with consequent errors. Secondly, counting to long times involves the use of a long timescale voltage ramp in the TAC, with consequent loss of precision at earlier times where anisotropy changes are most evident. The most serious drawback however is that there *are* systems for which the anisotropy is not expected to decay to zero during the fluorescence lifetime of the probe, e.g., in ordered or semi-ordered media such as liquid crystals, polymers of synthetic and biological origin, lipid bilayers, micelles, etc. Since these are the types of system most amenable to study by these techniques, the *assumption* implicit in tail-

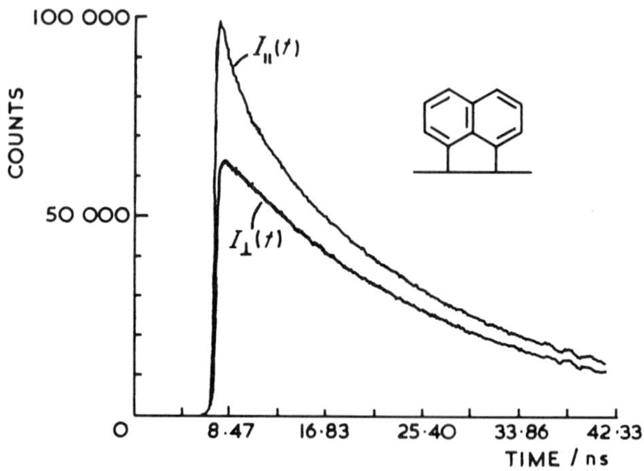

Figure 8.7 Decay of emission of acenaphthylene labelled polymethylmethacrylate in tetrahydrofuran. Laser excitation at 308 nm (after Drake *et al.*, 1982).

matching that r tends to zero is a dangerous one, unless checked carefully by other methods.

8.3.2(b) Leading edge matching

The initial anisotropy in a probe for which the emission and absorption transition dipoles have identical orientations with respect to the molecular framework is given by putting $\theta = 0$ in Equation 8.13, with the result that $r_0 = 2/5$. In any experimental system r_0 may be measured to be other than 2/5, and one of the causes of this is the G factor. If all other causes of deviations from the 2/5 result can with certainty be eliminated, then the G factor correction can be applied by weighting the $I_{\parallel}(t)$ and $I_{\perp}(t)$ decay curves such that the constructed anisotropy curve has $r_0 = 2/5$. This is an even more hazardous procedure than tail-matching, since it is not usually possible to be certain what the initial anisotropy should be. There are many probes for which the spectroscopy is complex, leading to a much reduced r_0. In addition, as the data in Table 8.1 illustrate, r_0 values can be different from 0.4 depending on the nature of the photoselection and the viewing geometry.

Moreover, leading edge matching requires manipulation of decay curves in the region where convolution with instrument response and excitation functions (see below) are of greatest importance.

Table 8.1 Polarization ratios and r_0 values in idealized photoelectron experiments (developed from Albrecht, 1970).

	Exciting light unpolarized				Exciting light vertically polarized			
	Head on viewing		90° viewing		Head on viewing		90° viewing	
Photoselection/observe	$(I_{\|\|}/I_{\perp})$	r_0	$(I_{\|\|}/I_{\perp})_0$	r_0	$(I_{\|\|}/I_{\perp})_0$	r_0	$(I_{\|\|}/I_{\perp})_0$	r_0
Single axis/same axis	1	0	2/1	0.2	3/1	0.4	3/1	0.4
Single axis/perp. axis	1	0	3/4	−0.1	1/2	−0.2	1/2	−0.2
In-plane/out-of-plane	1	0	3/4	−0.1	1/2	−0.2	1/2	−0.2
In-plane/out-of-plane	1	0	7/6	0.05	4/3	0.1	4/3	0.1

8.3.2(c) Use of a freely-rotating probe

Perhaps the simplest method of evaluating G factors is to use a probe fluorophore, the rotational relaxation time of which is very short compared with the decay time, and to carry out a continuous illumination experiment in the manner described in Section 8.2. Such a procedure carried out on our own experimental system using α-cyanonaphthalene in cyclohexane as calibrant, $\tau_F = 18.8$ ns, yielded a G factor of 1.03. Use of this to correct the $I_{\parallel}(t)$ and $I_{\perp}(t)$ curves shown in Fig. 8.7 showed quite clearly that there was *no* residual anisotropy in this sample, and hence that tail-matching would have been justified. The result of tail-matching, shown in Fig. 8.8, is identical to that resulting from application of the G factor correction, although extreme care is needed to ensure that this is so.

8.3.2(d) Synthesis of fluorescence decay curve

The total fluorescence decay of a fluorophore is given by Equation 8.5 as being $I_{\parallel}(t) + 2I_{\perp}(t)$ for plane polarized excitation. If $I_{\parallel}(t)$ and $I_{\perp}(t)$ are measured separately, they can be combined with a suitable weighting factor, such that the synthesized fluorescence decay curve $I_{\parallel}(t) + 2GI_{\perp}(t)$ matches exactly that measured under experimental conditions that eliminate effects of polarisation (e.g., Method 2, Section 8.2). This procedure evaluates non-unit

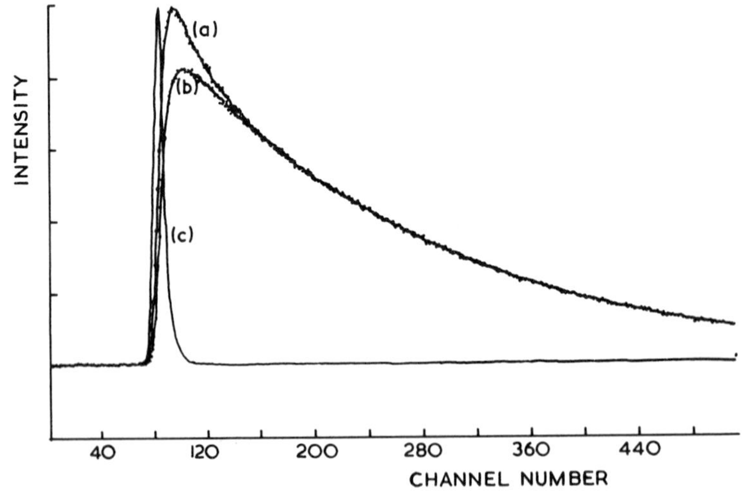

Figure 8.8 Illustration of tail-matching applied to $I_{\parallel}$(a) and $I_{\perp}$(b) curves for the sample described in Fig. 8.7. Points are the observed data and the lines are best fit curves. (c) Instrument response function. 0.082 ns per channel.

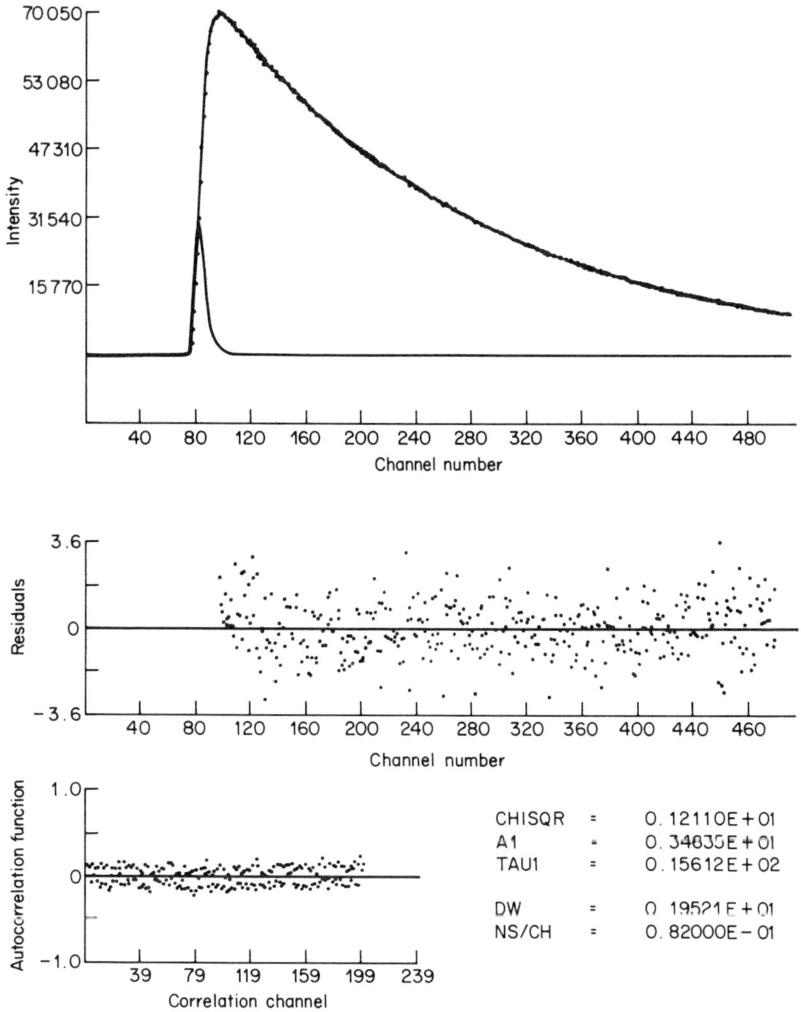

Figure 8.9 Synthesis and analysis of $I_{\|} + 2.06I_{\perp}$ for the sample described in Fig. 8.7. Upper section: data synthesized from observed $I_{\perp}$ and $I_{\|}$ (points), best fit curve (line) and instrument response function (small peak to the left). Middle section: plot of weighted residuals. Lower section: autocorrelation function of weighted residuals. Weighting factors given by Equation 8.36.

G factors in the system. Its application to a macromolecular sample is illustrated in Fig. 8.9.

A variation on this method is to use either Equation 8.3 or 8.4 to fit the $I_{\|}(t)$ curve or $I_{\perp}(t)$ curve, weighting such that the value of τ_F recovered matches that from "magic angle" excitation, although this procedure requires some assumption of the functional form (single component, multiple exponential) for τ_r. In our experience this method is quite successful; for when the double exponential fitting routine (e.g., that listed in Appendix 6.A2) is used to analyse $I_{||}(t)$ and $I_{\perp}(t)$ according to Equations 8.3 and 8.4 identical results are obtained, as indicated in Fig. 8.10. However, there may be

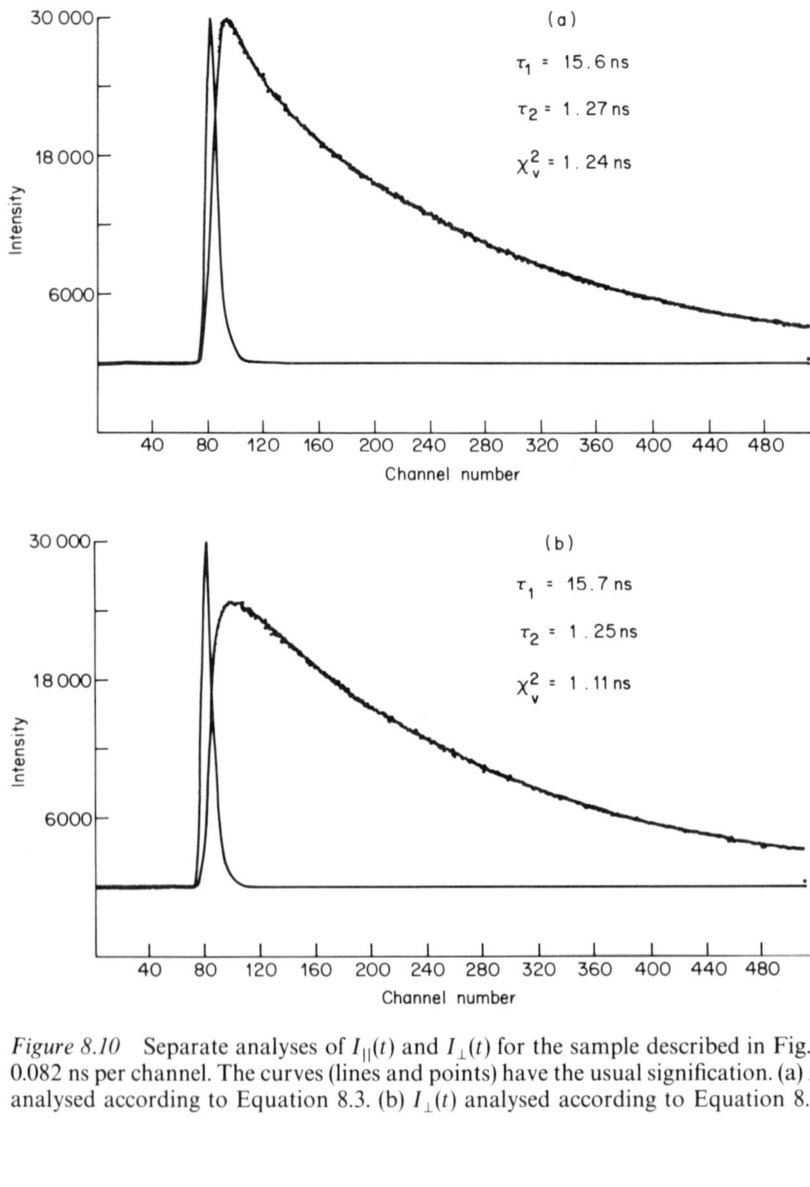

Figure 8.10 Separate analyses of $I_{||}(t)$ and $I_{\perp}(t)$ for the sample described in Fig. 8.7, 0.082 ns per channel. The curves (lines and points) have the usual signification. (a) $I_{||}(t)$ analysed according to Equation 8.3. (b) $I_{\perp}(t)$ analysed according to Equation 8.4.

uncertainties about the single value of τ_r resulting from this procedure, especially if the intrinsic fluorescence decay is non-exponential.

It is strongly recommended that combinations of all of the above methods are used vigilantly to ensure the reproducibility of data in these experiments.

8.4 Deconvolution Procedures

As we have stated above, much of the information concerning change in anisotropy with time will be contained in the early portions of fluorescence decay curves, where the effects of finite excitation profile and instrument response functions will be greatest. In such cases, some form of deconvolution will be essential in order to obtain reliable data, but the best procedure to adopt is still the subject of strenuous debate. It would of course be preferable if deconvolution were unnecessary, but this can only be realized in a small number of cases, using narrow temporal bandwidth excitation and very fast detection. Using the least-squares fitting routine favoured by us, the possibilities are outlined in Sections 8.4.1–8.4.4 with respect to a single polymer sample, and results are compared.

8.4.1 Create anisotropy r, from $I_{\|}(t)$, $I_{\perp}(t)$ without deconvolution

This procedure yields two components of τ_r, but with considerable question concerning the shorter of the two.

8.4.2 Analyse $I_{\|}(t)$ or $I_{\perp}(t)$ to obtain τ_F, τ_r from Equations 8.3 and 8.4

This was shown in Fig. 8.10. The CREWS program recovered τ_F accurately, but was able to recover, of course, only one τ_r value. Attempts to use a triple exponential fitting program to investigate other rotational components were unsuccessful. Clearly this method deconvolves out instrument response and finite excitation effects, and since complete decays are used the quality of data is such that confidence in values of τ_F and τ_r recovered should be high. However, for dual component anisotropies, successful recovery of the information requires a triple component fit to the data, since τ_F also appears as a term in the relevant equations, and even with excellent quality data, the occasions upon which six parameter fits can be applied with confidence are few.

8.4.3 Create difference, $[I_{\parallel}(t)-I_{\perp}(t)]$, then deconvolve

It can be seen from Equations 8.3 and 8.4 that

$$I_{\parallel}(t)-I_{\perp}(t)=3r_0\mathrm{e}^{-t/\tau_r}\mathrm{e}^{-t/\tau_F}. \tag{8.33}$$

Deconvolved curves can be fit to this function, and if τ_F is known from a separate experiment (e.g. magic angle viewing), τ_r can be recovered. For multiple components of τ_r, however, there are the same difficulties as in Method 8.4.2, but aggravated by the fact that *difference* curves have much poorer signal-to-noise ratios, and curve fitting procedures are even more subject to error. An example of deconvolving a difference curve of this type is illustrated in Fig. 8.11.

8.4.4 Deconvolve $I_{\parallel}(t)$, $I_{\perp}(t)$ using arbitrary function, create $r(t)$

This is the method we favour in principle. If $I_{\parallel}(t)$ and $I_{\perp}(t)$ are deconvolved, separately, in terms of *arbitrary* fitting functions, for example using the ten exponential program given in Appendix 7.A1, no functional form has to be assumed for either of these curves. Construction of the anisotropy from the resulting functions is straightforward and the time-dependent anisotropy curves, now free of convolution effects, can be analysed by curve fitting methods which in principle should yield data of greater reliability, since the expression for $r(t)$ does *not* contain terms in τ_F. The drawback is again that of relying upon a *difference* of two curves, with consequent loss of quality of data. This seems to be an unavoidable problem.

8.4.5 Weighting factors for anisotropy experiments

In fitting decays using the least-squares technique the fit is usually weighted according to Equation 8.34 (see also Equation 2.11):

$$w_i=\frac{1}{\sigma_i^2}=\frac{1}{I(t_i)}. \tag{8.34}$$

$I_{\parallel}(t)$ and $I_{\perp}(t)$ obey Poisson statistics and can be deconvolved in the usual way using this equation for weighting. If we wish to deconvolve $I_{\parallel}(t)-I_{\perp}(t)=d(t)$, $I_{\parallel}(t)+2I_{\perp}(t)=s(t)$ and $r(t)=d(t)/s(t)$ we must calculate the weighting factors by propagating the Poisson errors in the measured curves. It is easy to derive the following equations (Wahl, 1979):

$$\text{for } d(t_i),\qquad w_i=\frac{1}{I_{\parallel}(t_i)+I_{\perp}(t_i)} \tag{8.35}$$

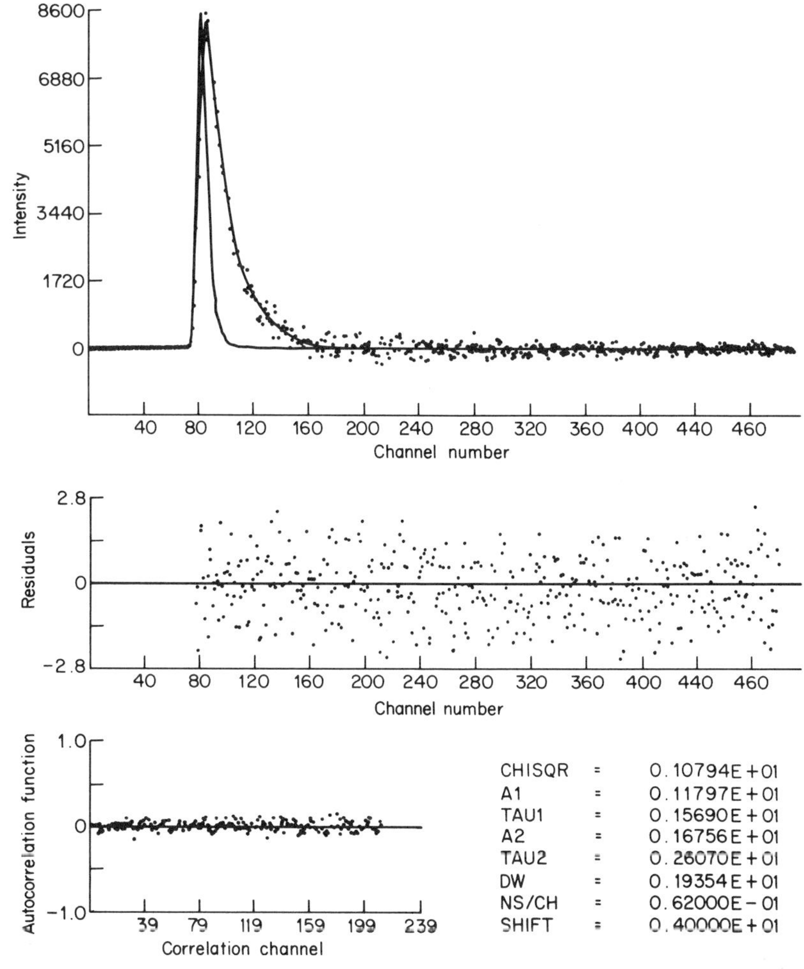

Figure 8.11 Analysis of the difference function $d(t) = I_{||}(t) - I_{\perp}(t)$ for the sample described in Fig. 8.7. Analysing function: Equation 8.33. Upper section: $d(t)$ calculated from observed data (points), best fit curve (line) and instrument response function (curve on l.h.s. of points). Middle section: weighted residuals. Lower section: autocorrelation function of weighted residuals. Correct weighting factors (Equation 8.35) were used.

for $s(t_i)$,
$$w_i = \frac{1}{I_{||}(t_i) + 4I_{\perp}(t_i)} \tag{8.36}$$

and for $r(t_i)$,
$$w_i = \frac{3[I_{||}(t_i) + 2I_{\perp}(t_i)]}{2 + r(t_i) + 5[r(t_i)]^2 - 2[r(t_i)]^3}. \tag{8.37}$$

Clearly $d(t_i)$, $s(t_i)$ and $r(t_i)$ are not Poisson distributed and use of incorrect weighting factors may lead to grossly incorrect results, as illustrated in Fig. 8.12 where analysis of the difference function $d(t)$ with (incorrect) weighting factors given by $w_i = 1/[I_{\parallel}(t_i) - I_{\perp}(t_i)]$ is shown. Analysis of this function with correct weighting factors is shown in Fig. 8.11 and it will be seen that the results are quite different. In particular the incorrect value for τ_r is almost an order of magnitude too small.

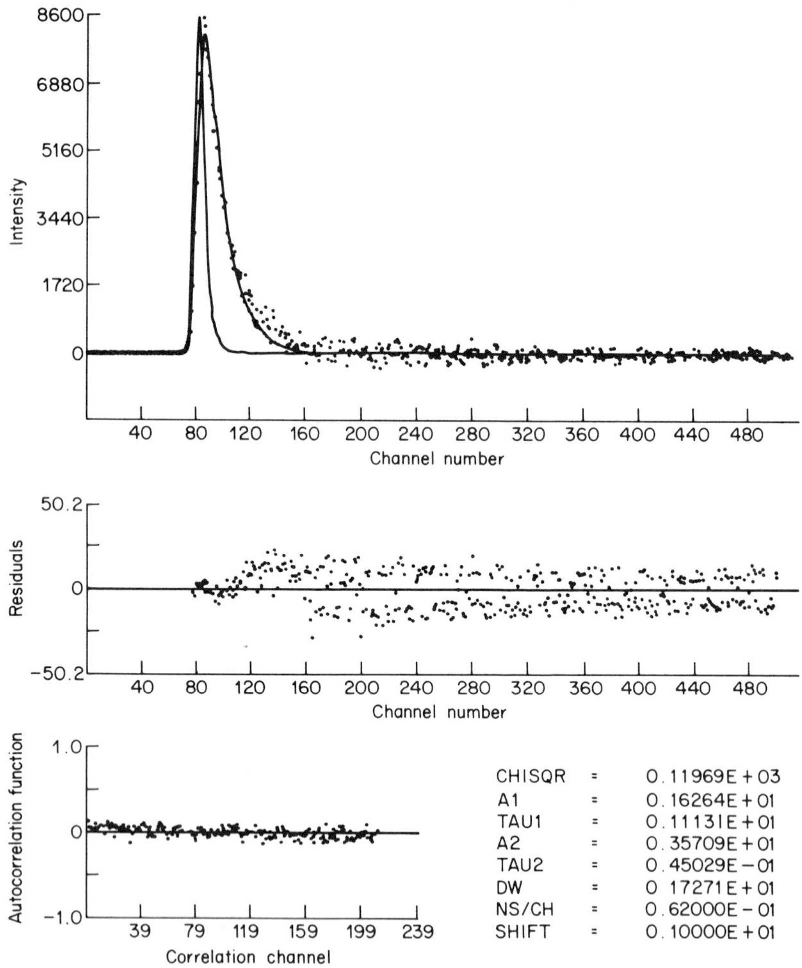

Figure 8.12 The difference function $d(t)$, described in Fig. 8.11 analysed using incorrect weighting factors (see text). Note the poor χ_v^2 value and the large discrepancy in τ values between this and the correct analysis (Fig. 8.11).

8.4.6 Fitting by means of Lagrange multipliers

Methods other than least-squares fitting routines have been attempted to recover $r(t)$. For example the use of a Lagrange multiplier technique (Andre *et al.*, 1982) is illustrated in Fig. 8.13, in which fluorescence from an anthra-

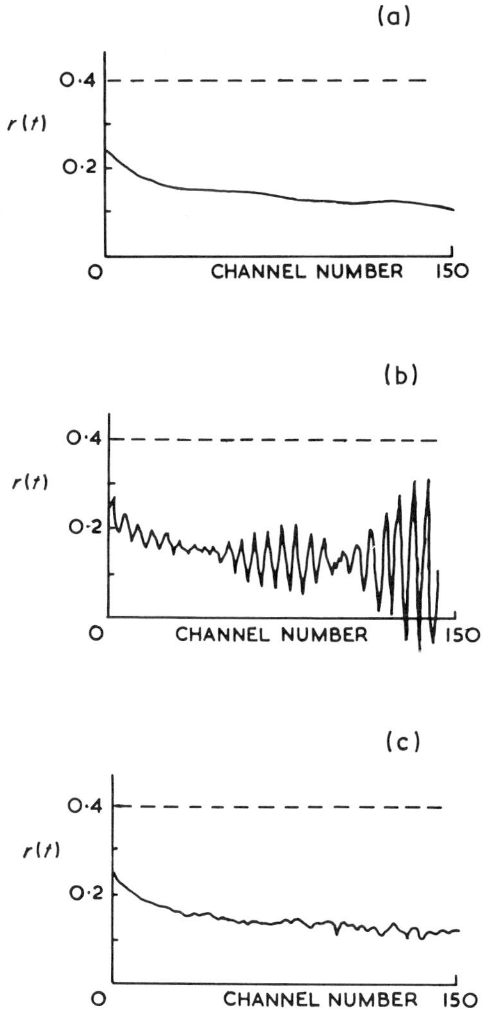

Figure 8.13 The decay of anisotropy of an anthracene probe inserted in a polybutadiene chain. Deconvolution of $[I_{||}(t)-I_{\perp}(t)]/[I_{||}(t)+2I_{\perp}(t)]$ using (a) exponential series, (b) Fourier transforms and (c) regularized Lagrange multipliers (after Andre *et al.*, 1982).

Table 8.2 Comparison of three "non *a-priori*" methods in analysing the total decay $[I_{||}(t)+2I_{\perp}(t)]$ of the probe anthracene in a polybutadiene chain (after Andre *et al.*, 1982).

Method	χ^2/χ^2 Lagrange
Exponential series	5.2
Fourier transform	2.3
Lagrange	1

cene fluorophore in the centre of a poly(butadiene) chain is analysed by least-squares fitting (Section 6.5.8), fast Fourier transform (Section 6.5.6), and Lagrange multiplier (Section 6.5.8) methods. Table 8.2 shows that the χ^2 values from this technique are superior to those from the other methods. Although this is still a good indicator of the relative quality of fitting, absolute values of χ^2 have no meaning in these cases, since *differences* of two curves are used in the construction of $r(t)$.

The discussions above will have served to illustrate that reliable values of $r(t)$ are difficult to obtain experimentally. It is pertinent to explore, however, how $r(t)$ should in principle be related to the motion of molecules. This is now discussed.

8.5 Theory of Time-dependent Fluorescence Depolarization

Following Tao (1969), the problem can be stated as follows. In Fig. 8.14, x, y, and z are the space-fixed coordinate system in the laboratory, with the sample placed at the origin. The exciting light travels along the x-axis, polarized with the electric vector e along the z-axis. The fluorescence, polarized along the z-axis, ($I_{||}$) and the x-axis, ($I_{\perp}$) will be observed along the y-axis. $\bar{\mu}$ is the transition dipole of the molecule, and θ, ϕ here, in contrast to the use of θ earlier in this chapter, define the orientation of $\bar{\mu}$ with respect to the space-fixed axes. What is sought is an expression for $r(t)$. The derivations will not be given here, but the reader is referred to a recent article by Lipari and Szabo (1980) and to earlier work by Belford *et al.* (1972), Tao (1969), Lombardi and Dafforn (1966), and Weber (1971). There was originally some dispute about the solution for $r(t)$, but the correct form for the emission anisotropy for a general rotating body was derived by Belford *et al.* (1972) and was found to contain five exponential terms:

$$\frac{5}{6}r(t)=\sum_{i=1}^{3} C_i\exp(-t/\tau_i)+[(F+G)/4]\exp(-6D-2\Delta)t$$

$$+[(F-G)/4]\exp[-(6D+2\Delta)t] \qquad (8.38)$$

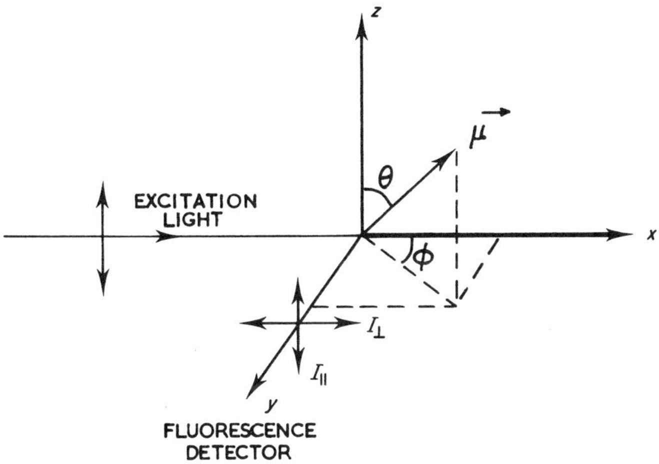

Figure 8.14 Experimental geometry for time-resolved fluorescence depolarization measurements.

where $D=(D_1+D_2+D_3)/3$, the mean rotational diffusion constant, given that D_1, D_2 and D_3 are the rotational diffusion constants around the three principal axes of the molecule labelled 1, 2 and 3 (to be distinguished from the laboratory axes labelled x, y and z in Fig. 8.14).

$$\Delta=(D_1^2+D_2^2+D_3^2-D_1D_2-D_1D_3-D_2D_3)^{1/2}.$$

$$C_i=\alpha_j\alpha_k\varepsilon_j\varepsilon_k \qquad \text{where } (i\,j\,k)=(1\,2\,3), (2\,3\,1) \text{ or } (3\,1\,2).$$

$\alpha_1, \alpha_2, \alpha_3$ are the direction cosines of the absorption dipole with respect to the

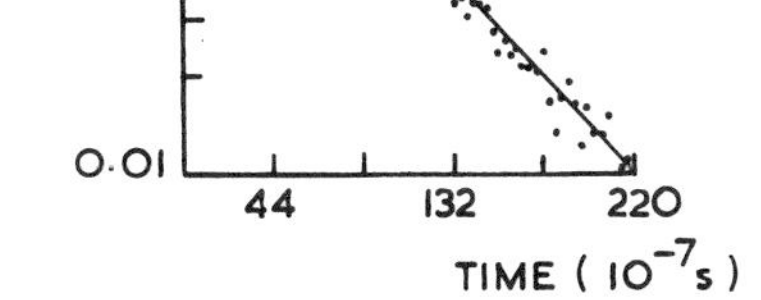

Figure 8.15 Decay of emission anisotropy simulated by computer (points) and compared with the prediction based on Equation 8.38 (line) (after Harvey and Cheung, 1972).

principal rotation axes. ε_1, ε_2 and ε_3 are, likewise, the direction cosines of the emission dipole.

$$\tau_i = (3D + 3D_i)^{-1}$$

$$F = \sum_{i=1}^{3} \alpha_i^2 \varepsilon_i^2 - 1/3$$

$$G\Delta = \sum_{i=1}^{3} D_i(\alpha_i^2 \varepsilon_i^2 + \alpha_j^2 \varepsilon_k^2 + \alpha_k^2 \varepsilon_j^2) - D \qquad (i \neq j \neq k \neq i)$$

It can be seen that Equation 8.38 represents $r(t)$ in terms of five exponential terms, which is rigorously correct. However, although each of these terms contributes to the time dependence of fluorescence, no more than three are independent and in most realistic experimental circumstances only two, or in extremely favourable cases, three (Small and Isenberg, 1977), decay rates will be distinguishable. Furthermore, as Belford *et al.* (1972) point out, any one of these experimental rates will usually be proportional to some weighted average of the five intrinsic decay rates represented by Equation 8.38. It should be stated that treatments by Lombardi and Dafforn (1966) and Weber (1971), which both predicted *two* decay times only, are not the consequence of any error, but reflect special symmetry restrictions in one case, discussed below, and the adoption of a six-position jump model in the other. In general α and ε values are not available, in which case weighting factors are not well known, and it has been suggested that the two decay times in Weber's model, being functions of α and ε, can properly be regarded as the two phenomenological decay rates which will ensue from any experiment. It should be further noted that the treatment by Tao (1969), which again predicts five exponential decay terms, is a special case, being valid only when the absorption and emission dipoles have coincidental directions ($\alpha_i = \varepsilon_i$ for all i), and also errs in that a simple one-to-one correspondence between the distribution of body orientations and the distribution of orientations of emitting dipoles is assumed. Computer simulated data have been used to check the validity of the general expression (Harvey and Cheung, 1972). In all cases good agreement was observed between the simulated data and theoretical curves. The results for one such experiment are shown in Fig. 8.15. It is possible in some cases to observe an initial increase in $r(t)$ as a consequence of the orientation of the absorption and emission dipoles.

There are several special cases resulting from the general expression. For a body that possesses an axis of symmetry, such as an ellipsoid of rotation, we may define

$$D_1 = D_2 = D_\perp$$

$$D_3 = D_{||}$$

The emission anisotropy in this case is given by (Tao, 1969):

$$r(t)=\frac{2}{5}\{A_1(\delta)\exp(-6D_\perp t)+A_2(\delta)\exp[-(5D_\perp+D_{||})t]\}$$
$$+A_3(\delta)\exp(2D_\perp+4D_{||}), \tag{8.39}$$

where

$$A_1(\delta)=(3/2\cos^2\delta-\tfrac{1}{2})^2,$$
$$A_2(\delta)=3\cos^2\delta\sin^2\delta,$$
$$A_3(\delta)=\tfrac{3}{4}\sin^4\delta,$$

and δ is the angle between the emission (and absorption) transition dipole and the axis of symmetry. For the special case in which δ is 90° (e.g. the dipole is oriented perpendicular to the axis of symmetry):

$$r(t)=\frac{1}{10}[\exp(-6D_\perp t)+3\exp[-(2D_\perp+4D_{||})t]] \tag{8.40}$$

and for the dipole parallel to the axis ($\delta=0$) the equation for $r(t)$ is reduced to a single exponential term:

$$r(t)=\frac{2}{5}\exp(-6D_\perp t). \tag{8.41}$$

For a rotating spherical body, the correlation function of order n is given by (Williams, 1978):

$$\langle P_n[\hat{\mu}(0)\cdot\hat{\mu}(t)\rangle=\exp[-n(n+1)D_r t)]. \tag{8.42}$$

Hence, for fluorescence depolarization measurements in which the second order correlation function is desired

$$r(t)=\frac{2}{5}\exp(-6D_r t). \tag{8.43}$$

Here D_r is the rotational diffusion constant and is given by the Einstein law as

$$6D_r=\frac{1}{\tau_r}=\frac{kT}{V\eta}, \tag{8.44}$$

where V is the volume of the sphere and η is the viscosity of the solvent. Some confusion exists in the definition of the rotational relaxation time τ_r; many reports preferring $\rho_0(=3\tau_r)$.

The three decay constants appearing in the expression for $r(t)$ for an ellipsoid of rotation have been calculated and are shown in Table 8.3. As may be seen, the three relaxation times diverge rapidly with increasing axial ratio

Table 8.3 Rotational diffusion decay constants for prolate and oblate ellipsoids

ρ	$1/\rho$	$\tau_{\|\|}/\tau_r$	τ_1/τ_r	τ_2/τ_r	τ_3/τ_r
Prolate ellipsoids					
1		1.0000	1.0000	1.0000	1.0000
2		0.8067	1.5049	1.3152	0.9543
3		0.7480	2.3408	1.7276	0.9674
4		0.7210	3.3956	2.0984	0.9777
5		0.7061	4.6405	2.4060	0.9842
6		0.6968	6.0616	2.6548	0.9884
7		0.6906	7.6506	2.8549	0.9911
8		0.6862	9.4008	3.0162	0.9930
9		0.6829	11.3080	3.1472	0.9944
10		0.6805	13.3680	3.2545	0.9954
15		0.6739	25.8607	3.5774	0.9979
20		0.6712	41.8199	3.7280	0.9988
Oblate ellipsoids					
1		1.0000	1.0000	1.0000	1.0000
2		1.4100	1.1316	1.1701	1.3032
3		1.8284	1.4645	1.5147	1.6885
4		2.2495	1.8431	1.9003	2.0955
5		2.6718	2.2398	2.3018	2.5104
6		3.0948	2.6455	2.7111	2.9290
7		3.5181	3.0564	3.1247	3.3495
8		3.9418	3.4705	3.5411	3.7711
9		4.3655	3.8869	3.9593	4.1934
10		4.7894	4.3049	4.3787	4.6162
15		6.9099	6.4073	6.4859	6.7338
20		9.0312	8.5193	8.6005	8.8538

ρ is the axial ratio (ratio of longitudinal semi-axis to equatorial semi-axis), $\tau_{||}=(6D_{||})^{-1}$, $\tau_1=(6D_{\perp})^{-1}$, $\tau_2=(5D_{\perp}+D_{||})^{-1}$, $\tau_3=(2D_{\perp}+4D_{||})^{-1}$, $\tau_r=(6D_r)^{-1}$ is the relaxation time under the spherical model. Adapted from Tao (1969).

for a prolate ellipsoid. However, for an oblate ellipsoid the deviations are small even for high axial ratios and experimentally it may prove difficult to resolve more than a single, mean relaxation time for $r(t)$ in this case. Thus, three situations exist in which the emission anisotropy may decay exponentially and it is not possible, therefore, to distinguish: (1) a true spherical geometry; (2) oblate geometry; (3) preferential orientation of the fluorescence probe with respect to the principal axis. Hence, it should be stressed, a single exponentially decaying emission anisotropy does *not* correspond uniquely to a spherical rotating body. On the other hand, the observation of non-exponential behaviour of $r(t)$ indicates deviations from spherical behaviour if the probe is known to have parallel absorption and emission dipoles. If the latter condition is not true, a spherical rotating molecule will also give rise to multiple component anisotropies.

8.6 Form of Time-dependent Anisotropy for Restricted Motion

The form for the emission anisotropy in several special cases is of interest. The decay of the anisotropy for a chromophore rotating between two barriers has been derived by Wahl (1975). The problem posed by a fluorophore with restricted rotation in an ordered lipid bilayer has been discussed by Kinoshita *et al.* (1977). If r_0 is the initial anisotropy created in the system at time 0 and r_∞ is the final limiting value after long times, than, as opposed to an isotropic, low viscosity solvent, r_∞/r_0 will not be zero, but will be a measure of the degree of confinement of the orientation of the label imposed by the architecture of the membrane. The decay in $r(t)$ between these two limits should reflect the degree of "wobbling" motions possible for the fluorophore. The case is illustrated in Fig. 8.16, where θ here is the orientation angle of a cylindrical probe with respect to the axis of a bilayer. Kinoshita *et al.*'s analysis leads to:

$$\frac{r(t)}{r(0)} = \langle P_2[\hat{\mu}(0) \cdot \hat{\mu}(t)) \rangle = \sum_i A_i \exp^{(-D_w t/\sigma_i)}, \tag{8.45}$$

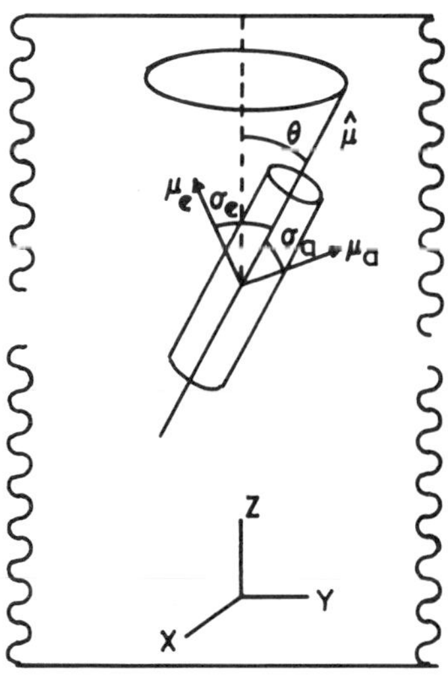

Figure 8.16 Schematic outline of a cylindrical fluorescent probe embedded in a lipid bilayer. θ is the (variable) orientation angle of the probe with respect to the axis of the bilayer and θ_0 (not shown) is the maximum value of θ (after Lipari and Szabo, 1980).

where D_w is the "wobbling" diffusion constant, which will be of magnitude greater than that for an unrestricted motion. The coefficients A_i and σ_i cannot be expressed as closed analytical functions of θ, defined in Fig. 8.16, but Kinoshita *et al.* have displayed these graphically, and in addition, have presented an approximate expression for $r(t)/r(0)$, Equation 8.46:

$$\left[\frac{r(t)}{r(0)}\right]_{\text{approx.}} = A_\infty + (1 - A_\infty)\exp(-D_w t/\langle\sigma\rangle), \tag{8.46}$$

where $A_\infty = r(\infty)/r(t)$ and is given by

$$\begin{aligned}\frac{r(\infty)}{r(0)} &= \langle P_2(\cos\theta)\rangle^2 \\ &= [(\tfrac{1}{2}\cos\theta_0(1+\cos\theta_0)]^2.\end{aligned} \tag{8.47}$$

θ_0 is the semi-angle of the cone in which the probe is free to wobble, and

$$\langle\sigma\rangle = \sum_{i\neq\infty} A_i\sigma_i/(1 - A_\infty). \tag{8.48}$$

This treatment is approximate, but Lipari and Szabo (1980) have shown that an exact solution is possible. The result is then

$$\begin{aligned}D_w\tau_{\text{eff}}(1 - A_\infty) = &-x_0^2(1+x_0)^2\{\log[(1+x_0)/2] \\ &+ (1-x_0)/2\}/[2(1-x_0)] \\ &+ (1-x_0)(6+8x_0-x_0^2-12x_0^3-7x_0^4)/24\end{aligned} \tag{8.49}$$

where $x_0 = \cos\theta_0$. Hence measurement of A_∞ yields a value for the "wobble" angle θ_0, and even if $r(t)/r(0)$ is not adequately represented by a single exponential, Equation 8.49 allows extraction of D_w from the decay of the emission anisotropy, since $\tau_{\text{eff.}}(1 - A_\infty)$ is exactly the area under $[r(t) - r(\infty)/r(0)]$. These expressions are valid *only* if either the absorption, μ_a, or emission, μ_e, dipole lies along the C_∞ axis μ of the probe. However, Lipari and Szabo (1980) have given an approximate solution for the general case, which will be discussed presently. For the moment, again assuming μ_a or μ_e coincides with the μ axis, the case of the probe molecule attached to a macromolecule can be considered. In this case, the approximate Equation 8.46 becomes:

$$\begin{aligned}\frac{r(t)}{r(0)_{\text{approx.}}} &= A_\infty\exp(-t/\tau_M) \\ &+ (1 - A_\infty)\exp[-t(\tau_M^{-1} + \tau_{\text{eff.}}^{-1})],\end{aligned} \tag{8.50}$$

where $\tau_M = D_M/6$, D_M being the rotational diffusion coefficient of the macromolecule.

This is exactly the form of the expression used empirically by Munro *et al.* (1979) to analyse data on the anisotropy of tryptophan fluorescence of a variety of proteins, when $\tau_M \gg \tau_{eff.}$ (a good approximation in many cases). The expression used by these authors to relate their effective correlation time to D, however, did not have a sound theoretical basis, and a reasonably accurate approximation is

$$\tau_{eff.} = 7\theta_0^2/24D_w \tag{8.51}$$

even at large cone angles. However, it seems clear that tryptophan is a poor probe in that the conditions inherent in the use of Equation 8.50 that the absorption or emission dipole is parallel to the axis which wobbles, are certainly not satisfied.

For the case where neither μ_a or μ_e are coincident with the C_∞ axis, an approximate solution has been given (Lipari and Szabo, 1980). It is

$$\left[\frac{r(t)}{r(0)}\right]_{\text{approx.}} = \frac{e^{-t/\tau_M}}{P_2(\cos\delta)} \sum_{m=-2}^{2} e^{-m^2 t/6\tau'_e}[S^2 + (1-S^2)e^{-t(1-m^2/6)/\tau''_e}] \times d_{m0}(\theta_e) d_{m0}(\theta_a) \cos m\phi_{ae}, \tag{8.52}$$

where θ_e and θ_a are the angles between μ_a and μ_e and μ respectively, δ is as before the angle between μ_a and μ_e, ϕ_{ae} is the difference between their azimuthal angles, τ''_e is an effective correlation time for the wobbling of the $\hat{\mu}$ axis, and $d^2_{Mo}(\beta)$ are reduced Wigner rotation matrices, that is

$$d^{(2)}_{\infty}(\beta) = (3\cos^2\beta - 1)/2 \tag{8.53}$$

$$d \pm 10(\beta) = \frac{4}{4}(3/2)^{1/2} \sin\beta \cos\beta \tag{8.54}$$

$$d \pm 20(\beta) = (3/8)^{1/2} \sin\beta^2 \beta \tag{8.55}$$

The sum over m in Equation 8.52 contains only three unique terms. As required, when θ_a or θ_e equals zero, Equation 8.52 reduces to Equation 8.50, with $\tau''_e = \tau_{eff.}$, and when $\theta_0 = \pi$, i.e., the motion is unrestricted, and $\tau_M = \infty$, it reduces to the result for a freely diffusing cylinder with $\tau'_e = 1/6D_2$ and $\tau''_e = 1/6Dx$.

There have been many other models of varying complexity which have been used to interpret time-dependent rotational anisotropy $r(t)$. We mention further here only the rather restricted model which attempts to mimic segmental motion of synthetic macromolecules (Valeur and Monnerie, 1976) by placing the chains upon a diamond lattice, and permitting only three and four centre motions upon this tetrahedral lattice, Fig. 8.17. The emission anisotropy in this case is given by

$$r(t) = r_0 \exp(-t/\rho)\,\text{erfc}\,(t/\rho)^{1/2}, \tag{8.56}$$

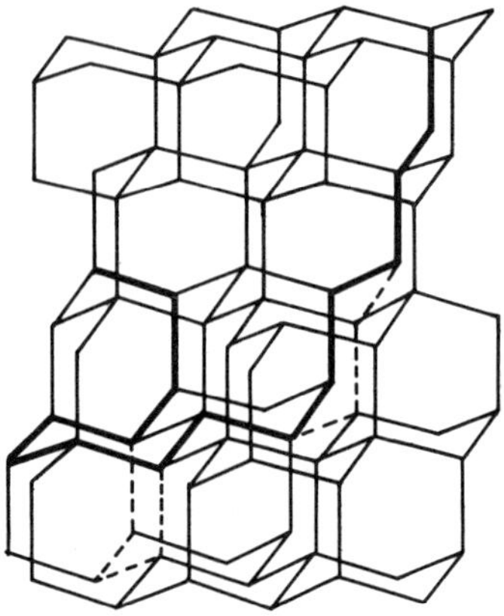

Figure 8.17 Schematic representation of a chain on a tetrahedral lattice (after Valeur and Monnerie, 1976).

where ρ is an effective relaxation time, and erfc is the error function complement. When valence angles and internal rotation angles are allowed to vary from an ideal lattice, and additional exponential term is introduced (Dubois-Violette *et al.*, 1969), and the expression for $r(t)$ becomes

$$r(t) = r_0 \exp(-t/\sigma) \exp(-t/\sigma) \operatorname{erfc}(t/\rho)^{1/2}, \tag{8.57}$$

where σ is a relaxation time reflecting the relaxation with respect to the ideal lattice.

Experimental measurements have in general yielded at most two, and often only one component decays in the analysis of $r(t)$. It is very clear from the discussions above that interpretation of $r(t)$ is far from straightforward, and that meaningful progress will only be made if experimental determinations of $r(t)$ can be made much more precisely than has in general been the case to date.

References

Albrecht, A. C. (1970). *Prog. React. Kinet.* **5**, 301–334.

Andre, J. C., Viovy, J. L., Bouchy, M., Vincent, L. M. and Valeur, B. (1982). In "Deconvolution and Reconvolution of Analytical Signals" (Bouchy, M., ed.), pp. 223–246. E.N.S.I.C.–I.N.P.L., Nancy.

Azumi, T. and McGlynn (1962). *J. Chem. Phys.* **37**, 2413.

Belford, G. G., Belford, R. L. and Weber, G. (1972). *Proc. Natl. Acad. Sci. USA* **69**, 1392–0000.

Drake, R., Phillips, D., Roberts, A. J. and Soutar, J. (1982). In "Photophysics of

Synthetic Polymers" (Phillips, D. and Roberts, A. J., eds), pp. 150–160. Science Reviews, Northwood.
Dubois-Violette, E., Geny, F., Monnerie, L. and Parodi, O. (1969). *J. Chem. Phys.* **66**, 1865–1876.
Ghiggino, K. P., Roberts, A. J. and Phillips, D. (1981). *Adv. Polym. Sci.* **40**, 69–167.
Harvey, S. C. and Cheung, H. C. (1972). *Proc. Natl. Acad. Sci. USA* **69**, 3670–3672.
Kinoshita, K., Kawato, S. and Ikegami, A. (1977). *Biophys. J.* **20**, 289–305.
Lipari, G. and Szabo, A. (1980). *Biophys. J.* **30**, 489–506.
Lombardi, J. R. and Dafforn, G. A. (1966). *J. Chem. Phys.* **44**, 3882–3887.
Mielenz, K. D., Cehelink, E. D. and McKenzie, R. L. (1976). *J. Chem. Phys.* **62**, 477–489.
Munro, I., Pecht, I. and Stryer, L. (1979). *Proc. Natl. Acad. Sci. USA* **76**, 56–63.
Small, E. W. and Isenberg, I. (1977). *J. Chem. Phys.* **66**, 3347–3351.
Spencer, R. D. and Weber, G. (1969). *Ann. N.Y. Acad. Sci.* **158**, 361–384.
Spencer, R. D. and Weber, G. (1970). *J. Chem. Phys.* **52**, 1654–1663.
Tao, T. (1969). *Biopolym.* **8**, 609–632.
Tschanz, H. P. and Binkert, Th. (1976). *J. Phys. E* **9**, 1131–1136.
Valeur, B. and Monnerie, L. (1976). *J. Polym. Sci. Polym. Phys.* **14**, 11–40.
Wahl, Ph. (1975). *Chem. Phys.* **7**, 210–219.
Wahl, Ph. (1979). *Biophys. Chem.* **10**, 91–104.
Weber, G. (1971). *J. Chem. Phys.* **55**, 2399–2407.
Williams, G. (1978). *Chem. Soc. Rev.* **7**, 89–131.

Subject Index